ENGINEERING UNCERTAINTY

AND RISK ANALYSIS

A Balanced Approach to Probability, Statistics, Stochastic
Modeling, and Stochastic Differential Equations

Sergio E. Serrano, Ph.D.

Second Edition, Completely Revised

 HydroScience Inc. Ambler, Pennsylvania

ENGINEERING UNCERTAINTY AND RISK ANALYSIS.
A Balanced Approach to Probability, Statistics, Stochastic Modeling,
    and Stochastic Differential Equations.
Second Edition, Completely Revised
By Sergio E. Serrano, Ph.D.

Published by:

HydroScience Inc.
1217 Charter Lane
Ambler, PA 19002
U.S.A.
Email: hydroscience@earthlink.net
http://home.earthlink.net/~hydroscience
SAN 299-3074

Library of Congress Cataloging-in-Publication Data

Serrano, Sergio E.
  Engineering uncertainty and risk analysis : a balanced approach to probability, statistics, stochastic modeling, and stochastic differential equations / Sergio E. Serrano. -- 2nd ed., completely rev.
      p. cm.
  Includes bibliographical references and index.
  ISBN 978-0-9655643-1-1 (pbk. : alk. paper)
  1. Engineering mathematics. 2. Uncertainty--Mathematical models.
  3. Reliability (Engineering). 4. Risk assessment--Mathematics. I. Title.
  TA330.S45 2011
  620.001'51--dc22

                              2010050460

ISBN 978-0-9655643-1-1

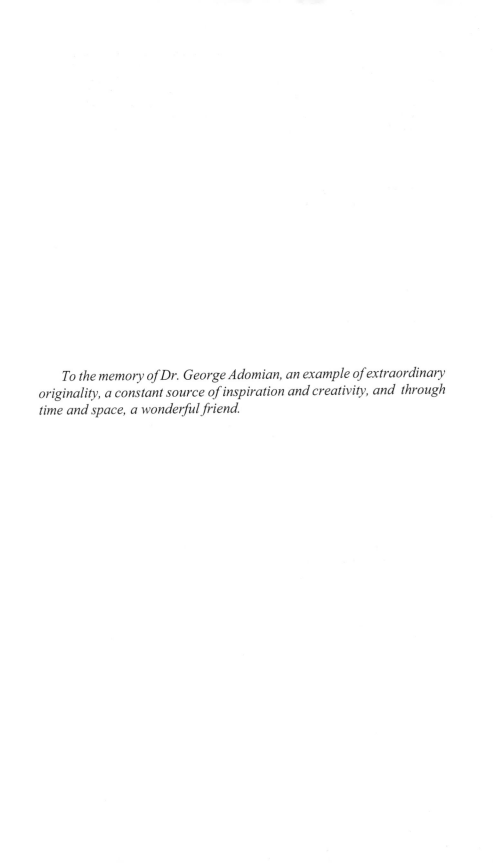

To the memory of Dr. George Adomian, an example of extraordinary originality, a constant source of inspiration and creativity, and through time and space, a wonderful friend.

*"Yet for all reasonable desires, for all honest endeavors, his benevolence hath established, in the nature of things, **a probability of success**. In all thy desires, let reason go along with thee, and fix not thy hopes beyond **the bounds of probability**; so shall success attend thy undertakings, thy heart shall not be vexed with disappointments."*

One of the greatest engineers of all times, Egyptian Pharaoh Amenhotep IV, 1360 B.C. From the book "Unto Thee I Grant." Supreme Grand Lodge of the Ancient and Mystical Order Rosae Crucis, AMORC. San Jose, California, 1995.

CONTENTS

PREFACE

Motivated by the positive response from the probability and statistics community, I release this second edition of *Engineering Uncertainty and Risk Analysis*. Based on suggestions and constructive criticism received from reviewers, consulting professionals, and many students, this new edition has been completely revised, enlarged, and updated. Many improvements and new developments are included, while preserving the original philosophy behind the text. Dozens of new examples in numerical applications and graphical visualization have been added.

This work attempts to offer the engineering community an integrated, balanced, and clear presentation of probability, statistics, stochastic modeling, and stochastic differential equations. There are excellent books in print that specialize in one or more of these subjects (i.e., Johnson, 2000; Walpole, et al, 1998; Rosenkrantz, 1997; Devore, 1995; Milton and Arnold, 1995; Montgomery and Runger, 1994). Depending on the authors' preference, most books emphasize either probability or statistics. The subjects of Monte Carlo simulations, stochastic processes, stochastic time series, and stochastic differential equations, which are essential pillars of modern risk analysis and engineering design, are regarded as "complicated" or "advanced" and are usually left for more advanced graduate courses. Thus, the average undergraduate engineering curriculum leaves the student prepared with only a small fraction of the subject matter of uncertainty analysis. The latest rule of the Accreditation Board for Engineering and Technology (ABET), requires that probability, statistics, modeling, and the design of experiments be an integral part of an engineering undergraduate curriculum.

Engineering Uncertainty and Risk Analysis is a response to the need for an integrated, harmonious, treatment of the above subjects. Besides an integrated coverage representative of the field of uncertainty analysis, the aim of this book is to demonstrate to the reader that the fundamental principles are inherently simple, and that probability, statistics, and stochastic modeling are practical and extremely useful in everyday engineering analysis or design. The focus is on engineering applications, rather than on lengthy theoretical digressions. The essential theory is illustrated with a multitude of numerical and computer examples of relevance in engineering practice, instead of elusive descriptions with games of chance (e.g., dice, cards, and colored balls). The subject of applied random differential equations, which appear in everyday engineering modeling, are presented for the first time at this level as a natural extension of the regular courses of college differential equations. Simple new methods to solve nonlinear differential equations analytically are presented with examples. This edition includes the modeling of engineering boundary-value initial-value problems in higher dimensions and new simple analytical procedures for irregularly-shaped domains, which frequently appear in field applications.

Computationally, this book departs substantially from the traditional textbooks of engineering statistics. I do not believe in introducing the first-time student to statistics software. Only a strict minority of them actually specializes in statistical analysis so as to justify the later purchase and application of the said software. Most engineers would eventually deal with probability or statistics as an integral part of projects where the uncertainty component is important. For this reason, I believe it is wiser to teach the average student how to apply the computational tools *he or she already has* in probabilistic or statistical calculations. *Engineering Uncertainty and Risk Analysis* emphasizes practical engineering applications, numerical calculations, and modeling via modern mathematics software, such as Maple™, rather than traditional computer languages, such as Fortran or C. Recent advances in computing technology are such that many modeling applications which previously required specialized software –such as flow lines and contour plotting, numerical integration, differential equations, and statistical analysis – are now easily done via standard mathematics software. Mathematics software provides the desired graphically-oriented, interactive, programming environment. Throughout an engineering curriculum of today, the student gradually learns to master a word processor, a spreadsheet, and mathematics software. All of the three applications are indispensable. A word processor satisfies the document preparation needs; a spreadsheet channels the basic engineering and graphics calculations; and a mathematics software satisfies the programming, matrix, additional graphics, and symbolic algebra needs. In this book all of the programs illustrated are in Maple (Versions 5 or later), since it is one of the best, one of the oldest, and one with the highest number of users worldwide. However, the reader may easily convert the commands to the mathematics software of his choice. It is strongly recommended that the reader does *all* of the computer exercises in the order presented in the chapters, especially if he is not familiar with Maple. *The book is self contained with respect to Maple and no prior knowledge of it is necessary*. Exercises are designed to gradually introduce the reader to the basic commands and operations in Maple, and to eventually master the more advanced features in the later chapters.

This book was written based on the believe that (1) every concept in uncertainty analysis may be explained in a simple fashion without losing scientific rigor and without resorting to detailed mathematical, statistical, or numerical theory; and (2) uncertainty analysis has practical applications in many engineering disciplines. My main objectives have been to offer clarity, rather than comprehensiveness, and to create a balance between probability, statistics, stochastic modeling, and random equations in engineering. I strongly believe that the fundamental laws behind uncertain phenomena are simple and that a good, solid, understanding of the fundamental principles is preferable to an extensive, and superficial, coverage of available methods. Learning is accomplished after a feeling of familiarity and confidence with the concepts overtakes the student. At that point the student acknowledges the *simplicity* of the subject and the desire to study more advanced concepts. I believe this is possible with undergraduate classes in probability and statistics.

In agreement with the above principles, *Engineering Uncertainty and Risk Analysis* implements many new features. The material in each chapter has been chosen so as to embark the reader in a gradual exposure to the same fundamental principles. This repetitive exposure to the same principles on a different level, or on a different application, pretends to accomplish the gradual familiarity required in a learning process. Consistency in notation is maintained throughout the text. The examples emphasize the engineering units of dimensions. There is a clear correspondence between the theoretical principles, the illustrative examples, and the corresponding proposed exercises. For the first 11 chapters, a minimum mathematical background is required from the reader. College algebra and integral calculus, such as that offered in most universities is sufficient. For the advanced (graduate) student, or the professional engineer interested in stochastic processes and stochastic differential equations, a knowledge of deterministic differential equations is necessary. However, even in Chapters 12 and 13 an effort has been made to demystify these topics and to emphasize engineering applications.

Chapter 1 introduces the philosophy and fundamental concepts behind uncertainty analysis in the context of engineering experimentation and design. The differences between the scientific method and the engineering method, the role of errors in design and modeling, and the origins of uncertainty are introduced to the reader along with a practical computer simulation exercise. The exercise aims at the observation of the effect of different levels of uncertainty and the need to characterize it and quantify it objectively. In addition, the exercise intends to immediately stimulate the reader to the practice of stochastic modeling.

Chapter 2 discusses the concepts of probability and frequency. Many engineering examples illustrate the quantifying of probabilistic events, independence, conditioning, and the fundamental operations with probability sets.

Chapter 3 introduces the concept of a random variable as a direct analog of regular deterministic functions. The most important discrete and continuous random variables in engineering are presented along with many examples. The mass and density functions are shown as intuitive means to quantify uncertainty. Computer algebra is offered as an excellent tool to perform the mathematical operations amongst the various variables and functions.

Chapter 4 presents the fundamentals of modeling of random systems. The principles behind random numbers generation and their transformation for Monte Carlo simulations are presented as an interesting and simple way to test engineering systems subject to uncertainty. Also, in this chapter, the essential differences between analytical and numerical modeling are illustrated. The method of decomposition is presented as a new way to solve nonlinear equations.

In Chapter 5, the concepts of single random variables in Chapter 3 are extended to systems with two or more random variables. Again, the concepts of the joint mass and the joint probability density functions are illustrated with several examples. Computer algebra is used for the various mathematical operations of differentiation and integration required in the derivation of moments.

Chapter 6 presents the basic concepts of statistics. The differences between population parameters and sample statistics are clearly stated. The problem of estimation of parameters is approached from the point of view of point estimation, interval estimation, and hypothesis testing. Depending on whether or not some of the parameters are known in actual practice, the various possible situations have been classified in cases with specific conditions. I have endeavor to create a systematic process where guessing of the method to use is minimized. Several engineering examples are shown.

Chapter 7 deals with the practical aspects of fitting probabilistic models to experimental data. Several methods are discussed along with numerical and computer examples in engineering applications. A practical procedure to build an empirical distribution from data; the application of statistical goodness-of-fit tests; and simple programs to produce probability paper of various distributions are discussed.

In Chapter 8 the reader is introduced to the basic linear regression techniques between two random variables. The concepts of covariance and linear correlation introduced in Chapter 5 are extended to the practical scenario of finding a relationship between two variables. Linear least square models, and nonlinear models reducible to linear, are discussed along with several manual and computer exercises.

In Chapter 9, the principles of single random variables in Chapter 3 are extended to the development of practical models of reliability of engineering systems. Time reliability, reliability of systems in series and of systems in parallel, and practical models of failure rates, along with simple programs to produce failure probability paper, are illustrated.

The practical procedures involved in the design of engineering experiments are presented in Chapter 10. Various approaches to the problem of estimating the mean from limited samples and that of finding the number of samples needed to assure a given project specification are discussed.

In Chapter 11, the fundamental methods for experiments and tests with two or more populations are presented. The concepts behind single-factor analysis of variance (ANOVA) are shown as a natural extension of hypothesis tests of single populations introduced in Chapter 6. Most of the mathematical derivations are omitted and the focus is centered on intuitive understanding and on simple practical applications the engineer may use immediately. As usual, references to more advanced work are given.

Chapter 12 presents an introduction to stochastic processes as a direct extension to the concepts of single random variables in Chapters 3 and 4. Again, the focus is on practical applications and on simple computer applications, rather than on lengthy theoretical descriptions. The visualization of random processes is illustrated via the construction of simple Maple programs. Spectral analysis, the analysis of simple time series, and some typical random processes (such as the Random Walk, the Brownian Motion, and the White Noise processes) are described, all with examples.

Since most engineering equations are differential equations of various kinds, and since one or more of its components are subject to uncertainty, I strongly believe all engineers should be capable of analyzing, even at some elemental level, stochastic differential equations. Yet, the study of applied random differential equations is rarely a component of engineering curricula, partly because it is regarded as a complex topic. This book attempts to provide such missing material. Under the belief that the fundamental aspects are simple, practical, and intelligible, Chapter 13 shows a unique treatment of applied random differential equations in engineering. We begin with basic notions on stochastic continuity, stochastic differentiation, and stochastic integration to arrive at a description of the origin of random differential equations. Equations are classified according to the way in which uncertainty enters the engineering system: the initial condition, the forcing function, the coefficients, the boundary conditions (for partial differential equations). An integrated method of solution is offered that naturally connects the methods of deterministic differential equations covered in most courses of college differential equations. Thus, an engineer who has studied the first 12 chapters of this book, and has a working knowledge of differential equations, will have no problem extending these concepts to cases involving random differential equations.

A unique feature of this book is that for the first time at this level it introduces simple new methods to solve linear and nonlinear differential equations analytically. The method of decomposition of Adomian is shown for ordinary or partial differential equations in one or more dimensions, steady state or transient, in regular or irregularly-shaped domains, and deterministic or stochastic conditions. With the advent of decomposition, it is demonstrated that the only thing an engineer needs to easily obtain an analytical solution in such cases is integral calculus. The result is a model easy to generalize in terms of the physical quantities, with dependent variables that are continuous in space and time –which permits the direct and continuous calculation of fluxes and gradients–, without the instability issues and computer memory requirements of traditional numerical techniques, and without the complexity required by traditional analytical methods. In Chapter 13 we describe *systematic* analytical decomposition solutions to multidimensional boundary-value problems in regular or irregular domains. For the case of stochastic partial differential equations, it is shown that decomposition does not require the usual restrictive assumptions of "smallness" (small perturbation schemes). In the examples shown (37 in total), I have strived to show the procedure to derive the solution, its mean, and its correlation function *step by step*. Several short programs and exercises are given to produce contours, three-dimensional plots, and velocity vectors, which no longer require specialized software.

In Chapter 14 is a short conclusion. From a higher perspective, and enriched with the knowledge acquired in the past 13 chapters, we revisit the fundamental ideas of uncertainty analysis in Chapter 1 and invite the reader to a more advanced level. This book was written for the undergraduate student of engineering. The first 11 chapters should be part of a one-semester 3-credit course in any engineering discipline. It is suggested that *all* of the proposed problems be done as part of regular homework, rather than a selected few from each chapter. Answers to all proposed exercises are given in Appendix B. Chapters 12 and 13 are intended for an advanced undergraduate or a graduate course in stochastic modeling. In such case, the first 11 chapters serve as an introductory review. Because of its integrated focus, and its emphasis on clarity, the book may also be used by the consulting professional or research engineer, wishing to familiarize himself or herself with the field of uncertainty and risk analysis. Many references to more advanced work are suggested to those readers interested in a more detailed study of a particular topic.

Sergio E. Serrano
Philadelphia, December 2010

1 ENGINEERING UNCERTAINTY ANALYSIS

1.1 PROBABILISTIC ANALYSIS OF ENGINEERING SYSTEMS

The Engineering Method

A heuristic definition of engineering disciplines may conceive them as the application of the scientific method to the improvement of humankind. The engineer is an individual that uses the scientific method for objectives that fall well beyond the mere understanding of a phenomenon. While the scientist studies a physical or natural phenomenon for the purpose of understanding and ultimately the formulation of a theory of knowledge, the engineer has a pragmatic objective in mind. The nature of this objective depends on the particular engineering discipline and the purpose of the project in question. Whether we are involved in the process of vehicle design, telecommunications, food production, building construction, or chemical manufacturing, the engineer has a concrete application in mind with a clear economic or social benefit.

While the objectives separate the scientist from the engineer, not always with a clear divide, the method unites them. The scientific method constitutes the fundamental tool. The application of this method has resulted in the impressive scientific and technological advance of the last few centuries. For a brief but accurate definition of this method one might say that *the scientific method is a procedure to acquire knowledge by means of objective experimentation and inductive inference.* It is the combination of *experimentation* with *inductive inference* that makes the scientific method different from traditional knowledge methods. This is not to say that traditional methods are always invalid. For instance, the ancient Greek philosophers used a combination of inductive and deductive reasoning to formulate their theories of the universe. Many of those ancient theories are now being confirmed via experimentation. The fact that the scientific theories may be *proved* is a fundamental distinction with traditional methods.

In summary, the scientific method may comprise three stages: (1) observations of a particular phenomenon are secured; (2) a hypothesis relating the observations is formulated; and (3) the hypothesis is tested by further designedly critical observations.

As an example, consider the case of some water resources engineer working with the objective of solving the present and future demand of water in a small community. She or he will gather observations and surveys on the current demand of water extrapolated by future estimates of population growth in the locality, as well as historical records of the flow rate in the nearby stream. This constitutes the phase (1) of the method. Next, the engineer will perform an analysis of the streamflow regime as compared to the water demand at the end of the service life of the project, say 50 years. For the times of the year when demand exceeds the natural supply, water coming from a reservoir should balance the deficit. This is the phase (2) of the project, which will conclude with the dimensional definition and specifications of the reservoir, water conveyance systems, water treatment plant, and service reservoirs (i.e., the engineering design). The final stage in the engineering project, the phase (3) of the method, consists in the construction of the water resources facility according to the engineering specifications. The success of the project is measured by an assessment of the ability of the project to satisfy the original objectives.

Similar examples could be mentioned about other engineering disciplines where the scientific method is applied in various forms. For this reason, engineering academic training begins with several required courses in science, such as physics, chemistry, and mathematics. These fundamental courses expose the engineer to the essential scientific laws and concepts with one fundamental assumption behind them: *the universe is governed by deterministic, immutable, and perfectly predictable laws.* However, is the universe truly deterministic? Can phenomena in nature be accurately predicted? We do not pretend to lose ourselves in philosophical discussions, but we would like to offer some insight into the fundamental reason for embarking ourselves in the study of probability and statistics.

Let us expand the previous questions more specifically. Can the engineer who works in flood forecasting predict a storm pattern with certainty? Can the structural engineer predict the loads an element being designed to withstand earthquakes will have? Can the quality control engineer predict with accuracy the number of defective parts in the next shift? Can the maintenance engineer predict the exact date of failure of a critical machine? Can the environmental engineer know with certainty where a contaminant spilled on the ground will go in the subsurface? The answer to these questions is, in most cases, negative. The fact remains that the engineer needs to work with inaccurate, or *uncertain*, data. Fundamental data for engineering projects is often insufficient and subject to various degrees of errors. The insufficiency in the data may be motivated by limitations of the project budget. The errors relate to human faults and the resolution of measurement devices.

If the engineer cannot predict the numerical value of many of the fundamental variables required in the solution, or in the design, of an engineering problem, why is it that most of the education he or she receives in the basic sciences and in subsequent engineering subjects ignores this fact? Except for courses specifically devoted to probability and statistics (usually one course), the vast majority of the engineering curricula adheres to the fundamental assumption that the universe is governed by deterministic and perfectly predictable laws. Indeed, these laws may be easily verified under ideal conditions in controlled laboratory settings. Under those circumstances, the errors and uncertainty in general are minimized and can be neglected. However, the vast majority of real-life engineering applications do not enjoy such an ideal scenario. Thus, the graduating engineer who has not received a proper training in probability and statistics not only has an unreal (i.e., idealistic) conception of the behavior of engineering systems, but also lacks the necessary skills to quantify and design under uncertainty.

The Origins of Uncertainty

From the previous section, engineering applications are faced with the practical problem of designing solutions with data and system parameters subject to uncertainty. There are several sources of uncertainty. One of the most commons is the error in the measurement of data or system parameters. We have stated that engineering practice employs the scientific method based on experimentation. This experimentation rests on the observation of data via the use of measurement devices. These are usually developed to quantify a parameter indirectly by analogy with another variable, which reacts as a function of the former. For instance, the measurement of temperature is based on the quantitative evaluation of the expansion, or contraction, of a column of mercury. This indirect variable (i.e., the length of the column) must be calibrated so as to yield the correct temperature. In this process, the level of resolution of the measurement device (the thermometer) is fixed. If the thermometer is built to measure up to a resolution of $0.1°C$, variability in the temperature of orders of magnitude of $0.01°C$ will not be detected. The resulting error may or may not be significant in the overall application, depending on the system. In a large system that employs hundreds of measurements the errors may be additive. Other measurement errors may be systematic, such as those produced by faulty observations originated in human tasks. Errors in the transcription of data from one type of data base to another may be substantial.

Measurement errors may or may not be neglected depending on the system sensitivity to the errors in question. We speak of an engineering *system* as a device or a model used in the design or in the solution of a particular objective. The system may be the structural design procedure in a project or the mathematical model to forecast a pollutant concentration in a lake. Small errors in the input data or the parameters of a system may result in large variability (i.e., large errors) in the output or dependent system variables. If large errors in the input variables are reduced or dissipated to produce small variability (i.e., small errors) in the output or predicted variables, then the errors may be neglected. At this point, "small" and "large" are subjective statements. The purpose of this book is to show the reader how to objectively quantify this in order to make an intelligent decision.

Other sources of uncertainty are not related to the collection of data and their errors, but rather to the engineering system itself. The engineering system, or the model, is actually an approximation to the prototype. In other words, the system is a *simplified* conceptual device to study the behavior of a much more complex prototype. If the prototype were not more complex, the engineer would study the prototype itself, a task too difficult in most cases. Thus, the model is a simplified abstraction, after certain assumptions are adopted, and after variables too difficult to measure are eliminated from the picture. This elimination of variables produces errors in the predicted variables. For example, the common equations used for the calculation of consolidation in geotechnical engineering are the result of many simplifications that regard the soil as homogeneous in texture and hydraulic properties, infinite in length, and one dimensional. With these assumptions the corresponding differential equations are solved using mathematical methods. The result is a simple procedure to *estimate* the consolidation of a soil underlying a footing under given conditions. It is clear that variations in soil properties that depart from model assumptions will produce errors in the estimation of a dependent variable. The ambitious engineer may reduce the number of assumptions in an attempt to make the model "more realistic." The result of this exercise is, indeed, a more realistic model that inevitably involves more complicated mathematical methods. In addition the data requirements of a more general model are substantially greater. The additional data might not be available, or it may be too expensive for the project in question. Thus, common sense rules in engineering dictate that one must choose the *simplest* model that would satisfy the project objectives, given the technical and economical constraints. The price for this simplicity is errors and uncertainty that should be quantified.

Another source of uncertainty facing engineering systems is related to the environmental fluctuations not detected in our model. Steady systems are chosen in many engineering applications out of necessity (i.e., simplicity). This implies that many truly transient or seasonal systems are modeled as steady systems with an additional increase in errors.

We will mention another type of uncertainty often neglected in engineering. It relates to the errors generated when a system is solved numerically, rather than analytically. This has nothing to do with the assumptions in the model or its ability to reproduce a prototype, but rather with the mathematical method used to solve the corresponding system (differential) equations. Analytical solutions are the preferred, so called "exact," solutions. However, differential equations in heterogeneous and irregular systems are difficult to solve analytically. The engineer instead applies numerical methods to approximate the solution. These methods may be unstable and generate significant errors in spatial or temporal grids of a certain discretization interval. A carefully-designed numerical solution should minimize these errors. However, with the advent of sophisticated graphics-oriented computer software, the applications engineer often does not know the solution method employed, much less the errors associated with it.

Deterministic Versus Probabilistic Analysis of Engineering Systems

From the previous discussion, we now have sufficient grounds to modify the fundamental assumption of a universe as guided by deterministic, perfectly predictable laws. We see that this assumption is incomplete since the existence of measurable, and sometimes significant, uncertainty causes us to question it. Let us expand the previous ideas with a simple law, such as Darcy's law:

$$u = -K\frac{\partial h}{\partial x} \tag{1.1}$$

where
u=specific discharge or Darcy velocity (*m/month*)
K=hydraulic conductivity (*m/month*)
h=hydraulic head (*m*)
x=distance (*m*)

Equation (1.1) governs the movement of groundwater flow in an aquifer (i.e., a porous geological unit saturated with water). This law is easily verifiable in the laboratory by taking a small sample of a homogeneous soil (i.e., constant K), and subjecting it to a known and constant hydraulic gradient, $\partial h/\partial x$. The hydraulic gradient may be approximated with finite differences as $\partial h/\partial x \approx (h_2 - h_1)/\Delta x$, where h_1, h_2 are the heads measured at two points in the soil separated by a distance Δx. After measuring the effluent fluid velocity, one easily confirms the validity of the law.

However, applying this law to field scale aquifers poses a problem of practical significance. The engineer must measure the hydraulic heads in wells located at specified distances. For economic and practical reasons, the separation between wells, Δx, is substantially greater in field applications. This results in a crude linearized approximation of a truly nonlinear gradient. In addition, the estimation of hydraulic heads is subject to the resolution of the stage meter or the analogic transducer employed, as well as other measurement errors discussed in the previous section. Thus, the errors in the measurement of the hydraulic head and the approximation of the gradient may be regarded as *errors in the input data.*

Next, we have the difficulty of estimating the system parameter, in other words, the hydraulic conductivity K. For a soil packed in a cylinder of $5cm$ in diameter and $10cm$ in length, the hydraulic conductivity is easily determined by well-known laboratory procedures. At that scale, the measured value of K is the bulk *average* of the small, microscopic, variability in soil texture. However, in an aquifer of $1000m$ in length, one encounters lenses of clay interlaced by regions of sandy loam, each with drastic changes in their individual values of K, even if the overall *average* texture is classified as sandy loam. The engineer then takes many soil samples and measures the individual values of K. Intuitively, he knows that the higher the number of samples, the more variability will be detected (i.e., a better resolution and accuracy), and the more expensive the project. A compromise between cost and accuracy is finalized and then the problem of errors in measurement of the individual samples needs to be addressed. Thus, the engineer is aware that the estimation of the system parameter has an uncertainty level that depends on the number of samples taken as well as the errors in measurement.

Having collected many samples, the engineer notes that each one has a different value, even though the same protocol was used in the measurement of K. Which value of K should he use? Intuitively, the engineer decides to take an *average* of the K values. After applying equation (1.1) he finds that the calculated value of the velocity, u, is only a mass *average* of the velocity. In verifying his calculations, the engineer measures the effluent velocity in the field only to find that the field-measured u is different from the calculated one. Upon verification of his measurements and calculations, and after concluding they are correct to the best of his efforts, the distressed engineer may conclude that Darcy's law is invalid at the field scale. However, is this really so? Recognizing the existence of possible errors in the determination of the field output velocity, he then decides to repeat its measurement several times. Once again, he notes that each field-measured value of u is different, even though an identical procedure has been used in its determination. Upon calculation of the *average u,* he observes that the average field-based velocity approaches in magnitude to the average model-based one. This coincidence between averages of prototype and average of model variables becomes more accentuated as the number of samples increases. Thus, for a single realization under controlled conditions (i.e., laboratory conditions), Darcy's law is valid within certain (small, unknown, or neglected) errors and certain values of the parameters. For the general case (i.e., the field conditions), Darcy's law is valid in probabilistic form with a high degree of certainty if the number of samples is large enough.

The foregoing example illustrates a typical problem in engineering analysis. Many engineering systems are composed in essence of fundamental laws, which have been proved under specific, ideal, conditions. This is the case of Darcy's law, which has shown to be valid under controlled laboratory conditions for small soil samples and under circumstances where the errors of measurement are so small that they may be neglected. However, when applied to general field-scale conditions the errors in measurement may not be negligible. On the contrary, the measurement errors might be so large that the ability of the system to predict the dependent variable is substantially diminished. In addition, under general conditions, the system parameters may not, and should not, be defined as fixed, constant, or perfectly predictable variables. System variability or heterogeneity, as in the case of the field-scale value of K, is better represented via a set of *repetitive* measurements that characterize such uncertain changes. Thus, under general conditions, the system is subject to uncertainty and the dependent variable has a degree of uncertainty that is functionally related to that of the system. Whether or not this uncertainty is negligible, depends on the level of uncertainty and on the system itself. Some dissipative systems reduce the size of uncertainty in the input variables as it is transformed into an output. Some systems are very unstable and amplify the uncertainty. The engineer must determine the *size* of the uncertainty and perform a sensitivity analysis of the output to various levels of uncertainty in the input. Only a quantitative assessment of the effect of uncertainty on the errors in the predicted variables may constitute an objective basis to neglect certain unimportant variables in order to concentrate on the most affecting ones. Figure 1.1

illustrates the distinction of a deterministic and a random or stochastic system. A deterministic system is fed by unique, perfectly measurable input(s). The input is transformed into unique, perfectly predictable, output(s) via the system equation, which has unique, errorless, parameters. The uncertain or stochastic system, on the other hand, constitutes the general case where the system input(s), the parameters, and thus the output(s) are subject to uncertainty.

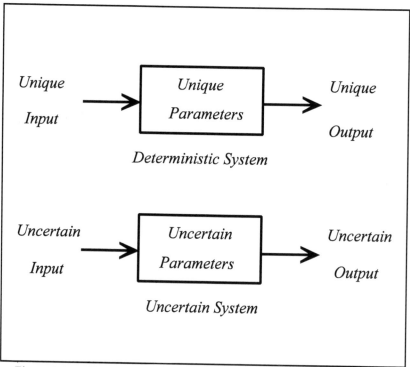

Figure 1.1: Difference between a Deterministic and an Uncertain System

This discussion may be extended to larger, more complex, systems. In the Darcy's law example, the *mathematical model* (or *engineering system*, or *system transfer function*) is described by equation (1.1). The *input variable* or *independent variable* is the hydraulic gradient. The *system parameter* is represented by the hydraulic conductivity K. The *system output* (or *dependent variable*) is the velocity u. In general, an engineering system is composed of several (differential) equations coupled in a specific manner, the input is composed of several variables, the system parameters may be many, and the system output may be made of several variables. Uncertainty in one or several of the system components may propagate, amplify, or dissipate, depending on the system properties. Finally, an engineering system does not have to be made of equations representing physical laws. They could be empirical or approximate equations. The variables involved do not have to be physical quantities. They could be equipment, materials, human resources, money, or intangible benefits.

Let us try to summarize the previous thoughts with some fundamental assumptions:

1. The universe is ruled by fundamental, immutable, laws. For a single realization under specific, controlled, circumstances, these fundamental laws predict a deterministic phenomenon within certain errors and in certain ranges of the dominant parameters. For many realizations, under general conditions, natural and physical laws predict the averages of many trials of a random phenomenon with a high degree of certainty if the number of trials is large enough.

2. There is a difference between a *law* or a scientific concept, and physical reality. The former is a human abstraction, a simplification, of a natural phenomenon. Similarly, there is a difference between an engineering *model* and a *prototype*. The former is a simplification resulting from neglecting certain variables and processes of a much more complex prototype.

3. A *deterministic* engineering system is a special (ideal) case of a more general *stochastic* one. In the former the errors and uncertainty are unknown or neglected. In the latter the uncertainty is acknowledged and quantified.

4. An engineering system subject to errors in measurement in the input variables or in the parameters produces an output subject to uncertainty. Output uncertainty depends on the size of input uncertainty and the system (differential) equation.

5. Uncertainty is always present in an engineering system. The appropriate way to eliminate or neglect variables in engineering analysis is by studying the output sensitivity to the uncertainty level in the said variables. If the magnitude of output uncertainty is acceptable, according to a desired resolution and project specifications, the engineer may neglect them.

We realize some of our readers will have reservations about some of the above assumptions. Many engineers believe probabilistic and statistical analyses in engineering are unnecessary. The argument between deterministic and stochastic analysis relates more to individual beliefs and attitudes than to a scientific rigor on the nature of uncertainty. Some argue that probabilistic analysis is only necessary because of our ignorance. However, from the previous discussion, uncertainty may arise not only from our ignorance, but also from errors in measurement and from deliberate simplifications in the process of building a workable model. Nature does not care whether or not we accept the existence of uncertainty. Uncertainty is always present whether the engineer is willing to acknowledge it or not. The quantitative evaluation of uncertainty is more expensive since it involves repetitive experimentation, but we believe it results in a more honest analysis or design. In other words, it is more honest to tell the customer "I have several uncertainties, but according to my calculations the concentration of radon in your building has an expected value of 175 pCi/L with a standard deviation of $\pm$ 25 pCi/L," than to state "the concentration of radon is 158 pCi/L.." The customer would appreciate the former statement more than if he is given the latter, only to find that subsequent measurements yield completely different results.

One recent objection to probabilistic analysis has been raised by the results of fractal analysis of engineering systems. Seemingly simple nonlinear systems subject to an initial condition that is changed every time the system is run generate an output which seems undistinguishable from a random process. However the output process is not random since it is perfectly predictable by the system. This feature has been used by critics of stochastic analysis to argue that natural processes interpreted as random are in fact deterministic. This would strengthen the idea that stochastic analysis is only necessary because of our ignorance (i.e., we would not need it if we knew the system). However, on close examination of these "chaotic" systems, we find that the key element is the continuous changing of the initial condition of a nonlinear system. Nonlinear systems are extremely sensitive to the initial condition. By repetitive changing of the initial condition, the resulting output shows the typical chaotic behavior. Thus, even if we know the system well, slight deviations in the initial condition generate large variability in the output simply because the system is nonlinear. If the said deviations are natural unpredicted environmental fluctuations, the predicted output is still stochastic.

The preceding discussion does not pretend to solve the historical philosophical discussion between deterministic and stochastic analysis. Rather it invites the reader to think about the issue and to assume a position of practical significance. This position focuses on the recognition that uncertainty exists and must be quantified. The engineer designing a structure to withstand earthquakes or a reservoir to withstand floods is faced with the practical dilemma of what input loads to use in the design. Ignoring uncertainty may be catastrophic at worst, or too expensive at best (i.e., over designing). This book intends to help the engineer deal with the issues of analysis and design under uncertainty in an objective way. For additional discussion on the meaning of probability the reader is referred to Papoulis (1984).

This leads us to the last objection against stochastic analysis: some engineers regard probabilistic data as theoretical, impractical, and useless. Having recognized that we cannot predict the output, in general, in a deterministic way, is it useful to predict it in a stochastic way? Since we are unable to predict if and when a system will fail, is it useful to know the *risk* or the chances of failure? We firmly believe the manager will prefer to know the chances of failure to nothing at all. Knowing there is an 80% probability of failure of a component after five years of operation will indeed help the manager to plan a budget, a schedule of maintenance, and the likely date of replacement. This ultimately relates to the minimization of damages and the maximization of profits. What is an acceptable or an unacceptable risk is a different matter, but the point is clear: probabilistic data is necessary, useful, and practical.

1.2 METHOD OF UNCERTAINTY ANALYSIS

Having recognized the existence and importance of uncertainty in engineering systems, we need to define a procedure to follow in order to quantify uncertainty. Our scenario is an engineering system which has been built in agreement with a set of objectives. For the time being, we limit our discussion to the evaluation of uncertainty of a specific system component. The traditional deterministic approach would be to measure its value once and to assume this number to be true, unique, and errorless. If, on the other hand, we desire to quantify the uncertainty associated with its value, we propose the following steps:

1. Repetitive measurement, or experimentation, *under identical conditions*.

2. Evaluation of Statistical Properties.

3. Building of a probabilistic model.

4. Model application in simulation, prediction, design, or risk assessment.

The first step implies the collection of data. Unlike deterministic data collection, we are now required to repeat our observation of the same parameter or phenomenon over and over. Unable to predict a parameter or system behavior, we decide to poke it many times and observe its behavior in order to find a pattern. The key phrase here is *under identical conditions.* This means that the same ASTM or EPA measurement, testing or laboratory protocol, with identical instrumentation, procedure, etc., must be used each time. By repeating our observations under identical conditions, we are eliminating other deterministic effects that might affect the system. We want to measure variability due to chance and in that effort intervening deterministic variables must be eliminated. For instance, if one repeats the procedure with different instrument types, we are measuring the effect of different instrument resolutions besides that of uncertainty. The meaning of "repetitive experimentation" differs according to the system or process in question. In many instances the "experiment" is simply the observation of a phenomenon with time. For example, repetitive "experimentation" in flood uncertainty means going back through the records and determining the maximum flood peak magnitude per year over many years of records. "Identical conditions" in this case means collecting data at a unique streamflow station. The data should have been collected by only one institution, which hopefully followed a uniform measurement procedure. By taking the maximum value per year (i.e., maximum values separated some months in time), we are minimizing other deterministic effects, such as the fact that values closely spaced in time respond strongly to direct variations in rainfall, etc. In other words, the data collected appears to be a set of random maximum flows.

As an alternative example, the engineer collecting information at a busy traffic intersection would measure the number of vehicles of certain kind at the peak hour during a normal week day. By proceeding to measure the number of vehicles at the peak hour only, the engineer is minimizing the effect other deterministic variables that may be affecting the data. If she chose to measure at different hours of the day, the data is biased by the (deterministic) fact that certain hours of the day exhibit more traffic volume than others. We will return to this issue, which is often overlooked in the design of experiments for data collection. What the engineer must keep in mind is that only *part* of the system component is random. The correct experiment measures this part only after eliminating all other intervening variables. The correct engineering model is composed of one deterministic part corresponding to the fundamental laws governing the processes under study and one stochastic or random part accounting for the associated uncertainty.

Once random data has been collected, the engineer proceeds to conduct step two in uncertainty analysis: evaluate the data statistical properties. These are mathematical measures, called *statistics*, which summarize and describe the uncertain data. For example, the most common statistics are the mean, or the average, and the standard deviation. The former describes the most commonly occurring value; the latter describes the average deviation from the mean. The statistics summarize the data: instead of having a collection of say 100 data points, each different in magnitude, now we have two numbers that summarize the uncertainty properties of the data. There are many statistical measures to calculate from a data set, each with a specific meaning and purpose. These analyses constitute the object of the field of statistics. This is a field of *deductive* thinking: based on a series of experimental data, one attempts to conclude its characteristics and properties.

Step three in uncertainty analysis consists in conceiving a general mathematical law, or probabilistic model, that conceptually and quantitatively describes the uncertain behavior of the system parameter or phenomenon in question. Based on the statistical analysis of step two, the engineer attempts to fit a probabilistic model that in theory describes the data. In this task, she finds a variety of existing models which engineers have found to be good models for similar sets of data. At this stage of the analysis, the data might be characterized as realizations from a random variable with a specific functional form. Under unusual circumstances (e.g., in the research of uncommon phenomena), the engineer may find that none of the existing models reasonably describes the data. In this case, she needs to find a mathematical function that achieves this purpose. In summary, step three constitutes the object of the field of probability. This is a field if *inductive* thinking: based on statistical properties and data from a phenomenon, one attempts to infer the conceptual law (i.e., the principle) that could generate all possible sets of data of such phenomenon.

From above, we may affirm that the field of uncertainty analysis constitutes an ideal example of the scientific method: based on experimental data of a phenomenon (step 1), we apply deductive thinking to conclude the statistical properties of the data (step 2). Next, we apply inductive thinking to infer and generalize a probabilistic law that describes the uncertain behavior of the phenomenon (step 3). The last step of uncertainty analysis (step 4), is the application of the knowledge gained in the previous steps. This is the realization of the objective of most engineering projects, that is, the reason why we embark ourselves in uncertainty analysis.

Step four takes many forms depending on the engineering project. For instance, for an environmental engineer interested in the possible consequences of a contaminant release in a lake, the application of the model may consist in the numerical estimation of the risk associated with a particular contaminant concentration. In other words, the engineer will compute the probability associated with a concentration that exceeds the maximum allowable EPA drinking-water norm. The traffic engineer, on the other hand, may use his model to forecast future probable realizations of traffic volume in order to test a projected intersection. In any case, the risk values or realizations generated help the engineer or the manager take appropriate decisions. In certain circumstances (e.g., water resources' systems), probabilistic values are specifically used in the design of engineering structures or engineering solutions: the structure is designed to withstand an adverse load or a condition associated with a given low probability of occurrence. Given a probability of occurrence, the engineer must find the numerical value of the load or condition corresponding to the given risk.

The next chapters of this book are devoted to a detailed description of an integrated (probability and statistics) method of uncertainty analysis introduced here. The following examples have the purpose of illustrating some of the concepts discussed in this chapter, as well as introducing the reader to the applications of mathematics software in uncertainty analysis, a tool we will use throughout this book. The readers not familiar with algebra software, and especially those not familiar with computer programming, should pay particular attention to these examples.

Example 1.1: Effect Input and Parameter Measurement Errors on System Output. A Computer Simulation

Consider a groundwater system governed by equation (1.1). Assume the hydraulic gradient is calculated from head measurements at two wells, h_1 and h_2, separated by a distance $\Delta x = 100m$. Based on several samples, the mean hydraulic head at wells 1 and 2 is $125m$ and $122m$, respectively. The mean hydraulic conductivity is $50m/month$. Investigate the effect of errors with a standard deviation of 1% of the mean head and of 5% of the mean conductivity. Specifically, estimate the standard deviation in the specific discharge. Assume the errors follow a Normal distribution.

Solution

We use Maple software for the simulations of measurement errors. The Maple worksheet has a prompt ">" after which the user types a command followed by a semicolon. When the Enter key is pressed, Maple executes the command and displays the output. Alternatively, if the user wants Maple to execute the command without displaying the output (as when generating thousands of random numbers), a colon must be typed at the end of the command, instead of a semicolon. The worksheet below illustrates the simulation of errors. Each command is preceded by an explanation comment.

[Start the statistics module.
[> with(stats):
[Define variables. Comments after the pound sign are ignored by Maple.
  Note a variable is defined by a name (a combination of letters and characters without spaces),
  followed by a colon, an equal sign, and its value (or its equation).
[> deltax:=100:   #Distance between wells.
[> h1_mean:=125:   #Mean head at well 1.
[> h2_mean:=122:   #Mean head at well 2.
[> K_mean:=50:   #Mean conductivity.
[> N:=20:   #Number of data points.
[> Sh1:=125*0.01:   #Standard deviation of head at well 1.
[> Sh2:=122*0.01:   #Standard deviation of head at well 2.
[> Sk:=50*0.05:   #Standard deviation of conductivity.
[Using the random[ ] command, generate N Normally-distributed random numbers
  representing head measurements at well 1. The subcommand normald[h1_mean,Sh1]
  requests Normal numbers with a mean and standard deviation specified by the values
  of the variables h1_mean and Sh1.
[> random[normald[h1_mean,Sh1]](N):
[Store these heads in an array of numbers [, , ,...] named h1_data.
[> h1_data:=[%]:   #The % means the "previous" command.
[With one command only, generate head measurements at well 2 and store them in an
  array caled h2_data. Make sure the number of opening brackets, [, match those closing,].
[> h2_data:=[random[normald[h2_mean,Sh2]]](N):
[Generate N samples of conductivity data.
[> K_data:=[random[normald[K_mean,Sk]]](N):
[Calculate the hydraulic gradient data by subtracting the array h1_data
  from the array h2_data and dividing by deltax.
[> dh_dx:=(h2_data-h1_data)/deltax:
[Using the seq(,) command, sequentially apply equation (1.1) to calculate the output
  velocity samples, the subindex i represents the position of the data value within an array.
  For example dh_dx[3]=-.1461151153.
[> seq(-K_data[i]*dh_dx[i],i=1..N):
[Store in an array.
[> u_data:=[%]:
[Calculate the mean velocity using the describe[ ] command.
[> describe[mean](u_data);

                      1.149354900

[Finally calculate the velocity standard deviation using the describe[ ] command.
[> describe[standarddeviation](u_data);

                      0.8536856761

Several notes are necessary at this point.

1. Except for the last two, all commands end with a colon, in which case, the command is executed but the output is not displayed. This saves space, but the reader is strongly advised to end all commands with semicolons the first time (see Problem 1.4). That way the user may check the output from each command for correctness.

2. Command syntax is very important. Each command must be typed exactly. Maple distinguishes between uppercase and lowercase letters. We recommend the reader type these commands and execute the entire worksheet. In addition, the reader should read the online description of the new commands, such as random[ ] and seq(,). This could be done by typing a question mark followed by the command name after the prompt. For example, type "?random" after the ">" prompt and press enter. Alternatively, one may use the help menu on top of the worksheet.

3. Generation of random numbers from a particular probability distribution (in this example the Normal distribution) constitutes the heart of Monte Carlo simulation models in uncertainty analysis. These computer-generated random numbers represent realizations of the uncertain phenomenon in question. In this example the random numbers represent sample measurement errors.

4. Arrays are used to store sets of data of one kind under a given name. For instance, the array of values $X = [0.33, 1.42, 3.22, -1.55]$ has four values. Using algebra notation, the individual values X_i, $i=1, 2, 3, 4$ are $X_1=0.33$, $X_2=1.42$, etc. Similarly, in Maple notation $X[1]=0.33$, $X[2]=1.42$, etc.

5. Since h1_data and h2_data are one-dimensional arrays, in the calculation of the hydraulic gradient Maple subtracts h2_data[1] from h1_data[1]; the subtraction is divided by the constant deltax and the result is stored as dh_dx[1]. This operation is repeated for the second term in each array, then the third, etc. For the reader familiar with linear algebra, this is simply the subtraction, term by term, of two one-dimensional matrices.

6. The calculation of the velocity values in the vector u_data, requires the product of K_data[1] times dh_dx[1]. The product is multiplied by -1 and the result is stored as u_data[1]. Next, K_data[2] is multiplied by dh_dx[2], etc. This term by term product is different

from the product of the two vectors as defined in linear algebra. Thus, the modeler needs to specify this procedure via a command that sweeps the arrays and performs the operations. This is achieved with the sequence command, " seq(,)," which is similar to a Do loop in the Fortran programming language, or a for command in the C language. The reader familiar with C should know that Maple allows the inclusion of portions of the worksheet to be written in C.

7. The describe[ ] command has many applications. In this example we use it to estimate the mean and standard deviation of an array of values, which provide us with a statistical description of the system output. In the next chapters we will return to the basic concepts behind moments (see Chapter 3).

Example 1.2: Comparison of System Input and Output Levels of Uncertainty

In Example 1.1, compare the levels of uncertainty of the input variables with respect to that of the output variable. Rationalize the effect of errors on system output.

Solution

To compare levels of uncertainty, we define a normalized measure of variability as the ratio of the standard deviation to the mean, called the *coefficient of variability*, C_v. For the hydraulic head data, the coefficient of variability is 1%, or 0.01; for the conductivity data, it is 5%. According to the results in Example 1.1, the average output velocity is about $1.149 m/month$ and its standard deviation is about $0.854 m/month$, which translates into a coefficient of variability, $C_v = 0.854/1.149 = 0.743$ or 74%.

Thus, in this system, seemingly small errors of measurement in the input variables (e.g., 1% in the head, and 5% in the conductivities) produce a large error in the output velocity. A 74% error in the output variable might substantially compromise our confidence in the ability of the system to forecast velocities. The output errors are substantially amplified. On close examination of the system equation (1.1) one finds the reason: the system equation has a derivative of an input variable. Differentiation, as opposed to integration, in the presence of uncertainty is an unstable procedure. The errors could increase to the point that individual estimates of the output variable are physically unrealistic. For instance, by increasing the errors in head to 5% (see Problem 1.6), the resulting coefficient of variability exceeds 100%, and some of the generated values of velocity would be negative since the sign of the derivative changes. The engineer unaware that this is an effect of uncertainty would incorrectly conclude a reverse direction in the flow.

Errors of 5%, or even 10% (see Problem 1.7) in conductivity do not seem to cause such a dramatic effect. Hence, to reduce uncertainty in the output, the engineer should focus on improving the measurement accuracy of heads. This is a perfect example of the application of uncertainty analysis in management decisions.

The previous exercises were conducted by generation of random realizations of a random process. We used a random number generator to simulate the said realization of measurements. We will return to this technique when we cover Monte Carlo simulation (see Chapter 4). We will also see that there are analytical techniques to estimate the output statistics without having to relay on thousands of numerical calculations.

QUESTIONS AND PROBLEMS

1.1 Consult a book on the scientific method or epistemology. Write a page (typewritten, single space) with a more detailed description of the scientific method to obtain knowledge. Mention other theories of knowledge and briefly describe their basis. Emphasize the fundamental difference between the different theories. Provide a complete bibliography.

1.2 Consult a book that discusses philosophical theories about the universe. Write a page describing the essence of the theories that conceive the universe as deterministic. Contrast these arguments with those that support a stochastic nature. Discuss the concepts of order, chaos, and fractal dynamics. Provide a complete bibliography.

1.3 Mention 5 applications where probability and statistics are necessary in your particular engineering field or major.

1.4 Rerun Example 1.1 using Maple, this time displaying each command output (i.e., changing the colon by a semicolon at the end of each command). If you do not have Maple, convert the program to a similar mathematics software to which you have access. If you obtain slightly different statistics (velocity mean and standard deviation), speculate on the possible reason.

1.5 Rerun Example 1.1 with 100 samples, instead of 20 and calculate the values of the velocity mean and standard deviation. Repeat for 1000, 2000, and 5000 samples (do not show the output). Write the results in a table. State how do the statistics change with sample size?

1.6 Rerun Example 1.1 with 2000 samples, increasing the coefficient of variability of the head to 5%. Show the velocity mean, standard deviation, and coefficient of variability. State your conclusions (i.e., explain the meaning of these numbers in this situation).

1.7 Rerun Example 1.1 with 2000 samples, increasing the coefficient of variability of conductivity to 10%, while keeping that of the head at 1%. Calculate the mean, standard deviation and coefficient of variability of the output velocity. State your conclusions.

"...We were asking ourselves: is it possible to draw in our spirit a true map of the universe, navigate according to this map towards definitive ends, and reach our chosen port? I believe our answer should be the following: human thinking cannot draw a precise map of the universe. It cannot set as goal the far and mythical shores of the kingdom of Utopia, but like the ancient navigators, it can bravely go from shipwreck to shipwreck, and from archipelago to archipelago, using the knowledge acquired by its ancestors, about the immutable constellations and about the capricious storms, completing this ancestral wisdom with actual experience, and observing the stars, the tides and the winds. This is sufficient, and the prudent Ulysses did not ask anything else from the gods."

André Maurois (1885-1967). "An Art of Living." *SpiralPress*, Ambler, PA, English translation edition by Sergio E. Serrano, 2007.

2 THE CONCEPT OF PROBABILITY

2.1 CHANCE EXPERIMENTS AND THEIR OUTCOMES

The fundamental concept of probability results from the idea of a *chance experiment*. This situation arises every time we face a phenomenon of an unpredictable nature. As explained in Chapter 1, this experiment may or may not be actually performed. The "experiment" may consist in the observation of the said phenomenon over a period of time. In order to fulfill the requirements of uncertainty analysis, this repetitive experimentation, or repetitive observation, must be performed under identical conditions so as to measure the manifestations of uncertainty, rather than other uncontrolled deterministic factors. For example, in the study of the performance of a screw attachment in the tail wing of a passenger airplane, the engineer needs to select all airplanes of the same class that have been equipped with the same part manufactured with the same specifications by (hopefully) a unique company. These selected planes should have been subjected to similar flying conditions and frequency, and similar maintenance protocols. By this systematic selection of samples, the effect of deterministic factors such as different design features, different shapes, and different materials are eliminated from the analysis. The experiment will then record the number of service hours of the part prior to failure. The numbers obtained are samples from the random variable "time to failure of a screw attachment in the tail wing of a passenger airplane."

Let us denote a chance experiment with boldface uppercase English letters, and its possible outcomes with Greek letters. For example, the experiment E_1 = number of vehicles per hour at a key traffic artery on a Monday at 8:00AM. More in detail, $E_1 = \{\xi_i, i=1, 2,..., N\}$, where $\xi_1 = 0$ vehicles, $\xi_2 = 1$ vehicle, $\xi_3 = 2$ vehicles, etc., and N is the theoretical maximum number of vehicles per hour the intersection can handle. As another example, consider the experiment E_2 = sunny or a cloudy day at a location on June 1st of a year. Prior to this, we declare a "sunny" day if the number of sunny hours, as measured by the heliometer was greater than 6, and "cloudy" otherwise. Thus, $E_2 = \{\xi_1 = Sunny, \xi_2 = Cloudy\}$. Unlike the previous experiment, this one has two possible outcomes only.

Consider now a slightly different experiment: E_3 = selecting 2 valves from a manufacturing line and assessing them as good, g, or bad, b. The quality engineer now takes 2 samples of the same production line. He then tests each one as to their ability to satisfy certain predetermined quality standards and judges them as either "good" (satisfactory), or "bad" (defective). The possible outcomes of this experiment are $E_3 = \{\xi_1 = gg, \xi_2 = gb, \xi_3 = bg, \xi_4 = bb\}$, where $\xi_1 = gg$ is the outcome first sample is good and second is good, $\xi_2 = gb$ the outcome first sample is good and second is bad, etc. Similarly, we could define the experiment $E_4 =$ selecting 3 valves and assessing them for quality. The possible outcomes are $E_4 = \{ggg, ggb, gbg, bgg, gbb, bgb, bbg, bbb\}$, where the outcome ggg implies that the first sample is good, the second sample is good, and the third sample is good, etc.

2.2 ESTIMATION OF THE NUMBER OF POSSIBLE OUTCOMES

Simple Enumeration

Defining the possible outcomes of an experiment is obvious when the sample size and the type of possible selections are small. In these cases a *tree diagram*, such as the one in the following example is useful.

Example 2.1: Inspection of Hydraulic Valves

In the previous experiment E_4 selecting 3 valves and assessing them for quality, construct a tree diagram representing the possible outcomes. Calculate the number of possible outcomes by simple enumeration.

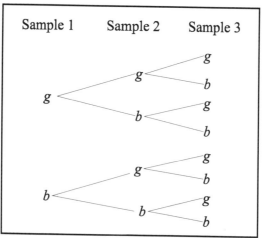

Figure 2.1: Tree Diagram of Example 2.1

Solution

We are selecting 3 samples. The first selection may be g or b. Each of these outcomes may be associated with 2 more samples, each having

2 possible outcomes. Graphically, we may represent the 3 samples in 3 columns (i.e., one per sample) and drawing lines between the outcomes in each column to represent the possible ways of arrangement (see Figure 2.1). A particular path in the tree indicates a possible outcome after 3 samples have been collected. The total number of possible outcomes, N, is obtained by counting the number of letters in the third column. Clearly, $N = 2 \times 2 \times 2 = 8$.

Example 2.2: Highway Base Compaction

A contractor has 3 types of machines, a, b, or c, to test the level of compaction of a highway base. When a pavement sample is taken, the machine measures its dry bulk density. The sample is judged as "compacted," c, or "non-compacted," n, depending on wether or not the sample density is greater than a set value. The sample is then subject to a measurement of water content, in which it is judged to be "dry," d, "medium," m, or "wet," w. If a test, T, consists in taking one sample with either machine, define the set of possible outcomes and calculate its number.

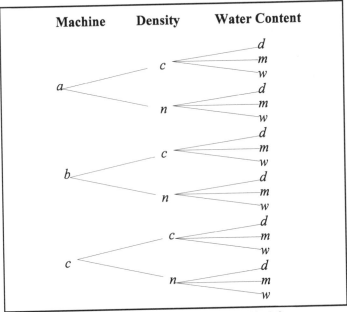

Figure 2.2: Tree Diagram of Example 2.2

Solution

Figure 2.2 shows the tree diagram of the experiment. The set of possible outcomes is $T=\{acd, acm, acw, and, anm, anw, bcd, bcm, bcw, bnd, bnm, bnw, ccd, ccm, ccw, cnd, cnm, cnw\}$. The number of possible outcomes is obtained by counting the elements in the set, but since we have 3 machines, 2 densities, and 3 water contents, it is easy to see that $N=3\times2\times3=18$.

Permutations: Sampling in a Specific Order Without Replacement

The basic principle of counting possible outcomes is this: if one experiment has n possible outcomes and another experiment has m possible outcomes, the total number of possible outcomes from the combination of the two experiments is $N=nm$. Generalizing to several experiments, the total number of possible outcomes is the product of the possible outcomes from each experiment. This result holds if the sampling at one time does not affect the possible outcome at the next. For example in the E_3 experiment of selecting 2 valves, the selection of the first valve as "good," g, will not deter the possibility that the selection of the second might be g as well. In other words, we are *sampling with replacement*. Example 2.3 illustrates the case of *sampling without replacement*, that is when we have a limited set of possible outcomes. By selecting one outcome at one time, we are withdrawing that outcome from the pool of possible outcomes at the next selection.

Example 2.3: Computer Server Login Passwords

The server of a computer network requires a 6-character login password. Valid characters are any of the letters of the alphabet or any digit from 0 to 9 as long as they are not repeated. How many passwords are possible?

Solution

As the user types one character he has 36 possibilities from which to choose (26 letters plus 10 single digits from 0 through 9). Once the user types a character (i.e., selects one), he will have 35 possibilities for the second slot, since characters may not be repeated. Thus, the number of possible ways to arrange 36 characters choosing 6 at a time is

$$\frac{36!}{(36-6)!} = 36\times35\times34\times33\times32\times31 = 1,402,410,240$$

Example 2.3 also illustrates the concept of sampling in a specific order. A *permutation* is a way in which we can arrange a set of objects when the *order is important* (e.g., in Example 2.3 the passwords *abcdef* and *acbdef* are different even though they both contain the same letters). If we have N objects, we can permute them in $N!$ ways. If we have a permutation of size $m<N$, where N is the total number of objects from which we make our selection, as in Example 2.1, then the number of possible permutations is:

$$P_m^N = \frac{N!}{(N-m)!} \qquad (2.1)$$

where
P_m^N =number of permutations of N objects choosing m at a time.

Combinations: Sampling without a Specific Order and without Replacement

A *combination* is a permutation in which the *order is not important*. The following example illustrates this concept.

Example 2.4: Pumps at an Underground Storage Tank

An underground gasoline tank at a service station has a network of 6 main outlet pumps. Under minimum demand, only 2 pumps are working. In how many ways can the pumps be arranged?

Solution

We are arranging sets of 2 pumps out of 6 in total. Lets call the pumps $p1, p2, p3, p4, p5$, and $p6$, respectively. By direct enumeration we can start arranging them in pairs as $\{p1p2, p1p3, p1p4, \dots \}$. Since the pumps may not be repeated, in the first selection we have 6 pumps to choose from and for the second selection we have 5 pumps to choose from. Thus, the number of ways to arrange them is 6×5 if the order were important. However, since order is not important, each group of 2 pumps would be repeated twice since $p1p2$ is the same as $p2p1$, $p1p3$ is the same as $p3p1$, etc. It follows that the number of combinations of 6 pumps taking 2 at a time is

$$\frac{6!}{(6-2)!\times2!} = 15$$

In general, the number of possible combinations of N objects choosing m at a time is given by

$$C_m^N = \binom{N}{m} = \frac{N!}{(N-m)!m!} \qquad (2.2)$$

where

C_m^N =number of combinations of N objects choosing m at a time.

Equation (2.2) is the *binomial coefficient* in algebra. A generalization of the binomial coefficient is the *multinomial coefficient*, which arises when arranging N objects into k classes of sizes m_1, m_2, ... , m_k, where $\Sigma m_i = N$. For the first class there are $C_{m_1}^N$ possible choices. For each choice in the first class there are $C_{m_2}^{N-m_1}$, and so on. By using the principle of multiplication, the total number of choices is

$$C_{m_1}^N C_{m_2}^{N-m_1} C_{m_3}^{N-m_1-m_2} ... C_{m_k}^{N-m_1-...-m_{k-1}} = \frac{N!}{m_1!m_2!m_3!...m_k!} \qquad (2.3)$$

Example 2.5: Human Resource Management at an Assembly Line

An assembly line is composed of 3 main tasks. Task 1 requires 7 workers, Task 2 requires 5 workers, and Task 3 requires 3 workers. How many possible ways can the manager assign the available 15 workers among the 3 tasks?

Solution

Using the multinomial coefficient, the total number of possible ways is

$$\frac{15!}{7!5!3!} = 43,243,200$$

2.3 SAMPLE SPACE AND EVENTS

For illustration purposes, it is convenient to represent chance experiments in terms of a *sample space, S*. The sample space constitutes the totality of all possible outcomes. In Figure 2.3 we represent as a rectangle the previous experiment E_3 = selecting 2 valves from a manufacturing line and assessing them as good, g, or bad, b. Each point in the sample space represents an outcome. A subset of the sample space is called an *event*. We will represent events as uppercase English letters. For example, we define the event E_1 of experiment E_3 as E_1 = at least one defective valve. The possible outcomes are $E_1 = \{gb, bg, bb\}$. In Figure 2.3 E_1 is represented as a triangle. These graphs are called *Venn diagrams*. In experiment E_4 =selecting 3 valves, we may define the event $E_2 = 2$ defective valves exactly. The event possible outcomes are $E_2 = \{gbb, bgb, bbg\}$. Alternatively, in the same experiment, we may define the event E_3 = at least 2 defective valves, and the event outcomes are E_3 = {*gbb, bgb, bbg, bbb*}. Note the difference between events E_2 and E_3. The language used in the definition of events is very important.

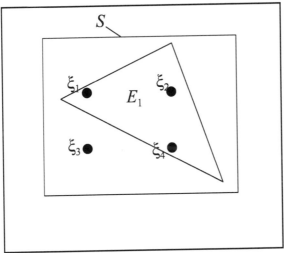

Figure 2.3: Venn Diagram of Outcomes
and Events of Experiment $\boldsymbol{E_3}$

Here we are concerned with the definition of the possible outcomes of a given experiment, and the possible outcomes of a given event. Why an experiment or an event is defined in a particular way is another matter. An experiment must be designed to satisfy a specific project objective known *a priori*.

2.4 QUANTITATIVE EVALUATION OF PROBABILITY

The Frequency Definition of Probability

In the previous sections, we stated that an experiment is constructed by defining the possible outcomes of a phenomenon subject to chance. The totality of possible outcomes constitutes the sample space, S. Thus, one needs to state as precisely as possible the phenomenon subject to chance. In the examples given above, we named an experiment with an uppercase bold letter and assigned to it a verbal description of the phenomenon or system being observed (e.g., number of vehicles at an intersection, the selection of 2 valves, the selection of 3 valves, the collection of 6 soil samples, etc.). Next, we specified the set of possible outcomes from the experiment and provided mathematical methods to calculate the number of outcomes according to the sampling procedure.

Once the experiment and the sample space had been defined, we proceeded to define certain events, or subsets of S, which had a particular meaning to the engineer (e.g., at least two parts are defective). We named those events with uppercase letters and defined the set of outcomes that satisfied their conditions.

The next step is to evaluate the probability that an event occurs when the experiment is executed *once*. For instance, in the experiment E_d= sunny or a cloudy day at a location on June 1st of a year, we may be interested in evaluating the probability that there will be a sunny day on June 1st, 2012. To that effect, define the event S_1 = day is sunny. Now the engineer observes the historical weather records for the past $N=25 years$ (i.e., repetitive experimentation) and notes that, out of those observed years, $n_f=15 years$ were sunny. Hence, 15 out of 25$years$ satisfied S_1 or 60%. If the engineer *assumes* that the number of years of observation, called the sample size N, is *representative* of the weather regime in the area, then she might conclude that the probability of a sunny day on June 1st of the year 2012 (or of any year for that matter) is $P(S_1)=0.6$ or 60%. This example summarizes several principles:

1. A probability is a number that estimates the chances of an event when the experiment is executed *once* in the future (e.g., the probability of a sunny day on a given day in the future).

2. To calculate the probability of an event, it is necessary to recourse to repetitive experimentation (e.g., the observation of the historical records), as described in Chapter 1.

3. There is an implicit *assumption* that the number of samples is representative, that is it characterizes the frequency behavior of the event. This is not the case if one takes an insufficient number of samples. As the sample size increases, the probability value approaches a constant.

4. There is also an assumption of stationarity involved in the calculation of probability. In other words, the engineer is assuming that the frequency behavior, as estimated by the observations of past records, is not going to change in the future. This assumption is incorrect if there are any changes in climate or other unaccounted deterministic factors.

Let us now state the fundamental concept behind the *relative frequency* definition of probability:

$$P(E)=\lim_{N\to\infty}\frac{n_f}{N}$$ (2.4)

where
  $P(E)$=probability of an event E
  n_f =number of favorable outcomes (i.e., satisfying E)
  N=number of trials in the experiment

The relative frequency definition of probability is intuitive, but not rigorous. In general, the experiment is not repeated an infinite number of times. Therefore, the limit at infinity is a conceptual generalization of the idea that as the sample size increases equation (2.4) approaches a constant, but it has no proof.

Example 2.6
  In experiment E_3=selecting 2 valves from a manufacturing line and assessing them as good, g, or bad, b, what is the probability of the event E_1=selecting at least one bad, b, valve?

Solution
  The Venn diagram of this event is illustrated in Figure 2.3. The possible outcomes of this experiment are E_3={gg, gb, bg, bb}. The possible outcomes of the event are E_1={gb, bg, bb}. Out of the N=4 possible outcomes, n_f=3 contains at least one bad valve. Thus, *assuming each outcome is equally likely to occur,* the probability of at least one bad valve is

$$P(E_1)=\frac{n_f}{N}=\frac{3}{4}=0.75 \quad or \quad 75\%$$

Example 2.7
  In the experiment E_4=selecting 3 valves from a manufacturing line and assessing them as good, g, or bad, b, what is the probability of the event E_2=2 defective valves exactly?

Solution
  The possible outcomes of this experiment are E_4={ggg, ggb, gbg, bgg, gbb, bgb, bbg, bbb}. The event outcomes are E_2={gbb, bgb, bbg}. Out of the N=8 possible outcomes, n_f=3 contain 2 bad valves exactly. Thus, *assuming each outcome is equally likely to occur,* the probability of 2 bad valves is

$$P(E_2)=\frac{n_f}{N}=\frac{3}{8}=0.375 \quad or \quad 37.5\%$$

Example 2.8

Recall Example 2.3, in which the server of a computer network required a 6-character login password consisting of any of the letters of the alphabet or any digit from 0 to 9 as long as they were not repeated. If the user is allowed to type an incorrect password three times, after which the system will deny access, what is the probability that a user will guess a valid password if there are 2,000 registered users?

Solution

From Example 2.3, the number of possible passwords is $N=1,402,410,240$. Assuming the user will attempt 3 times after which he will give up, then the number of possible successes is the number of registered users times the number of attempts (i.e., sampling times), or $n_f=2,000\times3=6,000$. Thus, the probability of guessing a password is $6,000/1,402,410,240=0.00000428$. Clearly, this is a very unlikely event.

Fundamental Axioms of Probability

The preceding section introduces us to the axiomatic definitions of probability.

1. The probability of an event E of a sample space S is a positive number between 0 and 1:
$$0\le P(E)\le 1 \tag{2.5}$$

2. The probability of the sample space is 1 (the certain event):
$$P(S)=1 \tag{2.6}$$

3. If two events E_1 and E_2 have no common elements (i.e., they are "*mutually exclusive events*"), then the probability of E_1 or E_2 (or both) is given by
$$P(E_1 or E_2)=P(E_1 \cup E_2)=P(E_1)+P(E_2) \tag{2.7}$$

where $E_1 \cup E_2$ denotes the union of the sets E_1 and E_2.

Example 2.9

In the experiment E_4=selecting 3 valves (see section 2.1), calculate the probability of zero or one defective parts.

Solution

Recall that the possible outcomes of this experiment are E_4={ggg, ggb, gbg, bgg, gbb, bgb, bbg, bbb}. Define the event G=zero defective valves and the event B=one defective part. From the explicit enumeration of outcomes, G={ggg}, and B={ggb, gbg, bgg}. From equation (2.4), *and assuming each outcome is equally likely to occur*, $P(G)$=1/8, and $P(B)$=3/8. Since G and B are mutually exclusive events (i.e., they do not have common outcomes), then from equation (2.7), $P(G \cup B)$=1/8+3/8=1/2 or 0.5. Figure 2.4 shows the Venn diagram of this problem.

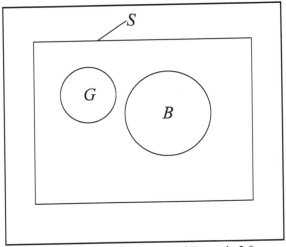

Figure 2.4: Venn Diagram of Example 2.9

Probability Axiom 3, as represented by equation (2.7), may be generalized to the case of n events. Thus, if n subset events E_1, E_2, ... , E_n of a sample space S, such that each pair E_i, E_j are mutually exclusive (i.e., have no common elements) for all $i \neq j$, then

$$P(E_1 orE_2 or...orE_n)=P(E_1 \cup E_2 \cup...\cup E_n)=\sum_{i=1}^{n} P(E_i) \qquad (2.8)$$

Sometimes events in a sample space are not mutually exclusive (i.e., they have common elements). The Venn diagram of this situation is depicted in Figure 2.5. Event A is a subset of S (or $A \subset S$, where $\subset$ denotes "subset"); event $B \subset S$; and the subset of common elements of A *and* B (or $A \cap B$, where $\cap$ denotes "intersection") is represented as the shaded common area between A and B. If we apply equation (2.7) to calculate the

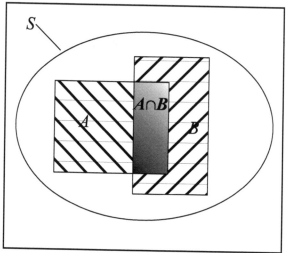

Figure 2.5: Venn Diagram of Events
with Common Elements

probability of A *or* B (or both), we see that the area $A \cap B$ is added twice. The correct calculation must subtract this area once. Thus, for events that have common outcomes, equation (2.7) becomes

$$P(A \, or \, B) = P(A \cup B) = P(A) + P(B) - P(A \cap B) \tag{2.9}$$

Example 2.10

In experiment E_4=selecting 3 valves (see section 2.1), determine the probability of "at least one defective valve or two defective valves exactly."

Solution

Recall that the possible outcomes of this experiment are E_4={ggg, ggb, gbg, bgg, gbb, bgb, bbg, bbb}. Let us call A the event "at least one defective valve" and B the event "two defective valves exactly." Knowing that we have N=8 possible outcomes, and assuming that each outcome is equally likely, we see that A={ggb, gbg, bgg, gbb, bgb, bbg, bbb} and $P(A)$=7/8. Similarly, B={gbb, bgb, bbg}, and $P(B)$=3/8. On the other hand, A and B have common elements:

$A \cap B = \{gbb, bgb, bbg\}$, with $P(A \cap B) = 3/8$. Thus, from equation (2.9), $P(A \cup B) = P(A) + P(B) - P(A \cap B) = 7/8 + 3/8 - 3/8 = 7/8$.

Example 2.11: Stock Planning at a Distribution Center.

An office-supply store has 3 types of computers for sale: laptop microcomputers, desktop models, and network stations. If there are 10 items of each computer type in stock, what is the probability of the event W that the next 5 orders are from the same computer type?

Solution

Here we have a more elaborate example of the rules of counting possible outcomes and the assumptions behind probability. First, we need to *assume* that the next 5 customers have no particular preference, or that there is no price incentive or any deterministic factor that would make one choice more likely than another. In other words, all the outcomes are *equally likely* to occur and the selection is at random. By formulating this assumption, the engineer becomes aware of the limitations of the model. If these assumptions are not realistic, the following calculations are only an approximation.

There are 30 computer units in total. The number of possible outcomes N is the number of combinations of 30 items choosing 5 at a time:

$$N = C_5^{30} = \frac{30!}{(30-5)!5!} = 142,506$$

Let us call the events L=selecting 5 laptop models; D=selecting 5 desktop models; and N_n= selecting 5 network models. The number of outcomes of L is simply the number of combinations of 10 items choosing 5 at a time. Thus, from equation (2.4), the probability of L is the ratio of the outcomes of L to the total number of outcomes:

$$P(L) = \frac{C_5^{10}}{N} = \frac{\dfrac{10!}{(10-5)!5!}}{142,506} = 0.00177$$

Since there 10 items of each computer type, then $P(D) = P(N_n) = P(L) = 0.00177$. In this case there are no common items to events L, D, and N_n. Therefore, equation (2.8) yields $P(W) = P(L) + P(D) + P(N_n) = 3 \times 0.00177 = 0.00531$.

Example 2.12

Recall from Example 2.11 that an office-supply store has 3 types of computers for sale: laptop microcomputers, desktop models, and network stations. There are 10 items of each computer type in stock, but 6 of the desktop computers can work as desktop and networking units. What is the probability of the event W that the next 5 orders are from the same computer type?

Solution

From equation (2.9), we need to subtract the common elements to events D and N_n. Thus,

$$P(W) = P(L) + P(D) + P(N_n) - P(D \cap N_n) = \frac{3 \times C_5^{10} - C_5^6}{C_5^{30}} = 0.00526$$

In the last two examples we had more than two events. For the case of three events, all of which having common outcomes, equation (2.9) becomes

$$P(A \cup B \cup C) = P(A) + P(B) + P(C) - P(A \cap B)$$

$$- P(A \cap C) - P(B \cap C) + P(A \cap B \cap C) \qquad (2.10)$$

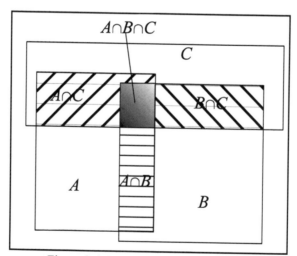

Figure 2.6: Venn Diagram of 3 Events
with Common Elements

Figure 2.6 shows the Venn diagram of three events A, B, and C, each with common elements represented by the hatched areas. Note the tripled shaded area $A \cap B \cap C$ has been subtracted 3 times in equation (2.10): once with $A \cap B$, once with $B \cap C$, and once with $A \cap C$. Therefore, to include these elements at least once, it must be added. A generalization of equation (2.10) to n sets is given in Ross (1998).

Basic Set Theory

Some fundamental set theory relationships may be presented now. They may be easily proved with the aid of Venn diagrams (see problems 2.15 and 2.16).

1. Two sets S_1 and S_2 are said to be *disjoint* if $S_1 \cap S_2 = \varnothing$, where $\varnothing$ denotes the "*empty set*" or the set without elements (without outcomes). For example, assume an experiment is such that $S = \{\xi \in \mathbb{R}\}$, where all the outcomes are represented by ξ; $\in$ denotes "*belongs to*"; and $\mathbb{R}$ denotes the set of all the real numbers. If $S_1 \subset S$, where $\subset$ denotes "*subset of*", is the subset corresponding to all negative numbers, and $S_2 \subset S$ is the subset of all the positive numbers, then $S_1 \cap S_2 = \varnothing$.

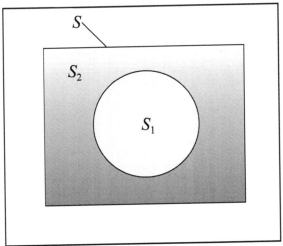

Figure 2.7: Venn Diagram of Complement Sets

2. The *complement* of a set A, denoted A^c (sometimes called "no A"), consists of all the elements in the sample space S of an experiment *not* in A. In the example of item 1 above, $S_1^c = S_2$, and $S_2^c = S_1$. Figure 2.7 shows the corresponding Venn diagram.

3. Certain relationships involving the sample space and the empty set

follow from the properties of union and intersection:

$$A \cup S = S, \quad A \cup \emptyset = A, \quad A \cap S = A, \quad A \cap \emptyset = \emptyset \tag{2.11}$$

4. If $B \subset A$, then

$$A \cup B = A, \quad A \cap B = B \tag{2.12}$$

5. If union is regarded as addition and intersection is regarded as multiplication, set operations follow the usual arithmetic commutative, associative, and distributive laws.

Commutative law: $A \cup B = B \cup A, \quad A \cap B = B \cap A \tag{2.13}$

Associative law: $(A \cup B) \cup C = A \cup (B \cup C), \quad (A \cap B) \cap C = A \cap (B \cap C) \tag{2.14}$

Distributive law: $(A \cup B) \cap C = (A \cap C) \cup (B \cap C) \tag{2.15}$

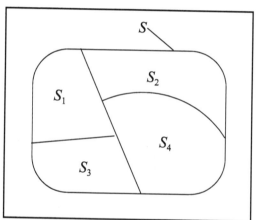

Figure 2.8: A Partition of a Sample Space

6. A *partition* of a sample space S is a collection of subsets S_1, S_2,..., S_n, such that all are disjoint and their union equals S. Mathematically,

$$S_i \cap S_j = \emptyset, \quad \forall i, j \mid i \neq j$$

$$S_1 \cup S_2 \cup \ldots \cup S_n = S \tag{2.16}$$

Where $\forall$ denotes *"for all"*; and $\mid$ denotes *"such that"* or *"given that."* Figure 2.8 shows the Venn diagram of a partition of S.

7. Two useful relationships called the *De Morgan's laws* are as follow:

$$(A \cup B)^c = A^c \cap B^c, \quad (A \cap B)^c = A^c \cup B^c \qquad (2.17)$$

Their validity may be easily demonstrated with Venn diagrams.

Additional Probability Relationships

With the probability axioms and the basic set theory of the previous sections, we now formulate additional probability relationships.

1. Since for any event A, $A \cup A^c = S$, and $P(S)=1$, then the probability of the complement of A is

$$P(A^c) = 1 - P(A) \qquad (2.18)$$

2. The probability of the empty set (i.e., the set with no outcomes) is $P(\varnothing) = 0$.

2.5 INDEPENDENCE AND CONDITIONING OF PROBABILISTIC EVENTS

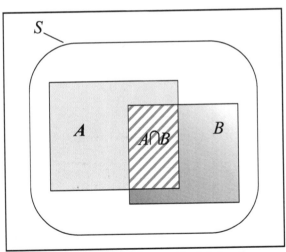

Figure 2.9: Venn Diagram Explaining the
Concept of Conditional Probability

In many circumstances, the probability of an event A is conditioned by the fact that a second event B has already occurred, or some partial information concerning the result of the experiment is already available. In these situations the number of possible outcomes in the experiment is reduced to that of B, and the number of favorable outcomes or successes is limited to the common outcomes between A and B. Figure 2.9 illustrates the Venn diagram of this concept. Mathematically,

$$P(A|B) = \frac{P(A \cap B)}{P(B)}, \quad P(B) > 0 \tag{2.19}$$

or equivalently,

$$P(A \cap B) = P(A|B)P(B) \tag{2.20}$$

Similarly,

$$P(B|A) = \frac{P(A \cap B)}{P(A)}, \quad P(A) > 0 \tag{2.21}$$

or equivalently,

$$P(A \cap B) = P(B|A)P(A) \tag{2.22}$$

Equations (2.20) and (2.22) are equivalent ways to express the probability of the event A *and* B. A generalization of these equations provides an expression for the probability of the intersection of an arbitrary number of events. This is referred to as the *multiplication rule*:

$$P(E_1 \cap E_2 \cap ... \cap E_n) = P(E_1)P(E_2|E_1)P(E_3|E_1 \cap E_2)$$
$$...P(E_n|E_1 \cap E_2 \cap ... \cap E_{n-1}) \tag{2.23}$$

Example 2.13

In the experiment E_4=selecting 3 valves (see section 2.1), the engineer found from workers reports that there is at least one defective valve somewhere in the assembly line in today's shift. If the experiment is executed, estimate the probability of finding 2 defective valves exactly.

Solution

Define the events A=at least one defective valve, and B=2 defective valves exactly. Recall that the number of possible outcomes in the experiment are E_4={ggg, ggb, gbg, bgg, gbb, bgb, bbg, bbb}. Thus, from direct enumeration, assuming each outcome is equally likely, A={ggb, gbg, bgg, gbb, bgb, bbg, bbb} and $P(A)$=7/8. Similarly, B={gbb, bgb, bbg} and $P(B)$=3/8. The common outcomes to A and B are $A \cap B$={gbb, bgb, bbg} and $P(A \cap B)$=3/8. From equation (2.21) the required probability is

$$P(B|A)=\frac{P(A\cap B)}{P(A)}=\frac{3/8}{7/8}=\frac{3}{7}$$

Statistical Independence

An important notion in probability is that of independence of events. If two events A and B have no common elements, and viewing intersection as multiplication, then the events are said to be *statistically independent* if

$$P(A\cap B)=P(A)P(B) \tag{2.24}$$

Substituting equation (2.24) into equations (2.19)and (2.21),

$$P(A|B)=P(A), \quad P(B|A)=P(B) \tag{2.25}$$

Thus, if A is independent of B, the probability of the occurrence of A is unchanged by information as to whether or not B has occurred.

Example 2.14

In Example 2.13, determine whether or not A and B are statistically independent.

Solution

In Example 2.13, $P(A)=7/8$, $P(B)=3/8$,, and $P(A\cap B)=3/8$. From equation (2.24), $P(A)P(B)=7/8\times3/8=21/64\neq3/8$. Therefore, A and B are not independent. We already know this, since from Example 2.13, A and B have common elements.

Example 2.15: Flooding in a Watershed

According to historical records, a watershed is flooded an average of 37 days per year; tornado storms touch down an average of 20 days per year; and simultaneous flooding and tornado conditions occur an average of 12 days per year. Assuming that storms are equally likely to occur any time during a year, estimate the probability of (1) a flood once a tornado has occurred; (2) a tornado once a flood has occurred; and (3) a flood or a tornado.

Solution

We first remark that these weather-related systems have certain deterministic conditions that affect their occurrence, such as the season in the year. Therefore they do not simply occur on any day as our naive probabilistic model implies. For instance, the fact that there is more likelihood of flooding during the rainy season, and more likelihood of wind-related storms during early Spring and late Fall would create errors in our calculations. The water resources engineer could improve the following predictions by separating these well-known deterministic effects from his calculations. For instance, by studying storms in the same season only the engineer assures that deterministic variables are removed and that only the effects of chance (random) are considered. Thus, the following calculations are first approximations only.

Let us define the events F=flood (i.e., water level greater than a certain threshold), and T=tornado (i.e., localized wind speeds greater than a certain threshold). From equations (2.4) and (2.18), the probabilities of flood and that of no flood in any day of the year are, respectively,

$$P(F)=\frac{37}{365}=0.101, \quad P(F^c)=1-P(F)=1-0.101=0.899$$

Similarly, the probabilities of tornado and that of no tornado are, respectively

$$P(T)=\frac{20}{365}=0.055, \quad P(T^c)=1-P(T)=1-0.055=0.945$$

Now, from equations (2.19), and (2.8),

(1) $P(F|T)=\dfrac{P(F\cap T)}{P(T)}=\dfrac{12/365}{20/365}=0.60$

(2) $P(T|F)=\dfrac{P(F\cap T)}{P(F)}=\dfrac{12/365}{37/365}=0.324$

(3) $P(F\cup T)=P(F)+P(T)-P(F\cap T)=0.101+0.055-\dfrac{12}{365}=0.123$

Example 2.16: Shortage of Water Supply

The water supply for a city C comes from 2 sources A and B. Pipelines carry the water to C (see Figure 2.10). Assume that either source can supply C with enough water. (1) Find an expression for shortage of water at C. (2) Find an expression for no shortage at C. (3) Find the probability of shortage if the probability of failure of pipe 1, 2, and 3 are assumed independent and equal to 0.01, 0.005, and 0.02, respectively.

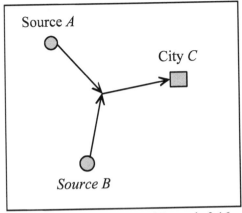

Figure 2.10: Illustration of Example 2.16

Solution

(1) Let us define S=shortage in city C, and E_i, i=1, 2, 3 the failure of pipe i. From the Venn diagram in Figure 2.11, shortage is described by the hatched area $S=(E_1 \cap E_2) \cup E_3$. In words, system failure occurs when pipe 1 *and* pipe 2 do not work, *or* when pipe 3 does not operate. Notice the correspondence of the word "and" to the symbol $\cap$ and the word "or" to the symbol $\cup$.

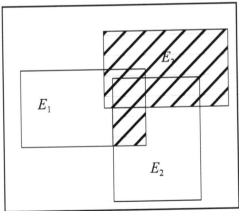

Figure 2.11: Venn Diagram of the Event
Shortage, S, in C for Example 2.16

(2) No shortage may be expressed as $S^c = [(E_1 \cap E_2) \cup E_3]^c$. Applying the DeMorgan's laws, equation (2.17) twice, one obtains

$$S^c = (E_1 \cap E_2)^c \cap E_3^c = (E_1^c \cup E_2^c) \cap E_3^c$$

(3) Presumably, the probabilities of failure of the individual pipes have been determined from past records of malfunction of pumping units, valves, unions, etc., as well as human factors such as improper maintenance, lack of personnel, etc. Since no additional information is given, we assume that the failure of any pipe is independent of that of the others. This would not be the case if the systems are electrically, mechanically, or administratively integrated. The following estimate could be improved if records of simultaneous failure of the system components were kept. They would serve as numerical evaluators of the conditional probabilities of failure of each pipe given that another has failed. If the events are independent, then from question (1), equation (2.8), and equation (2.23),

$$P(S)=P\big((E_1 \cap E_2)\cup E_3\big)=P(E_1 \cap E_2)+P(E_3)-P\big((E_1 \cap E_2)\cap E_3\big)$$

$$=P(E_1)P(E_2)+P(E_3)-P(E_1 \cap E_2 \cap E_3)$$

$$=P(E_1)P(E_2)+P(E_3)-P(E_1)P(E_2)P(E_3)$$

$$=0.01\times0.005+0.02-0.01\times0.005\times0.02=0.02$$

Thus, about 2% of the time there is shortage in C. In other words, 98% of the time the water supply is adequate (i.e., $1-P(S)$). If this service record is unacceptable, the engineer might want to investigate the ways to decrease the failure chances of pipe 3, which is the one with the highest probability of failure.

Example 2.17: Building Construction on Time

A company subdivides its building construction projects into 3 main tasks: excavation, foundation, and superstructure. After years of performance records in many projects, the management found that 80% of the time the excavation is completed on time; 70% of the time the foundation is completed on time; and 90% of the time the building superstructure is completed on time. The 3 operations are administered by 3 different independent contractors. Estimate the probability of the following events: (1) T_1 =project is finished on time; (2) T_2 =excavation is on time and at least one of the other tasks is not on time; (3) T_3 =only one of the 3 operations is on time.

Solution

Let us define the events E=excavation is finished on time; F=foundation is finished on time; and S=superstructure is finished on time. Since each operation is administered by different contractors, it is reasonable to assume that the events are independent.

(1) For the project to be completed on time, the excavation *and* the foundation *and* the superstructure must be completed on time.

$$P(T_1)=P(E \cap F \cap S)=P(E)P(F)P(S)=0.8 \times 0.7 \times 0.9 = 0.504$$

The interpretation of this result is an example of the practical use of probability in management decisions. Slightly over 50% of the time a project construction deadline is met. Clearly this is controlled by the foundation, which is the task with the lowest probability of completion on time. To improve the overall probability of project completion on time, the management should investigate what causes such a low performance in the foundation contractor, and implement solutions.

(2) We could phrase this question as "excavation is on time *and* either foundation *or* superstructure are not on time." Translating this statement into set operations,

$$P(T_2)=P\left(E \cap (F^c \cup S^c)\right)$$

Applying the distributive law of sets (see equation (2.15)), equation (2.9), and equation (2.22),

$$P(T_2)=P\left((E \cap (F^c \cup S^c)\right)$$

$$=P(E \cap F^c)+P(E \cap S^c)-P\left((E \cap F^c) \cap (E \cap S^c)\right)$$

$$=P(E)P(F^c)+P(E)P(S^c)-P(E \cap F^c \cap S^c)$$

$$=P(E)P(F^c)+P(E)P(S^c)-P(E)P(F^c)P(S^c)$$

$$=P(E)\left(P(F^c)+P(S^c)-P(F^c)P(S^c)\right)$$

$$=0.8 \times (0.3+0.1-0.3 \times 0.1)=0.296$$

(3) When only one of the 3 operations is completed on time, the others will not. One possibility of satisfying this condition is that the excavation is on time *and* the others not, $(E \cap F^c \cap S^c)$; a second possibility is that the foundation is on time *and* the others not, $(E^c \cap F \cap S^c)$; a third possibility is that superstructure is on time *and* the others not, $(E^c \cap F^c \cap S)$. Either the first, *or* the second, *or* the third possibility may occur. Hence, the required probability is

$$P(T_3)=P\left((E \cap F^c \cap S^c) \cup (E^c \cap F \cap S^c) \cup (E^c \cap F^c \cap S)\right)$$

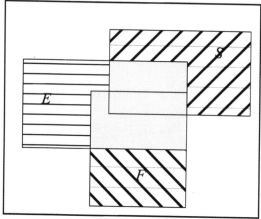

Figure 2.12: Event T_3 in Example 2.17

Notice how the words are translated into set operations: "not F" is the same as F^c; "and" is the same as $\cap$; "or" is the same as $\cup$. Also, the 3 possibilities, or sub-events, that make T_3 are mutually exclusive (see Figure 2.12). Thus,

$$P(T_3)=P(E\cap F^c\cap S^c)+P(E^c\cap F\cap S^c)+P(E^c\cap F^c\cap S)$$

$$=P(E)P(F^c)P(S^c)+P(E^c)P(F)P(S^c)+P(E^c)P(E^c)P(S)$$

$$=0.8\times0.3\times0.1+0.2\times0.7\times0.1+0.2\times0.3\times0.9=0.092$$

Example 2.18: Weld Flaws in a Structural Element.

A structural component has welds to be inspected for flaws (Figure 2.13). From past records, the company has found that the probability of detecting a flaw in a weld, $P(F_L)$, is proportional to its length L, that is $P(F_L)=0.2L$. If flaws are detected in section A_1, the probability of detecting them in section A_2 is 3 times the original value. If flaws are detected in section A, the probability of detecting them in section B is 2 times as much. Let F_{A_1}, F_{A_2}, F_A, F_B, and F be the events flaws are detected in A_1, A_2, A, B, and the whole structural element, respectively. Find the following probabilities: (1) $P(F_A)$; (2) $P(F)$; and (3) flaws are found only in A given that the component has flaws which are detected.

Solution

$$P(F_{A_1})=0.2\times1=0.2, \quad P(F_{A_2})=0.2\times1=0.2, \quad P(F_B)=0.2\times0.5=0.1$$

$$P(F_{A_2}|F_{A_1})=3\times0.2\times1=0.6, \quad P(F_B|F_A)=2\times0.2\times0.5=0.2$$

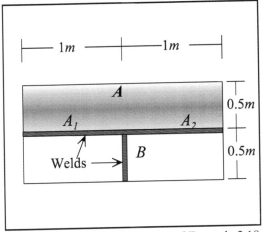

Figure 2.13: Structural Element of Example 2.18

(1) $P(F_A)=P(F_{A_1}\cup F_{A_2})=P(F_{A_1})+P(F_{A_2})-P(F_{A_1}\cap F_{A_2})$

$=P(F_{A_1})+P(F_{A_2})-P(F_{A_2}|F_{A_1})P(F_{A_1})=0.2+0.2-0.6\times0.2=0.28$

(2) $P(F)=P(F_A\cup F_B)=P(F_A)+P(F_B)-P(F_A\cap F_B)$

$=P(F_A)+P(F_B)-P(F_B|F_A)P(F_A)=0.28+0.1-0.2\times0.28=0.324$

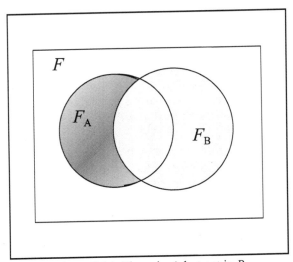

Figure 2.14: Flaws in A, but not in B

(3) The Venn diagram of flaws in A, but not in B, given that there are flaws is shown as a shaded area in Figure 2.14. From question (2) and Figure 2.14,

$$P(F_A \cap F_B^c | F) = \frac{P(F_A \cap F_B^c \cap F)}{P(F)} = \frac{P(F_A \cap F_B^c)}{P(F)}$$

$$= \frac{\text{(shaded area)}}{\text{(total area)}} = \frac{P(F_A) - P(F_A \cap F_B)}{P(F)} = \frac{P(F_A) - P(F_B | F_A)P(F_A)}{P(F)}$$

$$= \frac{0.28 - (2 \times 0.2 \times 0.5) \times 0.28}{0.324} = 0.692$$

Example 2.19: Gate Operation at a Spillway.

The spillway of a dam has 4 gates. Based on past records of gate performance, each gate is out of order an average of 36.5 days in a year. Calculate: (1) the probability that none of the gates is out of order in a day (any day); (2) the probability one gate is out of order in a day; (3) the probability that two of the gates are out of order; (4) the probability that three of the gates are out of order; (5) the probability that all of the gates are out of order; (6) the mathematical expectation in days that all of the gates are out of order; and (7) the probability that all the gates are out of order simultaneously with a storm having a probability of 1/100 (i.e., one in 100 years.

Solution

Let us define the events O_i, $i=1,...,$ 4 as gate i is out of order. Since no additional information is given, we assume that the event of a gate being out of order is independent of another one being out of order. This assumption is reasonable if each gate has an independent mechanical and electrical driving system, and if past records do not suggest the failure of one gate affecting that of the others.

(1) The probability of a gate (any gate) being out of order is 36.5/365=0.1. Table 2.1 describes the probabilities of each gate:

Table 2.1: Probabilities for each Gate in Example 2.19

Gate	1	2	3	4
$P(O_i)$	0.1	0.1	0.1	0.1
$P(O_i^c)$	0.9	0.9	0.9	0.9

The probability that *none* of the gates is out of order may be phrased as "gate 1 is *not* out of order *and* gate 2 is *not* out of order etc.":

$$P(O_1^c \cap O_2^c \cap O_3^c \cap O_4^c) = 0.9 \times 0.9 \times 0.9 \times 0.9 = 0.656$$

Over 65% of the time all of the gates are working properly.

(2) The probability of the first gate being out of order *(and* the others working properly) is

$$P(O_1 \cap O_2^c \cap O_3^c \cap O_4^c) = 0.1 \times 0.9 \times 0.9 \times 0.9 = 0.0729$$

However there are C_1^4 possible combinations of 4 gates choosing one at a time, each having a probability of occurrence of 0.0729. In other words, we could have "the first gate out of order (*and* the others not), *or* the second out of orders (*and* the others not), etc." Since the events are independent, and using equation (2.2) and equation (2.7), the probability of one gate being out of order is given by

$$\frac{4!}{1!3!} \times 0.0729 = 0.2916$$

(3) Reasoning as in question (2), the probability of 2 gates being out of order is

$$C_2^4 \times 0.1 \times 0.1 \times 0.9 \times 0.9 = 0.0486$$

(4) Similarly, the probability of 3 gates being out of order is

$$C_3^4 \times 0.1 \times 0.1 \times 0.1 \times 0.9 = 0.0036$$

(5) The probability of all gates being out of order is

$$C_4^4 \times 0.1 \times 0.1 \times 0.1 \times 0.1 = 0.0001$$

To check our calculations in questions (1) through (5), we know that the summation of all probabilities must equal 1 (see equation (2.6)): 0.656+0.2916+0.0486+0.0036+0.0001=1.0, which is correct.

(6) With a probability that all of the gates are out of order and a year of 365 days, the expected number of days of this event is $0.0001 \times 365 = 0.0365$. In other words, only a few minutes in a year the operator expects to have all of the gates out of order.

(7) Assuming that the event all of the gates are out of order *and* the event

a storm in one in 100 years are independent, the required probability is the product of the to event probabilities: $0.0001 \times (1/100) = 10^{-6}$. Thus, there is a chance of one in a million that this will occur.

Example 2.20: Computer Program of the Probability of Gate Failure

In Example 2.19, assume that the system has 12 gates, instead of 4. Using mathematics software, calculate the probability of 0, 1, 2, ..., and 12 gates being out of order, respectively.

Solution

The following Maple worksheet explains the solution. Each command is preceded by an explanation comment.

```
Define the probabilities of a gate being out of order, Oi,
and not being out of order, Oic.
> Oi:=0.1: Oic:=0.9:
```

```
(1) None of the Gates are out of order, W[1].
Note we store its value in position 1 of array W[ ].
> W[1]:=Oic^12;
```

$$W_1 := 0.2824295365$$

```
(2) One gate is out of order, W1. Note the use of the binomial (N,m) coefficient.
> W[2]:=binomial(12,1)*Oi*Oic^(11);
```

$$W_2 := 0.3765727153$$

```
(3) Use the sequence command, seq( , ) to calculate the probabilities of 2 and more gates
being out of order. The square brackets means that the output is sored in the array called W[ ].
Thefirst argument of the sequence command is the probability calculation,
or a generalization of question (2). The second argument in the sequence command indicates
the number of times to execute the first argument.
> W:=[ seq(binomial(12,i)*Oi^i*Oic^(12-i) , i=0..12 ) ];
```

$W := [0.2824295365, 0.3765727153, 0.2301277705, 0.08523250758, 0.02130812690,$

$0.003788111448, 0.000491051484, 0.000046766808, 0.3247695\ 10^{-5}, 0.160380\ 10^{-6}, 0.5346\ 10^{-8},$

$0.108\ 10^{-9}, 0.1\ 10^{-11}]$

```
(4) Check the calculations with the sum( , ) command. The first argument of the sum command
is the term to be added (i.e., W[i]). The second argument of the sum command gives the range
of values of the subindex i .
> sum(W[i],i=1..13);
```

$$1.000000000$$

Examples 2.19 and 2.20 introduce the *Binomial* probability distribution. The event has two possible outcomes *in one trial*: gate is out of order or it is not out of order. The first outcome occurs with a probability O_i, and the second outcome occurs with a probability $O_i^c = 1 - O_i$. The experiment is repeated $N = 4$ times (e.g., 4 gates). The probability of k "successes" *and* $(N-k)$ "no successes" (e.g., k gates out of order and $N-k$ working properly) is $O_i^k (1-O_i)^{N-k}$. However, there are C_k^N

combinations of N gates choosing k at a time, each with the same probability. Hence the probability of k gates out of order is given by the formula

$$P(k) = C_k^N O_i^k (1 - O_i)^{N-k} \tag{2.26}$$

We will return to this important distribution when we discuss probability distributions in Chapter 3.

The Total Probability and Bayes' Theorems

Sometimes a probabilistic experiment is made of an event $A \subset S$, such that A is a subset of the sample space S, and S already has a *partition*. Recall that a partition of a sample space S is a collection of subsets $S_1, S_2, ..., S_n$ such that all are disjoint and their union equals S (see equation (2.16) and Figure 2.8). If in addition to a partition, the sample space has another event A, the resulting Venn diagram is in Figure 2.15.

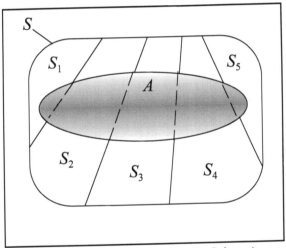

Figure 2.15 A Partition of S and a Subset A

Recalling that $A \cap S = A$ (see equation (2.11)), and using the distributive law of sets (see equation (2.15)),

$$A = A \cap S = A \cap (S_1 \cup S_2 \cup ... \cup S_m) = (A \cap S_1) \cup (A \cap S_2) \cup ... \cup (A \cap S_m)$$

where m is the number of subsets in the partition. Since events S_i are all mutually exclusive (i.e., disjoint), the events $A \cap S_i$ are also mutually exclusive. Then, from equation (2.8),

$$P(A) = P(A \cap S_1) + P(A \cap S_2) + ... + P(A \cap S_m)$$

However, from equation (2.20), $P(A \cap S_i) = P(A|S_i)P(S_i)$, for $i = 1, 2, ..., m$.

Hence, the probability of event A becomes

$$P(A) = \sum_{i=1}^{m} P(A \cap S_i) = \sum_{i=1}^{m} P(A | S_i) P(S_i) \tag{2.27}$$

which is called the *theorem of total probability*. Now, from equations (2.20) and (2.22) $P(A \cap S_i) = P(A | S_i) P(S_i) = P(S_i | A) P(A)$. Substituting equation (2.27) for $P(A)$,

$$P(S_i | A) = \frac{P(A | S_i) P(S_i)}{\displaystyle\sum_{j=1}^{m} P(A | S_j) P(S_j)}, \qquad 1 \le i \le m \tag{2.28}$$

which is called the *Bayes' theorem*. The Total probability theorem and the Bayes' theorems have numerous applications in science and engineering.

Example 2.21: Reliability of an Environmental Testing Program

Historical records indicate that an environmental engineering laboratory is 95% effective in detecting the presence of trichloroethylene, TCE (in concentrations greater than $1\mu g/L$) in a contaminated soil sample, when it is in fact present. However, due to sampling and protocol testing errors the laboratory has reported a "false positive" result in 1% of the samples tested. In other words, if a clean soil sample is tested the laboratory will indicate it is contaminated with TCE with a probability of 0.01. If it is known that 0.5% of the soil samples are indeed contaminated at a particular site, what is the probability that a sample is contaminated given that the said laboratory result is positive?

Solution

The partition of the sample space is made of $m=2$ events: C=soil is contaminated, and C^c=soil is not contaminated. Let us define A=the test result is positive. From the data, $P(A|C)=0.95$, $P(C)=0.005$, $P(A|C^c)=0.01$, and $P(C^c)=1-P(C)=0.995$. The desired probability is $P(C|A)$. Applying the Bayes' theorem equation (2.28),

$$P(C|A) = \frac{P(A|C)P(C)}{P(A|C)P(C) + P(A|C^c)P(C^c)} = \frac{0.95 \times 0.005}{0.95 \times 0.005 + 0.01 \times 0.995} = 0.323$$

Only 32% of the samples reported as contaminated actually are contaminated with TCE. As this result seems unexpectedly low, especially in the light of the good performance record of the laboratory, we present a more intuitive explanation: Since 0.5% of the soil samples are contaminated, then, on the average, 1 sample out of every 200 is polluted. The laboratory will correctly confirm the soil is contaminated with a probability of 0.95. On the average, out of 200 samples tested, the laboratory will correctly report that 0.95 samples are contaminated. On the other hand, out of 199 clean samples the laboratory will incorrectly report that 199×0.01 of these samples are contaminated. Thus, for every 0.95 contaminated samples that the laboratory correctly reports as contaminated, there are 199×0.01 clean samples that the laboratory incorrectly reports as contaminated. The proportion of time that the laboratory results are correct when they report contamination is

$$\frac{0.95}{0.95+199\times0.01}=0.323$$

PROBLEMS

2.1 One water sample is taken from a lake and tested for arsenic concentration. According to the measured concentration in $\mu g/L$, the water is classified as having a "low," "medium," or "high" level of concentration, where the limits for each of these classes have been determined previously by the water quality engineer. Define the set of possible outcomes of this experiment.

2.2 Two water samples are taken from a lake, tested for contaminant concentration, and classified as having a "low," "medium," or "high" level of concentration. Define the set of possible outcomes of this experiment by direct enumeration.

2.3 Three water samples are taken from a lake, tested for contaminant concentration, and classified as having a "low," "medium," or "high" level of concentration. Build the tree diagram and define the set of possible outcomes of this experiment by direct enumeration.

2.4 Recall that from Example 2.3, the server of a computer network requires a 6-character login password. Valid characters are any of the letters of the alphabet or any digit from 0 to 9. If the characters in the password may be repeated (i.e., sampling with replacement), how many passwords are possible?

2.5 The Kentucky lottery consists in correctly selecting 6 numbers from 1 to 42. (1) If the numbers have to be selected in the specific drawing order (i.e., in the same order drawn at the time of the game), how many possibilities does the player have? (2) If the correct numbers must be selected in any order, how many possibilities does the player have?

2.6 A freight company has a fleet of 50 trucks. During low demand, only 40 are used and the management wants a rotational system that assures regular maintenance. Estimate the number of 40-truck teams possible.

2.7 A freight company has a fleet of 50 trucks. The full fleet of 50 trucks is busy during the Christmas season. The management decided to allocate 20 trucks to serve the East of the country, 12 trucks to the Midwest, and 18 trucks to the West. Calculate the possible number of ways to arrange the trucks among the three service areas.

2.8 Recall that from Problem 2.1, one water sample was taken from a lake, tested for contaminant concentration, and classified as having a "low," "medium," or "high" level of concentration. Calculate the probability that the sample has a high concentration of arsenic.

2.9 Recall that from Problem 2.2, two water samples were taken from a lake, tested for contaminant concentration, and classified as having a "low," "medium," or "high" level of concentration. Determine the probability of the event M=no high concentration samples.

2.10 Recall that from Problem 2.3, three water samples are taken from a lake, tested for contaminant concentration, and classified as having a "low," "medium," or "high" level of concentration. Determine the probability of the event H=at least one sample with high concentration.

2.11 Recall that from Problem 2.3, three water samples are taken from a lake, tested for contaminant concentration, and classified as having a "low," "medium," or "high" level of concentration. Determine the probability of the event H_2=two high concentration samples, exactly.

2.12 Recall Examples 2.3 and 2.8, in which the server of a computer network required a 6-character login password consisting of any of the letters of the alphabet or any digit from 0 to 9 as long as they were not repeated. An obstinate user will attempt guessing a password 100 times. Calculate the probability that he guesses a correct password if there are 2,000 registered users?

2.13 Recall Example 2.11 in which an office-supply store had 3 types of computers for sale: laptop microcomputers, desktop models, and network stations. If there are 8 laptop units, 15 desktop units, and 6 network units, what is the probability, W, that the next 5 orders come from the same computer type?

2.14 Recall Example 2.11 in which an office-supply store had 3 types of computers for sale: laptop microcomputers, desktop models, and network stations. If there are 8 laptop units, 15 desktop units, and 6 network units, what is the probability, W, that the next 5 orders come from the same computer type if (1) 3 of the laptop computers can work as laptop and networking stations; and (2) 7 of the desktop computers can work as desktop and networking stations.

2.15 Show that relationships in equation (2.11) are valid using Venn diagrams.

2.16 Show that relationships in equation (2.12) are valid using Venn diagrams.

2.17 Recall Example 2.13 the experiment E_4=selecting 3 valves and determining if they are "good" or "bad." Determine the probability that 3 valves are defective given that there is at least one defective valve in the system.

2.18 Recall Example 2.16, in which the water supply for a city C came from 2 sources A and B. The probability of failure of pipe 1, 2, and 3 is equal to 0.01, 0.005, and 0.02, respectively. In addition, if the probability of failure of pipe 2 once pipe 1 has failed is 0.95, and the probability that pipe 3 fails once pipes 2 and 1 have both failed is 0.20 , calculate the probability of shortage of water in city C.

2.19 Recall Example 2.18 (Figure 2.13), in which a structural component had welds to be inspected for flaws. The probability of detecting a flaw $P(F_L)$=0.2L, where L is the weld length. If flaws are detected in section A_1, the probability of detecting flaws in section A_2 is 3 times the original value. If flaws are detected in section A, the probability of detecting flaws in section B is 2 times the original value. Let F_{A_1}, F_{A_2}, F_A, F_B, and F be the events flaws are detected in A_1, A_2, A, B, and the whole structural element, respectively. Determine the probability that flaws are found only in section B given that the component has flaws which are detected.

2.20 The spillway of a dam has 15 gates. Based on past records of gate performance, each gate is out of order an average of 20 days in a year. Write a Maple program to calculate the probability of 0, 1, 2, ... , and 15 gates being out of order, respectively. Check your calculations by adding all probabilities.

2.21 Reinforced concrete pipes are tested for leaks at a specified pressure prior to shipping. A pipe should fail (F) if a leak (L) is identified. Similarly, a pipe should not fail (F^c) if the test indicates no leak (L^c). However, it is possible that a pipe having no leak will fail. Previous records indicate that possible that a pipe having no leak will fail. Previous records indicate that $P(L^c)=0.95$, $P(F|L)=0.98$, and $P(F|L^c)=0.03$. (1) Use the total probability theorem to calculate the probability of failure, $P(F)$. (2) Use the Bayes' theorem to calculate the probability of a leak given that the pipe fails, $P(L|F)$, and the probability of no leaks given that the pipe does not fail, $P(L^c|F^c)$.

3 RANDOM VARIABLES
AND PROBABILITY DISTRIBUTIONS

3.1 RANDOM VARIABLE: A FUNCTION OF PROBABILITY

In Chapter 2 we defined a particular chance experiment with a written description of the mechanics or procedure to follow. For example, we defined a quality control experiment by stating it consisted in selecting at random a part from an assembly line and judging it as "good," g, or "bad," b. Instead of a written description, a more efficient manipulation of a chance experiment would transform each possible outcome into a predetermined number. Obvious advantages of this improvement are the ability to graph the numerical values of possible outcomes, and the mathematical manipulation of the new variable.

Thus, this new random variable is simply a mapping of each point in a set (i.e., the sample space S of possible outcomes) into another set (i.e., the space of real numbers $\mathbb{R}$). Recall that in mathematics this mapping of the elements of a set, or the "domain," into the elements of another subset of the real line, or the "range," is called a *function*. A function is a transformation of the values of a domain into those of a range. For example, the domain of the function $y=f(x)=x^2$ is the set of all the real numbers from minus infinity to plus infinity, $\mathbb{R}(-\infty, \infty)$. The range is the set of positive real numbers, $\mathbb{R}(0, \infty)$. The domain of the function $y=f(\theta)=\sin(\theta)$ is the set $\mathbb{R}(0, 2\pi)$ radians. The range is the set $\mathbb{R}(-1, 1)$. Similarly, the domain of a random variable $x=X(\xi)$ is the sample space S containing the possible outcomes ξ. The range may be a subset of the set $\mathbb{R}(0, \infty)$. Figure 3.1 shows the functional mapping, or transformation, of values in the domain, represented as points in the set, to points in the range. Points ξ_i in the domain sample space S represent the possible outcomes of a chance experiment. One may assign arbitrary values to each outcome. Points $x=X(\xi)$ in the range are the numerical values after a transformation of the domain outcome values. In Chapter 2 we denoted the individual sample outcomes with Greek lowercase letters. Now we denote the transformation, the function or rule of assignment of the outcome (e.g., $X(\xi)$), with uppercase English alphabet letters, and the value the random variable takes on (e.g., x) with lowercase letters.

As an example, consider again the experiment of selecting one part from an assembly line and judging it as good or bad, g or b, respectively. The domain sample space is $S=\{g, b\}$. Let us arbitrarily assign a value of 0 if the part is bad and 1 if it is good. The possible outcomes, or the elements of the domain set, are $\xi_1 = b$ and $\xi_2 = g$, or simply $S = \{g, b\}$. The values of the range are $x = X(\xi_1) = 0$ and $x = X(\xi_2) = 1$.

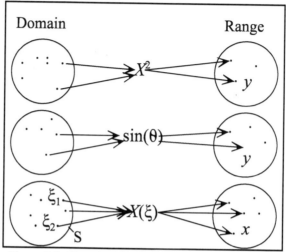

Figure 3.1: Functions Defined as Mapping of Points
in a Domain to Points in a Range

Now, suppose that in above example the experiment is repeated 3 times, instead of once. In other words, the engineer is selecting 3 parts from the line. From direct enumeration (see Chapter 2), we easily see that the sample space is S={ggg, ggb, gbg, bgg, gbb, bgb, bbg, bbb}, where ggb implies the first drawn part is good, the second is good, and the third is bad. Let us define the random variable X by assigning as before the value of 0 for a bad part, a value of 1 for a good part, and adding the numerical value of each of the 3 drawings. The values the random variable takes on are X={3, 2, 2, 2, 1, 1, 1, 0}. In this case, the values of the random variable are integers. For this reason it is classified as a *discrete random variable*. One could also have a *continuous random variable*, as in the case of the magnitude of atmospheric air pressure, deflection in a beam, or equipment amortization cost. In some special cases one could have a random variable that takes on integer values within a certain range and non-integer values within another. This is the case of a *mixed random variable*.

3.2 DISCRETE RANDOM VARIABLES

The Probability Mass Function

A random variable that takes on at most a countable number of possible values, m, is classified as a discrete random variable. For a discrete random variable X, we define the *probability mass function* of X as

$$p_X(x)=P(X=x) \tag{3.1}$$

where

$p_X(x)$=probability of the random variable (uppercase) X taking on values (lowercase) x

The usual rules of probability apply to the probability mass function:

1. $0 \leq p_X(x_i) \leq 10$, $i=1$, $2...$, and $p_X(x)=0$ for all other values of x.

2. $\sum_{i=1}^{\infty} p_X(x_i)=1$

A graphical representation of the probability mass function with respect to the values of a random variable is useful in applications of uncertainty analysis.

Example 3.1

In the example of selecting 3 parts from an assembly line (Section 3.1), assume that each outcome is equally likely. Calculate and plot the values of the probability mass function.

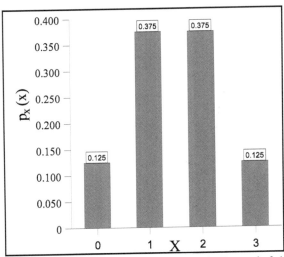

Figure 3.2: Probability Mass Function in Example 3.1

Solution

In section 3.1, we found that by assigning a value of 0 to a bad part, a value of 1 to a good part, and adding the results from each of the 3 drawings, the possible outcomes of the random variable were $X=\{3, 2, 2, 2, 1, 1, 1, 0\}$. Since by assumption each of the outcomes is equally likely to occur, then $p_X(3)=P(X=3)=1/8$. Since we have 3 outcomes with a value of 2, then $p_X(2)=P(X=2)=3/8$. Similarly, $p_X(1)=P(X=1)=3/8$, and $p_X(0)=P(X=0)=1/8$. Figure 3.2 shows a graph of the probability mass function with respect to the values of the random variable. Note that $X=1$ and $X=2$ are the most likely values to occur, whereas $X=0$ and $X=3$ are the least likely to occur. This is the most important advantage when graphing a probability mass function: in a simple picture the engineer has a quick overview of the "uncertainty behavior" of the variable in question.

Example 3.2

Let us consider a more realistic version of Example 3.1. Assume that from past experience about 5% of the parts produced are defective. Plot the probability mass function.

Solution

In this case, the likelihood of drawing a good part is much greater than that of a bad one (i.e., outcomes are not equally likely to occur). Recall that the set of possible outcomes is S={ggg, ggb, gbg, bgg, gbb, bgb, bbg, bbb}, which translates into values of the random variable $X=\{3, 2, 2, 2, 1, 1, 1, 0\}$ after assigning a value of 0 to a bad part, a value of 1 to a good part, and adding the results from each of the 3 drawings. Since the drawings are independent from each other, then (see equations (2.18) and (2.24))

$$P(X=3)=P(g \cap g \cap g)=P(g)\times P(g)\times P(g)=(1-P(b))\times(1-P(b))\times(1-P(b))$$

$$=(1-0.05)\times(1-0.05)\times(1-0.05)=0.857$$

This is the probability of occurrence of the first value in the set of possible outcomes of X. Similarly,

$$P(X=2)=P(g \cap g \cap b)=P(g)\times P(g)\times P(b)=(1-P(b))\times(1-P(b))\times(P(b))$$

$$=(1-0.05)\times(1-0.05)\times0.05=0.045$$

However, there are 3 occurrences of $X=2$ (3 possible ways of arranging 3 parts choosing 2 good ones at a time), and $P(X=2)=3\times0.045=0.135$. By a similar analysis, we find $P(X=1)=0.007$, and $P(X=0)=0.0001$. We check our results by making sure that the summation of all probabilities in the sample space equals one: 0.857+0.135+0.007+0.001=1. Figure 3.3 shows a graph of the probability mass function. Clearly we see that the most likely occurrence is X=3, or drawing 3 good parts, and the least likely occurrence is X=0, or drawing 3 bad parts.

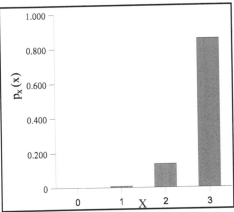

Figure 3.3: Probability Mass Function of
Example 3.2

The Cumulative Distribution Function

Another important function in probability is the *cumulative distribution function*, or simply the distribution function, defined as

$$F_X(x)=P(X \leq x) \qquad (3.2)$$

where
$F_X(x)$=probability a random variable (uppercase) X is less than or equal to a number (lowercase) x

For a discrete random variable, the distribution function may be expressed in terms of the probability mass function as

$$F_X(x)=\sum_{i=1}^{i=x} p_X(i) \qquad (3.3)$$

The distribution function represents the probability that the random variable does not exceed the value x. Many engineering design applications are based on the chances that a given variable does not exceed a critical number. The distribution function has the following properties:

1. $F_X(x)$ is a non-decreasing function. If $a<b$, then $F_X(a) \le F_X(b)$.

2. $\lim\limits_{x \to -\infty} F_X(x) = 0$.

3. $\lim\limits_{x \to \infty} F_X(x) = 1$.

4. The distribution function is continuous from the right. That is for any point of discontinuity a, $\lim\limits_{x \to a^+} F_X(x) = F_X(a)$.

5. The probability that the random variable lies in the interval $a<X \le b$ is

$$P(a<X \le b) = F_X(b) - F_X(a)$$

(3.4)

Example 3.3

In Example 3.1, which consisted in selecting 3 parts from an assembly line, (1) define the different values of the distribution function and plot it. (2) Estimate the probability that $2<X \le 3$.

Solution

(1) From Figure 3.2 and equation (3.3), $F_X(x)=0$, $X<0$; $F_X(x)=0.125$, $0 \le X<1$; $F_X(x)=0.125+0.375=0.5$, $1 \le X<2$; $F_X(x)=0.125+0.375+0.375 = 0.875$, $2 \le X<3$; and $F_X(x)=0.125+0.375+0.375+0.125=1$, $3 \le X$. Figure 3.4 show s a graph of the cumulative distribution function.

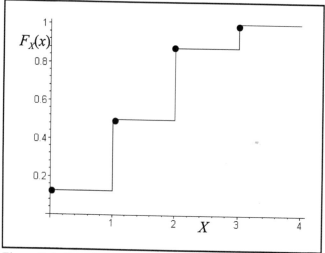

Figure 3.4: Cumulative Distribution Function of Example 3.3

(2) From equation (3.4), $P(2<X\leq3)=F_X(3)-F_X(2)=1.0-0.875=0.125$.

Expected Value of a Random Variable

One of the most important concepts in probability with extensive applications in engineering is that of the expectation of a random variable. The *expected value* is obtained after an operation consisting in calculating an average of the possible values a random variable can take on, each weighted by the probability of such value. For a discrete random variable, X, with a probability mass function, $p_X(x)$, the expected value is defined as

$$\mu=E\{X\}=\sum_{i=1}^{m} x_i\, p_X(x_i) \tag{3.5}$$

where

$E\{X\}$=the expected value of a random variable X.

x_i=i-th value a random variable X takes on

$p_X(x_i)$=the probability mass function that X takes on the value x_i

m=the number of values X takes on (not to be confused with N, which is the total number of possible outcomes or data points)

The expectation operator has the same useful properties of linear operators:

1. $E\{a\}=a$, a=constant.

2. $E\{aX\}=aE\{X\}$, a=constant.

3. If $g_1(X)$, $g_2(X)$,..., $g_n(X)$ are functions of a random variable X, then

$E\{g_1(X)+g_2(X)+...+g_n(X)\}=E\{g_1(X)\}+E\{g_2(X)\}+...+E\{g_n(X)\}$

4. If two random variables X and Y are *independent* (i.e., if their events are independent; see equation (2.24)), then $E\{XY\}=E\{X\}\cdot E\{Y\}$.

The expectation operation applied to a random variable X yields its center of mass, or center of area, or the mean μ, or the *first moment about the origin*, of the probability mass function. It is important to remark that the mean may be estimated from equation (3.5) based on the values of the probability mass function if available. Alternatively, the engineer may have the "raw" data (i.e., the measured values of the random variable) stored in an array. The mean may be computed from the usual formula:

$$\mu = E\{X\} = \frac{1}{N}\sum_{i=1}^{N} x_i \qquad (3.6)$$

where

N=the total number of possible outcomes or data points

Equations (3.5) and (3.6) give the same result.

Example 3.4

In Example 3.1, which consisted in selecting 3 parts from an assembly line, calculate the expected value of the random variable.

Solution

From Example 3.1, or from Figure 3.2, and equation (3.5)

$$\mu = E\{X\} = 0 \times \frac{1}{8} + 1 \times \frac{3}{8} + 2 \times \frac{3}{8} + 3 \times \frac{1}{8} = \frac{3}{2} = 1.5$$

which is the most common value, or the ordinary average, of X.

The Variance of a Random Variable

If the expected value of a random variable X, $E\{X\}$, yields a weighted average of the possible values of X, the variance of X, $Var(X)$, yields the average variation of X around the mean μ:

$$Var(X) = \sigma^2 = E\{(X-\mu)^2\} \qquad (3.7)$$

Equation (3.7) indicates that the variance is the expected value of the square of the distance between the mean and each of the values of X. A more efficient formula is found by expanding equation (3.7) and using the properties of the expectation operator (see below equation (3.5)):

$$Var(X) = E\{(X-\mu)^2\} = \sum_x (x-\mu)^2 p_X(x)$$

$$= \sum_x (x^2 - 2\mu x + \mu^2) p_X(x) = \sum_x x^2 p_X(x) - 2\mu \sum_x x p_X(x) + \mu^2 \sum_x p_X(x)$$

$$= E\{X^2\} - 2\mu^2 + \mu^2 = E\{X^2\} - \mu^2$$

Thus,

$$Var(X) = \sigma^2 = E\{X^2\} - (E\{X\})^2 = E\{X^2\} - \mu^2 \qquad (3.8)$$

This equation states that the variance of a random variable X is equal to the expected value of X^2 minus the square of its expected value. Often, this is the easiest way to compute the variance. The variance yields the *second moment about the mean* of the probability mass function. Structural engineers are quite familiar with the concept of the *moment of inertia* of structural profiles. Similarly, the variance is the moment of inertia of the probability mass function. A more intuitive quantity is obtained by taking the square root of the variance. This quantity, called the *standard deviation* σ, represents the average spread of X with respect to the mean μ:

$$\sigma = \sqrt{Var(X)} \tag{3.9}$$

Equations (3.7) or (3.8) may be used when one possesses the values of the probability mass function. Alternatively, it is possible to compute the variance from the N data values, x_i, from the usual formula,

$$Var(X) = \sigma^2 = \frac{1}{N}\sum_{i=1}^{N}(x_i - \mu)^2 \tag{3.10}$$

Example 3.5: Number of Vehicles Exceeding Maximum Load

After 30 one-hour experiments, the number of vehicles in violation of the maximum load at a highway section was $W=\{14, 10, 12, 11, 10, 13, 9, 11, 12, 12, 11, 13, 12, 9, 11, 10, 12, 10, 11, 12, 12, 13, 14, 10, 11, 12, 11, 10, 11, 13\}$. (1) Calculate the mean, the variance and the standard deviation of the random variable. (2) Calculate and plot the values of the probability mass function and the cumulative distribution function. (3) Calculate the mean and variance based on the probability mass function.

Solution
(1) From equation (3.6), the mean is

$$\mu_W = \frac{14+10+12...+13}{30} = 11.400 vehicles$$

From equation (3.10)

$$\sigma_W^2 = \frac{(14-11.400)^2 + (10-11.400)^2 + ... + (13-11.400)^2}{30} = 1.707 vehicles^2$$

From equation (3.9),

$$\sigma_W = \sqrt{1.707} = 1.306 vehicles$$

(2) Counting the number of outcomes corresponding to each number of vehicles and using equation (2.4), $p_W(9)=P(W=9)=2/30=0.067$, $p_W(10)=P(W=10)=6/30=0.200$, $p_W(11)=P(W=11)=8/30=0.267$, $p_W(12)=P(W=12)=8/30=0.267$, $p_W(13)=P(W=13)=4/30=0.133$,

$p_W(14) = P(W=14) = 2/30 = 0.067$. Figure 3.5 shows a graph of the probability mass function. Note the most frequent number of vehicles is between 11 and 12. The mean of 11.4 is the precise value.

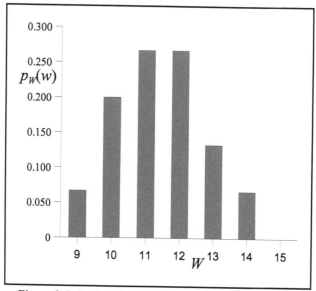

Figure 3.5: Probability Mass Function of Example 3.5

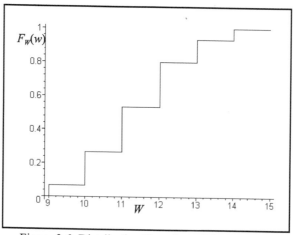

Figure 3.6: Distribution Function of Example 3.5

From equation (3.3), $F_W(9) = 0.067$, $F_W(10) = 0.267$, $F_W(11) = 0.534$, $F_W(12) = 0.801$, $F_W(13) = 0.934$, $F_W(14) = 1.0$. Figure 3.6 shows a graph of the cumulative distribution function.

(3) From equation (3.5),

$$E\{W\}=\mu_W=9\times0.067+10\times0.200+11\times0.267+12\times0.267$$

$$+13\times0.133+14\times0.067=11.411 vehicles$$

Note we have a roundoff error of 0.011. Now, from equation (3.8),

$$Var(W)=\sigma_W^2=9^2\times0.067+10^2\times0.200+11^2\times0.267+12^2\times0.267$$

$$+13^2\times0.133+14^2\times0.067-11.4^2=1.831 vehicles^2$$

The roundoff error in the variance is now 1.831-1.707=0.124. As we will see in our discussion of statistics, this discrepancy should decrease as the sample size, N, increases.

Example 3.6: Computer Calculation of Mean and Variance

In example 3.5, write a Maple program to calculate the mean, the variance and the standard deviation of the random variable number of vehicles, W.

Solution

The following Maple worksheet illustrates the solution.

```
[ Invoke the statistics sub-package.
[ > with(stats) :
[ Input data on the the random variable number of vehicles, W.
[ Press Shift-Enter simulataneously to create a new line.
[ > W:=[14,10,12,11,10,13,9,11,12,12,11,13,12,9,11,10,
    12,10,11,12,12,13,14,10,11,12,11,10,11,13]:
[ Use the describe[ ]( ) command to calculate the mean of W.
[ > describe[mean](W) ;
```

$$\frac{57}{5}$$

```
[ Evaluate the previous value in floating point using the command evalf( );
[ the % sign means "previous" result.
[ > evalf(%) ;
```

11.40000000

```
[ Calculate the variance of W using the describe[ ]( ) command.
[ > describe[variance](W) :
    evalf(%) ;
```

1.706666667

```
[ Take the square root of the above to calculate the standard deviation.
[ > sqrt(%) ;
```

1.306394530

Spreadsheet programs provide an alternative means of calculating the mean, the variance, and other statistics. For short lists of data, most hand calculators have intrinsic functions of these parameters.

The Binomial Random Variable

In Chapter 2, Examples 2.19 and 2.20, and the subsequent discussion culminating with equation (2.26), we defined the *Binomial random variable*. It results after an experiment is repeated N times. *Each outcome is independent* of the others and results in a "success" with a probability p, or in a "failure" with probability $(1-p)$. We seek to find the probability of k successes *and* $(N-k)$ failures after the N independent trials. Since the possible number of combinations of N objects choosing k at a time is C_k^N, and since each outcome has a probability $p^k(1-p)^{n-k}$, then *the probability of k successes in N trials* is given by the probability mass function of the Binomial random variable X as

$$P_X(k)=P(X=k)=C_k^N p^k(1-p)^{N-k}, \quad k=0,\ 1,\ 2...,\ N \qquad (3.11)$$

where

  p=probability of success *in one trial*
  N=number of *independent* trials
  k=number of successes

The graph of the Binomial probability mass function is symmetric with respect to its mean when the probability of success in one trial, p, is equal to the probability of failure in one trial (see Figure 3.2), and non-symmetric otherwise (see Figure 3.3). The cumulative distribution function of the Binomial random variable is, from equation (3.3)

$$F_X(k)=P(X\le k)=\sum_{i=0}^{k}P_X(i)=\sum_{i=0}^{k}C_i^N p^i(1-p)^{N-i} \qquad (3.12)$$

The mean of the Binomial random variable is, from equations (3.5) and (3.11),

$$E\{X\}=\mu_X\equiv Np \qquad (3.13)$$

The variance of the Binomial random variable is, from equations (3.8) and (3.11),

$$Var(X)=\sigma_X^2=Np(1-p) \qquad (3.14)$$

Example 3.7

Solve Example 3.2, which consisted of selecting 3 parts from an assembly line when 5% of the parts produced were defective, assuming the random variable follows a Binomial distribution.

Šolution
In a single drawing, a part could be good or bad. The probability of drawing a good part in one trial is $p=0.95$. The experiment consists in selecting $N=3$ parts. The probability of selecting $k=0$ good parts is, from equation (3.11),

$$p_X(0)=P(X=0)=C_0^3 0.95^0 \times (1-0.95)^{3-0}=0.000125$$

The probability of selecting 1 good part is,

$$p_X(1)=P(X=1)=C_1^3 0.95^1 \times (1-0.95)^{3-1}=0.007$$

Similarly, the probability of selecting 2 good parts and 3 good parts are, $p_X(2)=0.135375$, and $p_X(3)=0.857375$, respectively.

Example 3.8
Recall Example 2.19 (Chapter 2), in which the spillway of a dam had 4 gates and each one was out of order an average of 36.5 days in a year. Calculate the probability of (1) 2 gates being out of order, and (2) at least 2 gates being out of order in any day.

Solution
(1) Define the random variable O=number of gates being out of order. A gate (any gate) may be out of order in any day with a probability $p=0.1$. We have $N=4$ gates. Thus, the probability of 2 gates being out of order in any day is, from equation (3.11),

$$p_O(2)=P(O=2)=C_2^4 0.1^2 \times (1-0.1)^{4-2}=0.0486$$

(2) The probability of at least 2 gates being out of order is $P(O=2)+P(O=3)+P(O=4)$. This is equivalent to $P(O>1)=1-P(O\le 1)=1-F_O(1)$. Using the cumulative distribution function equation (3.12),

$$P(O>1)=1-F_O(1)=1-\sum_{i=0}^{1} p_O(i)=1-\sum_{i=0}^{k} C_i^4 p^i(1-p)^{4-i}$$

$$=1-\left(C_0^4 0.1^0 \times 0.9^4 + C_1^4 0.1^1 \times 0.9^3\right)=1-(0.6561+0.2916)=0.0523$$

Example 3.9
Recall Example 2.20 (Chapter 2), in which the spillway of a dam had 12 gates and each one was out of order an average of 36.5 days in a year.

Write a Maple program to calculate the probability of (1) 6 gates being out of order, and (2) at most 6 gates being out of order.

Solution

In this case we have $p=0.1$ (see Example 2.19), $N=12$, and $k=6$. The probability of 6 gates out of order is simply $p_o(6)$. The probability of at most 6 gates out of order could be expressed as $P(O \le 6) = F_o(6)$. We could use equations (3.11) and (3.12) for the solution of questions (1) and (2), respectively, with a hand calculator, but it would be easier if we use the intrinsic functions of Maple:

```
[ > with(stats): p:=0.1: N:=12: k:=6:
```
(1) Use the statevalf[,]( ) function to calculate values from a probability distribution. The first argument in the square brackets, pf, indicates that values from a probability mass function are requested. The second argument in the square brackets, binomiald[N, p], indicates that values of the Binomial distribution. Thus, the probability of k=6 gates out of order is
```
> statevalf[pf,binomiald[N,p]](k);
```
$$0.000491051484$$

(2) To calculate the probability of at most 6 gates out order, we use the cumulative distribution function. Thus, the first argument in the statevalf[,] is now dcdf, which stands for "discrete cumulative distribution function."
```
> statevalf[dcdf,binomiald[N,p]](k);
```
$$0.9999498197$$

The Geometric Random Variable

A modification of the Binomial random variable occurs when instead of inquiring about the probability of k successes in N trials, the user asks about the probability of the first success at the k-th trial. If the first success occurs at the k-th trial, then we must have $(k-1)$ failures prior to that. The probability of $(k-1)$ failures in a row is $(1-p)^{k-1}$, and the probability of $(k-1)$ failures followed by one success is $(1-p)^{k-1}p$. In general, *the probability of the first success at the k-th trial* is given by the *Geometric random variable* as

$$p_X(k) = P(X=k) = (1-p)^{k-1}p, \quad k=1, 2, 3,..., N \qquad (3.15)$$

where

p=probability of success *in one trial*
N=number of *independent* trials
k=trial number when *the first success* occurs

The mean and variance of the Geometric random variable are given by

$$E\{X\} = \mu_X = \frac{1}{p} \quad Var(X) = \sigma_X^2 = \frac{1-p}{p^2} \qquad (3.16)$$

Example 3.10: Telephone Line Congestion
The probability of a telephone line being congested in an hour is 0.25. (1) What is the probability that a new line becomes congested at the 4th hour of operation? (2) What is the expected number of hours one must wait before the line becomes congested?

Solution
(1) We assume that the event a line being congested, or not congested, in an hour is independent of its behavior the following hours. $p=0.25$, $k=4$. From equation (3.15), $P(X=4)=(1-0.25)^3 \times 0.25 = 0.106$. Note that the probability of a line becoming congested at the 4th hour is a much more restricted condition than that of a line being congested 1 hour out of 4 of operation, which would be, from the Binomial distribution, $C_1^4 0.25^1(1-0.25)^{4-1}=0.422$.

(2) From equation (3.16), $E\{X\}=1/p=1/0.25=4$ hours.

The Pascal Random Variable
A generalization of the Geometric random variable arises when we are interested in the probability of the k-th success at the Nth trial. This is the *Pascal random variable*. The first $(k-1)$ successes in any order in the first $(N-1)$ trials follow a Binomial distribution. At the Nth trial we must obtain the kth success. Thus, *the probability of the kth success at the Nth trial* is given by

$$p_X(k)=P(X=k)=C_{k-1}^{N-1}p^{k}(1-p)^{N-k} \qquad (3.17)$$

where
p=probability of success *in one trial*
N=number of *independent* trials *when the k-th success* occurs
k=number of successes

The mean and the variance of the Pascal random variable are given by

$$E\{X\}=\mu_X=\frac{k}{p}, \quad Var(X)=\sigma_X^2=\frac{k(1-p)}{p^2} \qquad (3.18)$$

Example 3.11
Recall Example 3.10, where the probability of a telephone line being congested in an hour was 0.25. (1) What is the probability that the third

hour in which there is congestion, will occur at the sixth hour? (2) Find the expected value, and its standard deviation, of the number of hours that will pass before congestion has occurred for 3 hours?

Solution

(1) From the Pascal random variable, equation (3.17), $k=3$, $N=6$, $p=0.25$, and

$$P_X(3)=P(X=3)=C_{3-1}^{6-1}p^3(1-p)^{6-3}=0.0659$$

(2) From equation (3.18),

$$E\{X\}=\frac{3}{0.25}=12 hours, \quad Var(X)=\frac{3(1-0.25)}{0.25^2}=36 hours^2, \quad \sigma=\sqrt{36}=6 hours$$

The Hypergeometric Random Variable

Our experiment now consists in sampling n times *without replacement* a sample space containing N elements. The total number of possible outcomes is the number of combinations of N elements choosing n at a time, C_n^N. Suppose, however, that out of the total number of elements, N, there are K which are of a "different" kind. *The probability of selecting k elements of the different kind* follows the *Hypergeometric random variable*. Thus, selecting k elements of the different kind *and* ($n-k$) of the regular kind is given by

$$P_X(k)=P(X=k)=\frac{C_k^K C_{n-k}^{N-K}}{C_n^N}, \quad k=0, 1, 2,..., n \qquad (3.19)$$

where

N=total number of elements in the sample space
n=number of trials (i.e., number of sampling times)
K=number of elements of a different kind
k=number of successes (number of elements of the different kind to draw)

The mean and the variance of the Hypergeometric random variable are given by

$$E\{X\}=\mu_X=\frac{nK}{N}, \quad Var(X)=\sigma_X^2=\frac{(N-n)(1-p)np}{N-1}, \quad p=\frac{K}{N} \qquad (3.20)$$

Example 3.12: Random Quality Inspection Knowing Defective Rates
A contractor buys iron bars in lots of 100 units. He routinely takes 10

samples at random from each lot and subjects them to a tension test. By policy, he rejects the lot if 3 of the 10 samples do not satisfy the specified tension strength. If the factory knows that 6% of the bars produced do not pass the test, (1) what is the probability that a lot will be rejected? (2) What is the expected number of defective parts to find in this experiment? (3) Solve question (1) using Maple.

Solution
(1) We have a case of $N=100$ units. $p=0.06=K/100$, or $K=6$, are defective. The engineer draws $n=10$ units without replacement. The lot is rejected if $k=3$ are defective. The probability of rejecting the lot is given by the Hypergeometric distribution, equation (3.19) as

$$p_X(3)=P(X=3)=\frac{C_3^6 C_{10-3}^{100-6}}{C_{10}^{100}}=0.0118$$

(2) From equation (3.20), $E\{X\}=10\times6/100=0.6$, or less than one bar is expected to be defective.

(3)

```
[ Initialize and enter data.
[ > restart: with(stats): N:=100: n:=10: K:=6: k:=3:
[ Use the statevalf[ ]( ) function to call the hypergeometric random variable.
[ > statevalf[pf,hypergeometric[K,N-K,n]](k);
                    0.01182632576
```

The Poisson Random Variable
Consider a Binomial random variable with p the probability of success in one trial and N the number of independent trials. If N is large and p so small that Np is a moderate size, then it can be shown that the Binomial random variable equation(3.11) approaches the *Poisson random variable* given by

$$p_X(k)=P(X=k)\approx\frac{\lambda^k}{k!}e^{-\lambda}, \quad \lambda=Np, \; k=0, 1, 2, 3,... \tag{3.21}$$

where
p=probability of success *in one trial*
N=number of independent trials
k=number of successes

When the above conditions are satisfied, the Poisson random variable is an efficient means of calculating the probability of k successes. In fact, the Poisson random variable remains a good approximation to the Binomial even when the trials are not independent, provided that their dependence is weak. In addition, the Poisson random variable has been found to describe many processes in engineering and the sciences. Typically, *arrivals* (e.g., of trucks, storms, earthquakes, electrons, customers, calls, etc.) may be described by the Poisson random variable (for a mathematical demonstration see Papoulis, 1984; and Ross, 1998). Thus, if one interprets p as the constant event arrival rate, ω, and N as an interval of time, t, it can be shown mathematically that the number of independent events occurring during a specified observation period follows a Poisson random variable:

$$p_X(k)=P(X=k)\approx\frac{(\omega t)^k}{k!}e^{-\omega t}, \quad k=0,\ 1,\ 2,\ 3,... \tag{3.22}$$

where

k=number of events arriving during a time interval t
ω=events arrival rate, or number of events during a recorded time T

The probability of k or less successes is described by the cumulative distribution function, equation (3.3), as

$$F_X(k)=P(X\le k)=\sum_{i=0}^{k}\frac{\lambda^i}{i!}e^{-\lambda}, \quad \lambda=Np,\ k=0,\ 1,\ 2,\ 3,... \tag{3.23}$$

The mean and the variance of the Poisson random variable are given by

$$E\{X\}=\mu_X=\lambda=Var(X)=\lambda \tag{3.24}$$

Example 3.13
Recall Example 3.9, in which the spillway of a dam had 12 gates and each one was out of order an average of 36.5 days in a year. Assuming the Poisson random variable is a good approximation, (1) calculate the probability that 6 gates are out of order; (2) calculate the probability that at most 6 gates are out of order; and (3) solve questions (1) and (2) using Maple.

Solution
(1) From Example 3.9, N=12, p=0.1, k=6, λ=12×0.1=1.2. From equation (3.21),

$$p_X(6)=P(X=6)\approx\frac{1.2^6}{6!}e^{-1.2}=0.00125$$

which still has an error of about one order of magnitude. The Poisson approximation will improve as N increases. (2) From equation 3.22,

$$F_X(6)=P(X\leq6)=\sum_{i=0}^{6}\frac{1.2^i}{i!}e^{-1.2}=0.9997$$

(3)

```
[ Initialize and enter data.
[ > restart: with(stats): lambda:=1.2: k:=6:
[ (1)
[ > statevalf[pf,poisson[lambda]](k);
                         0.001249112636
[ (2)
[ > statevalf[dcdf,poisson[lambda]](k);
                         0.9997488875
```

Example 3.14: Truck Arrivals at a Site

Thirty-five trucks of weight 10 or more tons are observed to cross a bridge in 5 hours. (1) Assuming that this is a typical day, what is the probability that 10 such trucks will arrive in 1 hour on the following day? (2) Calculate and graph the probability mass function.

Solution

(1) $\omega=35/5=7$ *trucks/hour*, $k=10$, $t=1$ hour. To use the Poisson random variable, we assume that the arrival rate is constant and that truck arrivals are independent. The engineer must confirm the latter with experimental observations. Do trucks belong to the same contractor? Are there peak hours and slow hours? If the arrivals are independent,

$$p_X(10)=P(X=10)=\frac{(7\times1)^{10}}{10!}e^{-7\times1}=0.071$$

(2) The following Maple worksheet illustrates the calculation and the probability mass function of truck arrivals.

```
[ Initialize and call the plotting tools module.
[ > restart: with(stats): with(plottools): omega:=7:
  Using the sequence command, seq( , ), calculate the coordinates [k, P(X=k)]
  of the probability mass (PMF) function for M values of k.
  Store data in an array called PMF
[ > M:=15:      #number of class intervals
    PMF:=[seq(statevalf[pf,poisson[omega]](k),k=0..M)]:
[ Draw a PMF as a series of vertical lines.
[ > seq( line([k,0],[k,PMF[k]]), k=1..M ):
    plots[display](%,thickness=12,labels=[` k `,`P(k)`],
    labeldirections=[HORIZONTAL,VERTICAL],
    labelfont=[TIMES,ROMAN,16]):
```

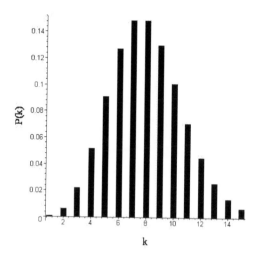

An Approximation to the Poisson Random Variable

In the previous section we saw that the Binomial random variable approaches a Poisson random variable when N is large and p is so small that Np is a moderate size. Similarly, when the parameter λ in the Poisson random variable is large, the Poisson distribution may be approximated as

$$p_X(k) = \frac{\lambda^k}{k!} e^{-\lambda} \approx \frac{e^{-(k-\lambda)^2/2\lambda}}{\sqrt{2\pi\lambda}}, \quad \lambda \gg 1 \tag{3.25}$$

This is called the *De Moivre-Laplace* theorem. It states that the Poisson distribution becomes a properly scaled and shifted Gaussian distribution (see section 3.3) when λ is large. When the latter condition is satisfied, it can also approximate the Binomial distribution if $\lambda = Np$.

3.3 CONTINUOUS RANDOM VARIABLES

The Probability Density Function

In section 3.1 we described discrete random variables as those whose set of possible outcomes is either finite or countably finite. Alternatively, we may have *continuous random variables*, whose set of possible outcomes is uncountable (e.g., ambient air temperature, time of arrival of a shipment, lake contaminant concentration, beam actual load, etc.). We defined the probability mass function of a discrete random variable as the probability that the random variable takes on a specific value (see equation (3.1)). For a continuous random variable X, we define the *probability density function* $f_X(x)$ as a nonnegative function defined for all real values $x \in (-\infty, \infty)$ such that the probability that X takes on values between two limits $a < b$ is given by

$$P(a \leq X \leq b) = \int_a^b f_X(x)dx \qquad (3.26)$$

where
   $f_X(x)$ = probability density of a random variable (uppercase) X taking
         on values (lowercase) x

The probability density function has the following properties:

1. $f_X(x) \geq 0, \ \forall \ x \in (-\infty, \infty)$.

2. $P(-\infty < X < \infty) = \int_{-\infty}^{\infty} f_X(x)dx = 1$.

   Thus, if a random variable X is discrete, its probability mass function is a bar diagram of positive numbers, each determining the probability that X takes on a particular value x. For a continuous random variable, the probability density function is a smooth curve of positive values corresponding to the possible number, x, the random variable, X, takes on. Areas under the probability density function represent probabilities that the random variable takes on values in between the limits of the integration. Unlike a discrete random variable, the probability that a continuous random variable takes on a fixed value is zero: if we let $a=b$ in equation (3.26),

$$P(X=a) = \int_a^a f_X(x)dx = 0 \qquad (3.27)$$

Example 3.15: Maximum Beam Deflection
   After several experiments, it was found that the maximum deflection, $D(mm)$, of a variably-loaded beam follows a probability density function given by

$$f_D(d) = w(10d - d^2), \qquad 0 \leq D \leq 10$$

where w is a positive constant. (1) Estimate the value of w such that the above function follows the properties of a probability density function. (2) Find the probability that the deflection is less than or equal to 5mm. (3) Find the probability that the deflection is in between 3 and 6mm. (4) Find the probability that the deflection exceeds 8mm.

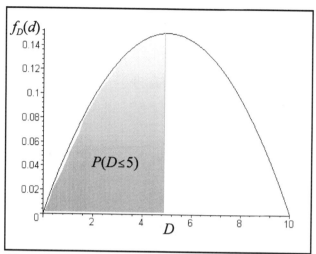

Figure 3.7: Probability Density Function of Example 3.15
Showing Area Corresponding to $P(D \leq 5)$

Solution

(1) From property 2 above, the total area under the probability density function must equal 1. This implies that

$$\int_0^{10} f_D(d) = 1 = \int_0^{10} w(10d - d^2)dd = w\left[5d^2 - \frac{d^3}{3}\right]_0^{10} \Rightarrow w = 0.006$$

(2) Figure 3.7 shows the probability density function with the area sought. From equation (3.26),

$$P(D \leq 5) = \int_0^5 f_D(d)dd = \int_0^5 0.006(10d - d^2)dd = 0.006\left[5d^2 - \frac{d^3}{3}\right]_0^5 = 0.5$$

(3) Similarly (see Figure 3.8),

$$P(3 \leq D \leq 6) = \int_3^6 f_D(d)dd = \int_3^6 0.006(10d - d^2)dd = 0.006\left[5d^2 - \frac{d^3}{3}\right]_3^6 = 0.432$$

(4) Finally (see Figure 3.9),

$$P(D > 8) = 1 - P(D \leq 8) = 1 - \int_0^8 f_D(d)dd = 1 - \int_0^8 0.006(10d - d^2)dd$$

$$= 1 - 0.006\left[5d^2 - \frac{d^3}{3}\right]_0^8 = 0.104$$

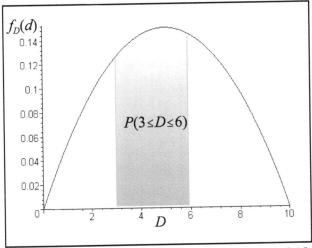

Figure 3.8: Probability Density Function of Example 3.15
Showing Area Corresponding to $P(3 \leq D \leq 6)$

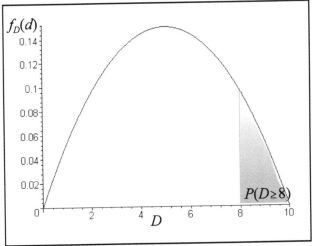

Figure 3.9: Probability Density Function of Example 3.15
Showing Area Corresponding to $P(D \geq 8)$

Example 3.16
Solve Example 3.15 using Maple.

Solution
The following Maple worksheet describes the solution.

> (1) Define the probability density function of beam deflection as a function mapping
> values of "d" into (->) its algebaric form. Note the syntaxis for the definition of a function.

```
> restart: fD:=d->w*(10*d-d^2);
```

$$fD := d \rightarrow w\,(10\,d - d^2)$$

> Using the int(,) command, integrate the probability density fucntion between 0 and 10.
> The first argument of int(,) states the function to be integrated (i.e., the integrand).
> The second argument of int(,) describes the limits of integration.

```
> int(fD(d),d=0..10)=1;
```

$$\frac{500\,w}{3} = 1$$

> Using the solve(,) command, solve the above equation for w. The first argument
> in solve(,) states the equation to solve. The second argument in solve(,)
> states the variable to solve for. Next, evaluate in floating point and equate it to w.

```
> solve(%,w): w:=evalf(%);
```

$$w := 0.006000000000$$

> (2) Integrate the probability mass function between 0 and 5.

```
> int(fD(d),d=0..5);
```

$$0.5000000000$$

> (3) Integrate the probability mass function between 3 and 6.

```
> int(fD(d),d=3..6);
```

$$0.4320000000$$

> (4) Integrate the probability mass function between 0 and 8, and subtract from 1.

```
> 1-int(fD(d),d=0..8);
```

$$0.1040000000$$

A visual observation of the graph of the probability density function, its shape, and tails reveals the "uncertain behavior" of the variable in question. In other words, it shows the ranges of values of the variable that are more likely to occur, and those less likely to occur.

The Cumulative Distribution Function

In Section 3.2, we introduced the cumulative distribution function of a discrete random variable as a "staircase" curve that yields the probability that the random variable takes on a value of less than or equal to its abscissa (see equation (3.2)). This translated into a summation of the probability mass function (equation(3.3)). For a continuous random variable, the *cumulative distribution function* becomes a smooth curve gradually increasing in magnitude from 0 to 1. Equation (3.3) now becomes an integral of the probability density function:

$$F_X(x) = P(X \le x) = \int_{-\infty}^{x} f_X(u)\,du \qquad (3.28)$$

Properties 1 through 5 (after equation(3.3)) are valid for continuous random variables. If the cumulative distribution function of a continuous random variable is obtained by integrating the probability density function, then the density function may be obtained by differentiating the cumulative distribution function. Thus, from equation (3.28),

$$f_X(x) = \frac{dF_X(x)}{dx} \qquad (3.29)$$

Example 3.17

In Example 3.15, (1) derive the cumulative distribution function of beam deflection. (2) Use the cumulative distribution function to calculate the probability that the deflection is less than 7mm, and the probability that the deflection is in between 7 and 8mm. (3) write a Maple program that solves items (1), (2), and plots the cumulative distribution function.

Solution
(1) From equation (3.28),

$$F_D(d) = P(D \le d) = \int_0^d f_D(u)\,du = \int_0^d w(10u - u^2)\,du = w\left[5u^2 - \frac{u^3}{3}\right]_0^d = w(5d^2 - \frac{d^3}{3}$$

(2) Now from equations (3.28) and (3.4), respectively,

$$P(D \le 7) = F_D(7) = 0.006 \times (5 \times 7^2 - \frac{7^3}{3}) = 0.784$$

$$P(7 \le D \le 8) = F_D(8) - F_D(7) = 0.006 \times (5 \times 8^2 - \frac{8^3}{3}) - 0.006 \times (5 \times 7^2 - \frac{7^3}{3}) = 0.112$$

(3)

```
 Define the probability density function of beam deflection as a function.
[ > restart: fD:=d->w* (10*d-d^2) :
[ Integrate the probability density function between 0 and d.
[ > int(fD(u) ,u=0..d) ;
```

$$-\frac{1}{3} w\, d^3 + 5\, w\, d^2$$

```
 Define the cumulative distribution function using the unapply( , ) command. The first
 argument of this command states the expression to use (i.e., the previous expression here).
 The second argument states the dependenet variable (i.e., d is the dependent variable in FD(d)).
[ > FD:=unapply(%,d) ;
```

$$FD := d \to -\frac{1}{3} w\, d^3 + 5\, w\, d^2$$

```
 Set the actual value of w and estimate P(D<7).
[ > w:=0.006: FD(7) ;
                        0.7840000000

[ Estimate P(7<D<8).
[ > FD(8)-FD(7) ;
                        0.1120000000

 > plot(FD(d) ,d=0..10,color=black,thickness=3,
      labels=[`D (mm)` , `FD(d)`],
      labeldirections=[HORIZONTAL,VERTICAL],
      labelfont=[TIMES,ROMAN,16]);
```

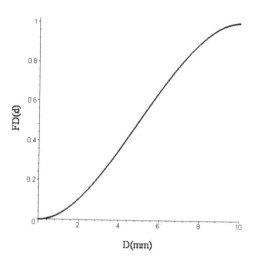

D(mm)

Mean, Variance, and Moments of Continuous Random Variables

In section 3.1, we defined the expected value of a discrete random variable as the first moment with respect to the origin of the probability mass function. For a continuous random variable, the probability mass function is replaced by the probability density function, and the summation in equation (3.5) is replaced by integration:

$$E\{X\}=\mu_X=\int_{-\infty}^{\infty}xf_X(x)dx \tag{3.30}$$

Clearly the expectation operator enjoys all of the associative and distributive properties of linear operators. Once again, the expected value of X represents the mean, or the average value, or the center of mass of the density function. Now using the definition of the expectation operator, equation (3.30), and the definition for the variance, equation (3.7), the variance of a continuous random variable is given by

$$Var(X)=E\{(X-\mu)^2\}=\sigma_X^2=\int_{-\infty}^{\infty}(x-\mu)^2f_X(x)dx \tag{3.31}$$

The variance represents the second moment with respect to the mean, or the moment of inertia, of the probability density function. Equation (3.31) is equivalent to equation (3.8) if one expands the square and splits the integrals. The variance is a measure of the square of the average deviation with respect to the mean. A more intuitive measure is the standard deviation, which is defined as the square root of the variance (see equation (3.9)). The standard deviation represents the average spread of the data around its mean. The greater its value, the greater the degree of uncertainty as represented by the random variability around a mean value. Conversely, the lesser the magnitude of the standard deviation, the lower the degree of uncertainty, or the greater the "determinism" of the data. A normalized version of the standard deviation, called the *coefficient of variability*, is often useful for comparing different populations of a random variable:

$$C_v = \frac{\sigma_X}{\mu_X}$$ (3.32)

where
  C_v=coefficient of variability of a random variable X
  σ_X=standard deviation of random variable X
  μ_X=mean of random variable X

The coefficient of variability is simply a ratio of the standard deviation to the mean. It may be expressed as a percentage. Using the definition for the expectation operator equation (3.30), one may calculate higher-order moments as

$$E\{X^n\} = \int_{-\infty}^{\infty} x^n f_X(x)dx, \quad E\{(X-\mu)^n\} = \int_{-\infty}^{\infty} (x-\mu)^n f_X(x)dx$$ (3.33)

where
  $E\{X^n\}=n^{th}$ order moment with respect to the origin of X
  $E\{(X-\mu)^n\}=n^{th}$ order moment with respect to the mean of X

When $n=1$, the first moment with respect to the origin represents the mean of the random variable X. When $n=2$, the second moment with respect to the mean represents the variance of X. When $n=3$, the third moment with respect to the mean represents the degree of *symmetry* with respect to the mean of a probability density function. A normalized parameter that measures symmetry is the *skewness coefficient*, γ, defined as the ratio of the third moment with respect to the mean to the standard deviation to the power of 3. Skewness is a measure of symmetry with respect to the mean. If it approaches zero, the two flanks of the probability density function on each side of the mean value are symmetric. A positive value in the skew coefficient means that the probability density function possesses a tail to the right of the mean. A negative value implies a tail extending to the left. Often, this is useful in fitting sample data to theoretical density functions whose skewness is known. Moments of order greater than 3 have specific meaning in statistical applications. In fact, moments' values of a sample from a random variable constitute a practical way to evaluate the uncertainty properties of a phenomenon under study. In many engineering applications with limited data, the only information available is a limited sample with its corresponding first and second-order moments. In cases where sufficient experimental information is available, the moments constitute the basis for the parameter estimation of the model chosen as probability density function.

Example 3.18

Calculate the mean, variance, standard deviation, coefficient of variability, and skewness coefficient of the random variable D in Example 3.15.

Solution

From equation (3.30), the mean deflection is given by

$$E\{D\}=\mu_D=\int_0^{10} d.f_D(d)dd=\int_0^{10} d.w(10d-d^2)dd=w\left[\frac{10d^3}{3}-\frac{d^4}{4}\right]_0^{10}=5mm$$

Thus, the abscissa $D=5$ is the mean deflection as can be seen in Figure 3.7. For the integrations of the second and third moments we use the following Maple program:

```
Define the parameter value and the density function.
  a and b are the lower and upper limits of the random variable.
> restart: w:=0.006: a:=0: b:=10: fD:=d->w*(10*d-d^2):
Calculate the mean from equation (3.30).
> mu:=int(d*fD(d),d=a..b);
                          μ := 5.
Calculate the variance in mm^2 from equariuon (3.31)
> sigma2:=int((d-mu)^2*fD(d),d=a..b);
                          σ2 := 5.
The standard deviation in mm from equaiton (3.9)
> sigma:=sqrt(sigma2);
                          σ := 2.236067977
The coefficient of variability from equation (3.32)
> Cv:=sigma/mu;
                          Cv := 0.4472135954
Skewness coefficient from equation (3.33) with n=3, divided by sigma^3
> gama:=int((d-mu)^3*fD(d),d=a..b)/sigma^3;
                          gama := 0.
```

The standard deviation, or average spread around the mean, is about 2.24 *mm*. The coefficient of variability is about 0.45. This implies that the standard deviation is about 45% of the mean. As expected, the skewness is zero, since D is symmetric with respect to the mean (see Figure 3.7).

The Exponential Random Variable

In section 3.2, we learned that the number of independent events, k, arriving during an interval of time, t, when the average arrival rate was ω is usually modeled by the Poisson random variable, equation (3.22). If the arrivals of events are modeled by the Poisson random variable, *the time interval between successive occurrences of events* is modeled by the Exponential random variable. To see this, let us call T a time interval with *no* occurrences of a given event (i.e., $k=0$), and t a time interval with at least *one* occurrence. The probability that $T \leq t$ is the cumulative distribution function of T, $F_T(t)$, which would represent the case when there is at least one Poisson event in the interval from 0 to t. From equation (3.22),

$$P(T \leq t) = F_T(t) = 1 - P(X=0) = 1 - e^{-\omega t} \qquad t \geq 0 \qquad (3.34)$$

To obtain the probability density function of the Exponential distribution we apply equation (3.29) (i.e., differentiate with respect to t):

$$f_T(t) = \omega e^{-\omega t} \qquad t \geq 0 \qquad (3.35)$$

where
  T=time interval with no arrivals of events
  ω=average rate of arrival of events

It is interesting to interpret the analytical form of the Exponential density function equation (3.35). Its maximum value, ω, occurs when $t=0$. As t increases, the probability density function decreases gradually to zero. In other words, maximum chances of observing no events occur for very short time intervals. As time increases, the chances of observing no events decrease. The mean and variance of the exponential random variable may be obtained by applying equations (3.30) and (3.31), respectively. Thus,

$$E\{T\} = \mu_T = \frac{1}{\omega}, \quad Var(T) = \sigma_T^2 = \frac{1}{\omega^2} \qquad (3.36)$$

Example 3.19

In Example 3.14, (1) calculate the probability that there is an interval of 2 hours without any truck arrival. (2) Calculate the probability that there is a time between a quarter of an hour to half an hour without any truck arrival. (3) Use Maple to solve the item (2) and plot the probability density function and the cumulative distribution function.

Solution

(1) From equation (3.27), $P(T=2)=0$ (i.e., the probability of a continuous random variable taking on a constant value is zero).

(2) From equations (3.4) and (3.34),

$$P(\frac{1}{4} \leq T \leq \frac{1}{2}) = F_T(\frac{1}{2}) - F_T(\frac{1}{4}) = (1 - e^{-7 \times 1/2}) - (1 - e^{-7 \times 1/4}) = 0.144$$

(3)

```
[ > restart: with(stats): omega:=7:
  (2) Using the statevalf[ , ](t), calculate the difference between the cumulative
  distribution function at t=1/2 and at t=1/4. The first argument of the
  statevalf[ , ](t) command requests the cumulative distribution function, "cdf."
  The second argument requests the Exponential distribution with parameter mu.
[ > statevalf[cdf,exponential[omega]](1/2)
      -statevalf[cdf,exponential[omega]](1/4);
                        0.1435765600
  Plot the probability density function with the plot( , , ) command. The first argument
  of this command indicates the function to plot. The second argument indicates the range
  of values for the dependent variable. The third or more argument(s) of the plot command
  are plotting options. Note that, in this case, the first argument of the statevalf[ , ](t)
  is the probabilty density fucntion, "pdf."
[ > plot( statevalf[pdf,exponential[omega]](t), t=0..1,
      color=black,thickness=2,labels=[`T`,`fT(t)`],
      labeldirections=[HORIZONTAL,VERTICAL],
      labelfont=[TIMES,ROMAN,16] );
```

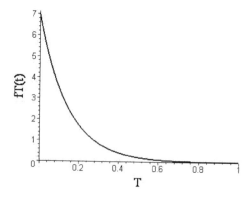

```
[ Plot the cumulative distribution function.
[ > plot( statevalf[cdf,exponential[omega]](t), t=0..1,
      color=black,thickness=2,labels=[`T`,`FT(t)`],
      labeldirections=[HORIZONTAL,VERTICAL],
      labelfont=[TIMES,ROMAN,16] );
```

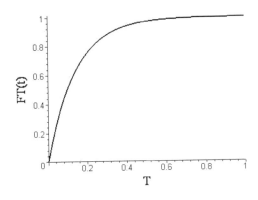

The Uniform Random Variable

A purely random phenomenon whose sample values oscillate between a lower and an upper limit is modeled as a *Uniform random variable*. The experiment underlying the Uniform random variable, U, produces values with an equal likelihood of occurrence between a lower and upper limits. The probability density function is given by

$$f_U(u) = \begin{cases} \dfrac{1}{b-a}, & a \leq U \leq b, \ a < b \\ 0 & otherwise \end{cases} \qquad (3.37)$$

where

a=lower limit of occurrence of the random variable U
b=upper limit of U

The probability density function of the uniform random variable is a horizontal line between a and b, and zero everywhere else. The horizontal line implies that any value is equally likely to occur, or alternatively, that there are no values with a greater chance of occurrence than others. Many games of chance are modeled by the Uniform random variable. The results from the tossing of a single *fair* die follow a discrete Uniform random variable between a=1 and b=6: in a single tossing, any number is equally likely to occur. Lottery games consisting of drawing numbers with replacement follow Uniform random variables, provided the drawing game is conducted honestly. Engineering systems whose underlying experiments generate real numbers with an equal chance of occurrence between two limits are modeled as continuous Uniform random variables. The Uniform distribution is also important in computer simulations of engineering systems governed by other probability distributions when the seed numbers are uniformly-distributed random numbers. We will return to this important application when we study Monte Carlo simulation methods.

By integration of equation (3.37), we find that the cumulative distribution function of the Uniform random variable is given by

$$F_U(u) = \begin{cases} 0, & U \le a \\ \dfrac{u-a}{b-a}, & a < U \le b \\ 1, & U > b \end{cases}$$

(3.38)

Taking first and second moments, the mean and variance of the Uniform random variable are given by

$$E\{U\} = \frac{a+b}{2}, \quad Var(U) = \frac{(b-a)^2}{12}$$

(3.39)

Example 3.20: Completion Date of a Project

An engineering project is expected to be completed anytime between September 1 and September 30 of a year. (1) Define the probability density and cumulative distribution functions of project completion date. (2) Calculate the probability that the project is completed prior to September 8. (3) Calculate the probability that the project is completed between the 14th and the 28th of September. (4) Write a Maple program that solves questions (1)-(3) and plots the density and cumulative distribution functions.

Solution

(1) Since any day during September is, by assumption, equally likely to be the completion date, the probability density function of the date of completion, C, is a Uniform random variable between $a=1$ and $b=30$ in September. From equations (3.37) and (3.38),

$$f_C(c) = \begin{cases} \dfrac{1}{29}, & 1 < C < 30 \\ 0, & otherwise \end{cases}, \quad F_C(c) = \begin{cases} 0, & C \le 1 \\ \dfrac{c-1}{29}, & 1 < C \le 30 \\ 1, & C > 1 \end{cases}$$

(2) $P(C \le 8) = F_C(8) = \dfrac{8-1}{29} = 0.241$

(3) $P(14 < C \le 28) = F_C(28) - F_C(14) = \dfrac{28-1}{29} - \dfrac{14-1}{29} = 0.483$

(4) Since the Uniform random variable has jump discontinuities, we characterize them mathematically as

$$f_C(c)=\frac{1}{29}H(c-a)H(b-c), \quad F_C(c)=\frac{c-1}{29}H(c-a)H(b-c)+H(c-b)$$

where
  $H(x)$=the *Heaviside Unit Step function* defined as

$$H(x)=\begin{Bmatrix} 0, & x<0 \\ 1, & x>0 \end{Bmatrix}$$

The following Maple program implements these expressions:

```
Define the Uniform density and cumulative distribution functions with the
"Heaviside" Unit Step function.
> restart: a:=1: b:=30:
  fC:=c->Heaviside(c-a)*Heaviside(b-c)/29:
  FC:=c->Heaviside(c-a)*Heaviside(b-c)*(c-1)/29
        +Heaviside(c-b):
  FC(8.);            #(1)
  FC(28.)-FC(14.);   #(2)
                              0.2413793103
                              0.4827586207
> plot(fC(c),c=0..b+1,y=0..0.04,color=black,thickness=2,
       labels=[`C`,`fC(c)`],
       labeldirections=[HORIZONTAL,VERTICAL],
       labelfont=[TIMES,ROMAN,16]);
  plot(FC(c),c=0..b+3,y=0..1.2,color=black,thickness=2,
       labels=[`C`,`FC(c)`],
       labeldirections=[HORIZONTAL,VERTICAL],
       labelfont=[TIMES,ROMAN,16]);
```

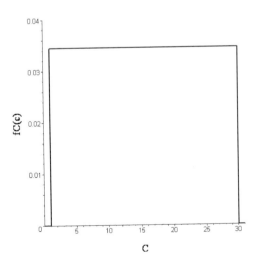

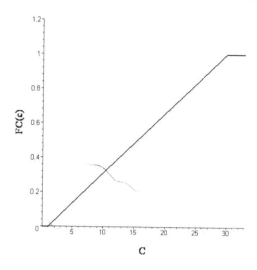

C

The Gaussian Random Variable

The *Gaussian random variable* received its name after the German mathematician Karl Friedrich Gauss (1777-1855) who was one of the first to use it to predict the location of astronomical objects. It was first used by the French mathematician Abraham De Moivre in 1733 as an approximation to the Binomial and Poisson distribution (see equation (3.25)). However, by the late 19th century most statisticians believed that the majority of random phenomena could be described by the typical bell-shaped Gaussian density. For this reason, the Gaussian random variable is also called the *Normal random variable*. It is the most important random variable in probability theory after the formulation of the central limit theorem (discussed later), which establishes a theoretical basis to the empirical observation that many random phenomena obey the Normal probability distribution. Informally stated, the random summation of independent, non-Normal, random variables results in a symmetrical, bell-shaped, no-skewness, Normal distribution. A *Standard Gaussian,* or *Standard Normal*, random variable has a probability density function given by

$$f_X(x) = \frac{e^{-\frac{x^2}{2}}}{\sqrt{2\pi}}, \quad -\infty < X < \infty \qquad (3.40)$$

Figure 3.10 shows the probability density function of the Standard Gaussian random variable. The mean is $\mu_X = 0$, and the variance is $\sigma_X^2 = 1$. We write $X \sim N(0, 1)$ to denote a random variable X follows a Normal distribution with a mean of zero and a variance of 1. The Standard

Gaussian density function is symmetric with respect to the line $X=0$. For this reason the skewness coefficient $\gamma_X=0$.

The cumulative distribution function is obtained by integrating the density function. However, an exact integral of equation (3.40) cannot be found. Thus, numerical approximations of this integral are part of the built-in functions in computer programs, or recorded in tables (see Table A.1, Appendix A). The following example illustrates the estimation of probabilities from a Standard Gaussian random variable.

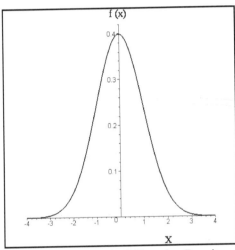

Figure 3.10: Standard Gaussian Density

Example 3.21: Calculation of Probabilities from a Standard Gaussian Random Variable

Suppose that after many trials the total error in degrees, X, at the end of a surveying polygonal line follows a Standard Normal distribution. Calculate the probability that in the next line (1) the error is less than 1.25 degrees; (2) the error is greater than 2 degrees; (3) the error is less than -2 degrees; (4) the error is between ±1 degree; and (5) the error is greater than -0.55 degrees. (6) What should be the maximum error if we want to have a probability of occurrence of 0.65? (7) Solve questions (1) through (6) using Maple.

Solution
All questions adapt and use Table A.1 (Appendix A) as follows:

(1) Directly from Table A.1., read the value of 1.20 in the left column; then move horizontally to the value in the column corresponding to 0.05.

Thus, $P(X \leq 1.25) = 0.8944$.

(2) Table A.1., gives the cumulative distribution function (i.e., $P(X \leq x) = F_X(x)$). To obtain probabilities of the complement, we use equation (2.18): $P(X>2) = 1 - P(X \leq 2) = 1 - 0.9772 = 0.0228$, where the value of 0.9772 was read in the table.

(3) Table A.1 gives cumulative areas for positive values of X. To estimate areas corresponding to negative abscissas, we use the symmetric properties of the Normal distribution (see Figure 3.11). Thus, $P(X<-2) = P(X>+2) = 1 - P(X \leq 2) = 1 - 0.9772 = 0.0228$, where the value of 0.9772 was read in the table.

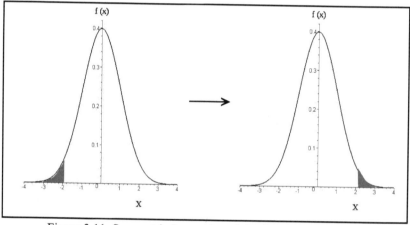

Figure 3.11: Symmetric Properties of the Gaussian Distribution

(4) From equation (3.26), and using again the symmetry properties of the Gaussian distribution,

$$P(-1 \leq X \leq 1) = P(X \leq 1) - P(X \leq -1) = P(X \leq 1) - [1 - P(X \leq 1)]$$

$$= 2P(X \leq 1) - 1 = 2 \times 0.8413 - 1 = 0.6823$$

(5) Similarly, $P(X>-0.55) = P(X \leq +0.55) = 0.7088$.

(6) Here we seek the value x such that $P(X \leq x) = 0.65$. Thus, we look up in the table the value that is closest to an area of 0.65, and read (i.e., interpolate) the corresponding abscissa $x = 0.384$.

(7)
The following Maple program calculates items (1)-(6):

[(1) P(X<1.25). Use the statevalf[,]( ) command.

```
> restart: with(stats):
  statevalf[cdf,normald](1.25);
```
$$0.8943502263$$

[(2) P(X>2).

```
> 1-statevalf[cdf,normald](2);
```
$$0.0227501319$$

[(3) P(X< - 2). Here we do not need to worry about symmetry. Maple yields the cumulative areas for positive as well as negative abscissas.

```
> statevalf[cdf,normald](-2);
```
$$0.02275013195$$

[(4) P(-1 <X< +1).

```
> statevalf[cdf,normald](1)
      -statevalf[cdf,normald](-1);
```
$$0.6826894922$$

[(5) P(X>-0.55).

```
> 1-statevalf[cdf,normald](-0.55);
```
$$0.7088403132$$

Find x such that P(X<x)=0.65. In the first argument of the statevalf[,]( ) command type "icdf" to request the inverse of the cumulative distribution function.

```
> statevalf[icdf,normald](0.65);
```
$$0.3853204664$$

The Standard Gaussian random variable, and its tables, may still be used in the general case of a Gaussian variable, say Y, that has a mean $\mu_Y \neq 0$ and a variance $\sigma_Y^2 \neq 1$ by the transformation

$$X = \frac{Y-\mu}{\sigma} \qquad (3.41)$$

X is a Standard Normal random variable with a density function given by equation (3.40), whereas Y is a Normal random variable with a density function given by

$$f_Y(y) = \frac{e^{-\frac{1}{2}(\frac{Y-\mu_Y}{\sigma_Y})^2}}{\sqrt{2\pi\sigma_Y^2}}, \qquad -\infty < Y < \infty \qquad (3.42)$$

Example 3.22: Lake Contaminant Concentration

After repeated measurements of the water in a lake a long time after a spill, it was concluded that the concentration C of trichloroethylene (TCE) in $\mu g/L$ followed a Normal distribution with a mean of $650 \, \mu g/L$, and a standard deviation of $150 \, \mu g/L$. If an inspector takes a sample at random, find the probability that (1) the concentration is less than 700 $\mu g/L$; (2) the concentration is less than $400 \, \mu g/L$; and (3) the concentration is between 450 and 850 $\mu g/L$. (4) Solve questions (1) through (3) using Maple.

Solution

After a long time, the dilution effects of dispersion and advection assure that the contaminant distribution is random. Thus, $C \sim N(650, 150^2)$. From equation (3.41), $X=(C-650)/150$, $X \sim N(0, 1)$.

(1) $X=(700-650)/150=0.3333$. Hence, $P(C \leq 700)=P(X \leq 0.3333)=0.6293$, where the value of 0.6293 was read in Table A.1.

(2) $X=(400-650)/150 = -1.6666$ Hence,

$$P(C \leq 400)=P(X \leq -1.6666)=1-P(X(\leq 1.6666)=1-0.9525=0.0475$$

(3) $X_1=(850-650)/150=1.3333$, $X_2=(450-650)/150=-1.3333$.

$$P(450 \leq C \leq 850)=P(-1.3333 \leq X \leq 1.3333)=P(X \leq 1.333)-[1-P(X \leq 1.3333)]$$

$$=2P(X \leq 1.3333)-1=2 \times 0.9082-1=0.8164$$

(4)

(1) P(C<700). With Maple we do not have to make a transformation into a Standard Gaussian variable. Simply indicate the mean and the standard deviation as a sub-command of the normald[,] argument. When this sub-command is omited, Maple assumes it is a Standard Gaussian variable.

```
> restart: with(stats):
  statevalf[cdf,normald[650,150]](700);
```
$$0.6305586598$$

(2) P(C<=400).

```
> statevalf[cdf,normald[650,150]](400);
```
$$0.04779035227$$

(3) P(400<=C<=850).

```
> statevalf[cdf,normald[650,150]](850)
     -statevalf[cdf,normald[650,150]](450);
```
$$0.8175775606$$

The Lognormal Random Variable

While the Normal random variable is very important in theoretical analysis and in some applications, many engineering random variables do not follow a Normal distribution. However, just as the summation of several independent random variables results in a Normal distribution, the product of several independent random variables results in an asymmetrical distribution with positive skewness (i.e., a density function with a right tail). If the logarithm of the product is taken, the result once again tends toward a symmetrical bell-shaped Normal distribution, indicating that the random variable is Lognormally distributed. In other words, if the original random variable Z is Lognormally distributed, then the transformed variable $Y = \ln(Z)$ is Normally distributed with a density function given by equation (3.42). In terms of Z, the *Lognormal random variable* has a density function given by

$$f_Z(z) = \frac{e^{-\frac{1}{2}\left(\frac{\ln(Z)-\mu_Y}{\sigma_Y}\right)^2}}{\sqrt{2\pi\sigma_Y^2 Z^2}}, \qquad Y=\ln(Z),\ 0<Z<\infty,\ -\infty\le Y<\infty \qquad (3.43)$$

where

μ_Y=mean of a transformed Normal variable $Y=\ln(Z)$

σ_Y^2=the variance of Y

Be careful here. μ_Y And σ_Y are not the mean and the standard deviation of Z but of $Y=\ln(Z)$. The mean and variance of the Lognormal variable Z are, respectively,

$$\mu_Z=e^{\mu_Y+\frac{\sigma_Y^2}{2}}, \qquad \sigma_Z^2=\mu_Z^2\left(e^{\sigma_Y^2}-1\right) \qquad (3.44)$$

where

μ_Z=mean of the Lognormal (i.e., untransformed) random variable Z

σ_Z^2=the variance of Z

These relationships provide a basis for finding the mean and variance of the original Lognormal variable, Z, if the corresponding mean and variance of the transformed Normal variable, Y, are known. In practice, the opposite is true; that is, the mean and variance of Z are known from the raw data, and equations (3.44) are solved simultaneously to estimate the mean and variance of Y to yield

$$\mu_Y=\ln(\mu_Z)-\frac{1}{2}\ln\left(\frac{\sigma_Z^2}{\mu_Z^2}+1\right), \qquad \sigma_Y^2=\ln\left(\frac{\sigma_Z^2}{\mu_Z^2}+1\right) \qquad (3.45)$$

Alternatively, one may take logarithms to the data to create the series of Y, from which the parameters are computed using the moments equations. Areas under the Lognormal distribution are easily found using the tables for the cumulative areas of the Normal distribution for the transformed variable, Y.

Many engineering processes may be conceived as the product of independent random variables. Engineering variables having values greater than zero and being theoretically unconstrained in the upper range are often Lognormal random variables. Several environmental engineering variables have been reported to follow a Lognormal distribution (Ott, 1995).

Example 3.23: Contaminant Concentration
Assume that in Example 3.22 a better model for C is a Lognormal distribution with a mean $\mu_C=650\mu g/L$ and a standard deviation $\sigma_C=150\mu g/L$. Since $C>0$, the lower bound of the Lognormal variable appears more realistic. (1) Using the Normal table, calculate the probability of having a concentration of less than $700\,\mu g/L$. (2) What should be the concentration in $\mu g/L$ such that we are 85% sure this value is not exceeded? (3) Write a Maple program to solve items (1) and (2).

Solution
(1) C follows a Lognormal distribution given by equation (3.43), whereas $Y=\ln(C)\sim N(\mu_Y,\ \sigma_Y^2)$. From equations (3.45), the mean and variance of the transformed variable Y are given by

$$\mu_Y=\ln(650)-\frac{1}{2}\ln\left(\frac{150^2}{650^2}+1\right)=6.451,\ \sigma_Y=\sqrt{\ln\left(\frac{150^2}{650^2}+1\right)}=0.2278$$

To transform Y into a Standard Normal variable $X\sim N(0, 1)$, we use equation (3.41):

$$X=\frac{Y-\mu_Y}{\sigma_Y}=\frac{\ln(C)-\mu_Y}{\sigma_Y}=\frac{\ln(700)-6.451}{0.2278}=0.4393$$

Now, from Table A.1,
$$P(C\le700)=P(Y\le\ln(700))=P(X\le0.4393)\approx0.6698$$

(2) We seek the value of c such that $P(C\le c)=P(\ln(C)\le y)=P(Y\le y)=P(X\le x)=0.85$. Interpolating the area of 0.85 in Table A.1, the corresponding abscissa is $X=1.036$. From equation (3.41),

$$X=\frac{Y-\mu_Y}{\sigma_Y}\Rightarrow Y=\ln(C)=X\sigma_Y+\mu_Y=1.036\times0.2278+6.451=6.687$$

Hence, $C=e^{6.687}=801.9\mu g/L$.

(3)

```
┌ Enter data and calculate muy and sigmay of the Y–ln(Z) variate from equations (3.45).
│ > restart: with(stats): muz:=650.: sigmaz:=150.:
│   muy:=ln(muz)-ln(sigmaz^2/muz^2+1)/2:
└   sigmay:=sqrt(ln(sigmaz^2/muz^2+1)):
│ (1) P(C<=700). Note that the parameters to use with the statevalf[ ]( ) command are
└ the mean, muy, and the standard deviation, sigmay, of the log-transformed variate Y=ln(Z).
┌ > statevalf[cdf,lognormal[muy,sigmay]](700);
└                          0.6697549618
┌ (2) Find c such that P(C<=c)–0.85.
│ > statevalf[icdf,lognormal[muy,sigmay]](0.85);
└                          802.0021280
```

The Gamma Random Variable

The *Gamma random variable* is a versatile model that describes a wide variety of engineering systems. Its mathematical form may represent different shapes in the probability density function:

$$f_X(x) = \frac{c^b X^{b-1} e^{-cX}}{\Gamma(b)}, \quad b,\ c,\ X > 0 \tag{3.46}$$

where

b=a shape parameter
c=a scale parameter
$\Gamma(b)$=the *gamma function* with the properties

$$\Gamma(b) = (b-1)!, \quad b=1,\ 2,\ 3,...(integer)$$

$$\Gamma(1) = \Gamma(2) = 1, \quad \Gamma\left(\frac{1}{2}\right) = \sqrt{\pi}$$

$$\Gamma(b+1) = b\Gamma(b) \tag{3.47}$$

$$\Gamma(b) = \int_0^\infty u^{b-1} e^{-u} du, \quad (in\ general)$$

The gamma function may be read in tables. More conveniently, both the gamma function and areas under the Gamma random variable may be obtained from intrinsic functions in computer programs and spread sheets.

Example 3.24

Using Maple, explore the effect of the parameter b on the shape of the Gamma density function, and the effect of the parameter c on the scale of the Gamma density function. To do this, plot the Gamma density for several values of each parameter.

Solution

Effecft of c: Keeping $c=1$ constant, plot the Gamma density function for $b=1$, $b=2$, and $b=3$, respectively. Note that the parameter c in equation (3.46) corresponds to $1/c$ in the Maple program (see Maple help for details). By changing the value of b, one changes the shape of the density fucntion.

```
> restart: with(stats):
  b:=1: c:=1:
  #End each graph command with a colon
  G1:=plot( statevalf[pdf,gamma[b,1/c]](X),X=0..6,
     linestyle=1,color=black,thickness=2,legend="b=1"):
  b:=2: G2:=plot(statevalf[pdf,gamma[b,1/c]](X),X=0..6,
     linestyle=3,color=black,thickness=2,legend="b=2"):
  b:=3: G3:=plot(statevalf[pdf,gamma[b,1/c]](X),X=0..6,
     linestyle=4,color=black,thickness=2,legend="b=3"):
  #Display all curves in one graph: end with a semicolon
  plots[display](G1,G2,G3,labels=["X","f (x)"],
     labeldirections=[HORIZONTAL,VERTICAL],
     labelfont=[TIMES,ROMAN,16]);
```

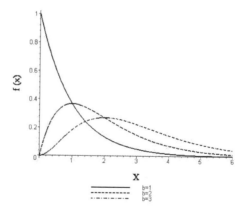

Now keep $b=3$ constant and plot the Gamma density function for $c=1$, $c=2$, and $c=3$, respectively. By changing the value of c, one changes the scale of the density funciton.

```
> b:=3: c:=1:
  G1:=plot(statevalf[pdf,gamma[b,1/c]](X),X=0..6,
     linestyle=1,color=black,thickness=2,legend="c=1"):

  c:=2: G2:=plot(statevalf[pdf,gamma[b,1/c]](X),X=0..6,
     linestyle=3,color=black,thickness=2,legend="c=2"):
  c:=3: G3:=plot(statevalf[pdf,gamma[b,1/c]](X),X=0..6,
     linestyle=4,color=black,thickness=2,legend="c=3"):
  plots[display](G1,G2,G3,labels=[`X`,`f (x)`],
     labeldirections=[HORIZONTAL,VERTICAL],
     labelfont=[TIMES,ROMAN,16]);
```

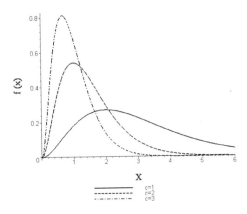

The mean and the variance of the Gamma random variable are, respectively,

$$E\{X\} = \frac{b}{c}, \quad Var(X) = \frac{b}{c^2} \tag{3.48}$$

Other Probability Distributions

Many mathematical functions have been adapted to fit probability density functions of observed phenomena. Occasionally, engineering and science journals announce a new random variable that exhibits special properties in its mean, variance, or skewness, such that the uncertain characteristics of a special system variable are properly described. In this section we present some random variables commonly encountered in engineering practice. We leave to the reader the investigation of the detailed properties of these and many other random variables.

The *Cauchy random variable* is a random variable with a density function given by (Ziemer, 1997)

$$f_X(x) = \frac{\alpha/\pi}{(X - a)^2 + \alpha^2}, \quad \alpha > 0 \tag{3.49}$$

where
a = the x location of the mean

The *Laplace random* variable has a density function made of a double-sided exponential and two parameters, α and a, which is the location of the mean:

$$f_X(x) = \frac{1}{2\alpha} e^{-\frac{|X-a|}{\alpha}}, \quad \alpha > 0 \qquad (3.50)$$

The *Rayleigh random variable* has a probability density function given by

$$f_X(x) = \frac{X}{\alpha^2} e^{-\frac{X^2}{2\alpha^2}}, \quad X > 0 \qquad (3.51)$$

The *Weibull random variable* is widely used in reliability studies. Its probability density function is given by

$$f_X(x) = \frac{m}{c} \left(\frac{X}{c} \right)^{m-1} e^{-(X/c)^m}, \quad X, m, c > 0 \qquad (3.52)$$

where

 m=a shape parameter
 c=a parameter called "the characteristic time"

Several distributions have been adapted to the modeling of extreme events, such as flood/drought discharges in rivers. The tails of probability density functions (i.e., the extremely high or extremely low occurrences), have been found to follow a distribution related to, although different from, the parent distribution: the *Gumbel random variable*, and the *Logpearson* family of random variables are good examples. The reader is referred to the standard hydrology texts for the details (i.e., Serrano, 2010; Haan, 1977). Other distributions have been adapted to the development of statistical inference theory, such as the *Chi Square random variable*, which is a special case of the Gamma random variable; the *Student's t random variable*; and the *F random variable*. We will return to these in the chapters on statistics.

PROBLEMS

3.1 Solve Example 3.2 if about 15% of the parts produced are bad.

3.2 Plot the cumulative distribution function of Problem 3.1.

3.3 Calculate the expected value of the random variable in Example 3.2.

3.4 Annual energy consumption level, E, at a facility has been classified into 4 different categories 1, 2, 3, and 4, respectively. Each category has been deemed as randomly occurring since it depends on management allocation of resources and operation efficiency in a complex manner. Records indicate that the consumption over the past 20 years was as follows: $E=\{3, 1, 3, 2, 4, 3, 1, 2, 2, 2, 1, 4, 3, 3, 1, 1, 3, 4, 3, 3\}$. Calculate the mean annual consumption level and its standard deviation.

3.5 In Problem 3.4, calculate and plot the probability mass function and the cumulative distribution function of E.

3.6 Table 3.1 shows the number of earthquakes per year, Q, with magnitude 7.0 or greater, as reported by the U.S. National Earthquake Information Center. (1) Using a spreadsheet program, calculate the mean annual number of earthquakes μ_Q and its standard deviation σ_Q. (2) Transfer the data to a Maple program and repeat question (1).

Table 3.1: Annual Number of Earthquakes

Year	Q	Year	Q	Year	Q	Year	Q	Year	Q
1900	13	1920	8	1940	23	1960	22	1980	18
1901	14	1921	11	1941	24	1961	18	1981	14
1902	8	1922	14	1942	27	1962	15	1982	10
1903	10	1923	23	1943	41	1963	20	1983	15
1904	16	1924	18	1944	31	1964	15	1984	8
1905	26	1925	17	1945	27	1965	22	1985	15
1906	32	1926	19	1946	35	1966	19	1986	6
1907	27	1927	20	1947	26	1967	16	1987	11
1908	18	1928	22	1948	28	1968	30	1988	8
1909	32	1929	19	1949	36	1969	27	1989	7
1910	36	1930	13	1950	39	1970	29	1990	13
1911	24	1931	26	1951	21	1971	23	1991	10
1912	22	1932	13	1952	17	1972	20	1992	23
1913	23	1933	14	1953	22	1973	16	1993	16
1914	22	1934	22	1954	17	1974	21	1994	15
1915	18	1935	24	1955	19	1975	21	1995	25
1916	25	1936	21	1956	15	1976	15	1996	22
1917	21	1937	22	1957	34	1977	16	1997	20
1918	21	1938	26	1958	10	1978	18	1998	16
1919	14	1939	21	1959	15	1979	15	1999	20

3.7 In Example 3.8, calculate the probability of 3 gates being out of order.

3.8 In Problem 3.6, assume that the probability of an earthquake anywhere in a day is $p=\mu_Q/365$, where μ_Q is the mean annual number of earthquakes. Assume that the daily occurrences of earthquakes in the world are independent and that in any day an earthquake occurs with probability p or does not occur with a probability $(1-p)$ (i.e., there is a maximum of one earthquake in a day). Calculate the probability of 2 earthquakes in one month.

3.9 In Problem 3.6, assume that the probability of an earthquake anywhere in a day is $p=\mu_Q/365$, where μ_Q is the mean annual number of earthquakes. What is the probability of having the first earthquake on the 15th day of the year?

3.10 In Problem 3.6, assume that the probability of an earthquake anywhere in a day is $p=\mu_Q/365$, where μ_Q is the mean annual number of earthquakes. What is the expected value, and its standard deviation, of the number of days one must wait until the first earthquake occurs in a year?

3.11 In Problem 3.6, assume that the probability of an earthquake anywhere in a day is $p=\mu_Q/365$, where μ_Q is the mean annual number of earthquakes. (1) what is the probability of having the second earthquake at the 30th day of the year? (2)What is the expected number of days to wait before 4 earthquakes have occurred?

3.12 A geotechnical company has taken 20 soil samples for an inspection visit. 5 of those were taken with a malfunctioning equipment, which produce samples that fail a triaxial test. The inspector will select 4 samples at random, send them to a laboratory and reject the company work if 2 of the samples fail the test. What is the probability that the job will be rejected?

3.13 In Problem 3.6, $p=\mu_Q/365$, where μ_Q is the mean annual number of earthquakes. It appears that a better model for the occurrence of earthquakes is the Poisson random variable. With this assumption, (1) Calculate the probability of 2 earthquakes in one month, and (2) find the probability of 6 earthquakes or less in 3 months.

3.14 A random variable has a probability density function given by
$$f_X(x)=a(x^2-x^3), \quad 0\leq X\leq 1$$

(1) Estimate the value of the parameter a. (2) Calculate the probability that X is less than or equal to 0.2. (3) Calculate the probability that X is in between 0.4 and 0.6. (4) Calculate the probability that X exceeds 0.9.

3.15 In Problem 3.14 calculate the mean, the variance, the standard deviation, the coefficient of variability, and the skewness coefficient of the random variable X.

3.16 In Problem 3.6, $p=\mu_Q/365$, where μ_Q is the mean annual number of earthquakes. Using the Poisson random variable, calculate the probability that there is a time of less than, or equal to, 1 month without an earthquake. Plot the probability density function and the cumulative distribution function using Maple.

3.17 In Example 3.20, calculate the probability that the project will be completed after September 18.

3.18 Annual product sales at an engineering company follow a Normal distribution with a mean of 350 (millions of dollars) and a standard deviation of 125. Calculate the probability that next year the sales (1) will be less than 400 (millions of dollars); (2) will exceed 300; (3) will be between 250 and 350. (4) How much should the sales be if the chances are 90% that this figure will not be exceeded?

3.19 In Problem 3.18, assume that the sales follow a Lognormal distribution. Solve questions (1) and (4) using the table in Appendix A. Show your procedure.

3.20 In Problem 3.18, assume that the sales follow a Lognormal distribution. Solve questions (1) through (4) using Maple.

3.21 (1) Using Maple, derive an expression for the cumulative distribution function of the Cauchy random variable. (2) Assume arbitrary values of the parameters and plot the density functions of the Cauchy, Laplace, Rayleigh, and Weibull random variables.

Detail of tiled walls of the palace of Alhambra, Granada, left by ancient Islamic civilizations occupying Spain. Prolonged contemplation of precise deterministic geometric patterns (i.e., "Mandalas") leads to random meditation.

4 SIMULATION OF RANDOM SYSTEMS

In this chapter we study the fundamental concepts behind the practical computer modeling of engineering systems affected by one or more *independent* random variables. Many engineering applications are concerned with the numerical simulation of systems or processes that receive random inputs, or have parameters, in the form of random variables. We describe briefly the problem of finding the probability density function of the output variable when the input has a known density function. Subsequently, we study the problem of synthesizing realizations of the output. This includes the essentials of Monte Carlo simulation, and the generation of random numbers from various density functions. A basic knowledge of linear systems theory is advisable at this point. There are many excellent texts on the subject (e.g., Ogata, 1998; Woods and Lawrence, 1997; Ziemer et al., 1993; Research and Education Association, 1982). We also introduce the method of decomposition of Adomian (1994) as a fundamental tool to solve analytically nonlinear system equations. We use decomposition to derive analytical forms of first and second-order statistics of the system output.

4.1 DERIVING A SYSTEM OUTPUT DENSITY FUNCTION

We now consider the practical situation of an engineering system, or process, receiving input from a variable, X, best described as a random variable (see Figure 4.1). The system, represented as a box, takes the input coming from each random realization of X and transforms it into a realization of the output variable, Y. The transformation of samples of X into samples of Y is achieved via the system operator function, $G(X)$. This function represents an intrinsic process or transformation that produces the output from the input and ultimately defines the engineering system. For example, the system may be as simple as one that transforms the minutes spent in a telephone conversation, X, into the amount of money to charge a customer. In this case the system function is simply $Y=aG(X)$, where a is the charge rate per minute. The system function is normally a more complicated expression. Recall that in Chapter 1 we used a differential equation to characterize a system (see equation(1.1)) that takes values of hydraulic head as input and transforms them into values of groundwater flow velocity. Many engineering systems have operator functions made of a solution to a differential equation. The differential equation may be empirical, as in the case of many electrical or physically based systems as in the case of continuum-mechanics systems.

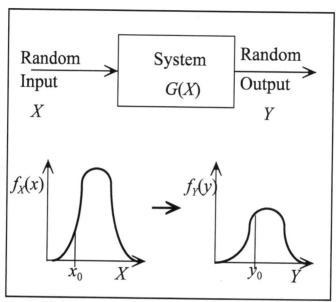

Figure 4.1: A System Subject to a Random Input

Since the system input is random in nature, the system output is also random. The input X is best characterized by the probability density function f sub $f_X(x)$. For a single realization of X, say a value x_0, the corresponding system output is $y_0=G(x_0)$. y_0 constitutes a realization (i.e., one sample) of the random variable Y. In general, for all possible samples of Y we have the probability density function of the output Y, $f_Y(y)$. This is the best description of the system output. Thus, the density function of the input is functionally related to that of the output. Ideally, the engineer seeks to identify the form and features of the output density function, given those of the inputs. We say *ideally* since in practice one does not have the input density, or if we do the difficulty in the derivation of the output density is proportional to the mathematical complexity of the function $G(X)$.

Let us derive the output density function in terms of the input density function. If $Y=G(X)$ and the output to x_0 is y_0, then from the definition of the cumulative distribution function we can say

$$F_Y(y_0)=P(Y\leq y_0)=P\big(G(X)\leq G(x_0)\big)=P(X\leq x_0)=F_X(x_0) \qquad (4.1)$$

Thus, the cumulative of Y evaluated at y_0 is equal to the cumulative of X evaluated at x_0. Differentiating with respect to y and using the chain rule of differentiation,

$$f_Y(y) = \frac{dF_Y(y)}{dy} = \frac{dF_X(x)}{dx} \cdot \left.\frac{dx}{dy}\right|_{x=G^{-1}(y)} = f_X(x) \cdot \left.\frac{dx}{dy}\right|_{x=G^{-1}(y)} \tag{4.2}$$

where the substitution $x = G^{-1}(y)$ is used to obtain a function of y after differentiating (i.e., we seek to express $f_Y(y)$ in terms of y), and the absolute value is needed to satisfy the condition that a density function must be positive. Equation (4.2) states that the density function of the output is equal to the density function of the input times dx/dy in terms of y. If there is more than one root to the inverse function $x = G^{-1}(y)$, then equation (4.2) becomes

$$f_Y(y) = \sum_{i=1}^{m} f_X(x) \cdot \left.\frac{dx_i}{dy}\right|_{x_i=G_i^{-1}(y)} \tag{4.3}$$

where
  m=number of roots in the inverse function

Example 4.1
   The probability density function of a system is Cauchy with $\alpha = 1$ and $a=0$. If the system function is given by $Y = G(X) = 4/X$, derive the output probability density function.

Solution
   From the system function we write the input in terms of the output:

$$x = \frac{4}{y}$$

Differentiating with respect to y,

$$\frac{dx}{dy} = -\frac{4}{y^2}$$

From equations (4.2) and (3.49),

$$f_Y(y) = f_X(x) \cdot \left.\frac{dx}{dy}\right|_{x=G^{-1}(y)} = \left.\frac{1/\pi}{x^2+1} \cdot \left[\frac{4}{y^2}\right]\right|_{x=\frac{4}{y}} = \frac{1/\pi}{\left(\frac{4}{y}\right)^2+1} \cdot \left(\frac{4}{y^2}\right) = \frac{4/\pi}{y^2+4^2}$$

which is also a Cauchy random variable.

Example 4.2
   A system is excited by uniformly-distributed random numbers

between 2 and 6. If the system function is given by $Y=G(X)=2X^2$, derive the output density function.

Solution

If $y<0$, $=2x$ has no real solutions and $f_Y(y)=0$. If $y>0$, $y=2x^2$ has two solutions:

$$x_1 = \sqrt{\frac{y}{2}}, \quad x_2 = -\sqrt{\frac{y}{2}} \quad \therefore \left|\frac{dx_i}{dy}\right| = \frac{1}{2\sqrt{2y}}, \quad i=1, 2$$

From equation (4.3),

$$f_Y(y) = \sum_{i=1}^{2} f_X(x) \cdot \left|\frac{dx_i}{dy}\right|_{x_i = G_i^{-1}(y)}$$

However, x_1 and x_2 produce the same result. Since the density of the input is (see equation (3.37))

$$f_X(x) = \frac{1}{4}, \quad 2 \le X \le 6$$

then the output density is given by

$$f_Y(y) = \frac{1}{4} \cdot \frac{1}{2\sqrt{2y}} = \frac{1}{\sqrt{128y}}$$

At the lower limit of the input, $X=2$ and $Y=8$; at the upper limit, $X=6$ and $Y=72$. Figure 4.2 shows a graph of the system input and output densities.

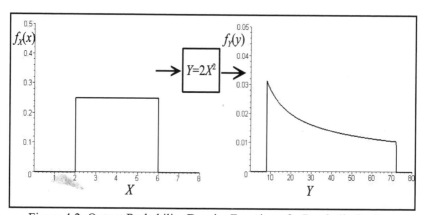

Figure 4.2: Output Probability Density Function of a Parabolic System
Forced by Uniform Random Numbers

Example 4.3

If the input density function is Standard Normal and the system function is $Y = ce^{-aX}$, X, c, $a > 0$, find the expression for the output density function.

Solution

The system function may be written as

$$x = -\frac{1}{a}\ln\left(\frac{y}{c}\right), \quad 0 < y < c$$

Differentiating with respect to y,

$$\frac{dx}{dy} = -\frac{1}{ay}$$

From equations (4.3) and (3.40),

$$f_Y(y) = \frac{e^{-\frac{x^2}{2}}}{\sqrt{2\pi}} \cdot \left[\frac{1}{ay}\right]_{x=-\frac{1}{a}\ln(y/c)} = \frac{e^{-\frac{1}{2a^2}\ln^2(y/c)}}{\sqrt{2\pi a^2 y^2}}$$

4.2 MONTE CARLO SIMULATION

From the previous examples, one can see that the derivation of the output density function may be difficult when the mathematical form of the system function is not a simple expression (i.e, when $= G^{-1}(y)$ is not simple). Alternatively, the derivation of the output density is inferred numerically via a procedure called *Monte Carlo simulation*. Using computer-generated random numbers that follow the input density function, each number is transformed into a sample output via the system operation function. When this procedure is repeated many times, one has a large sample of the output process, from which a density function may be fitted if desired. This procedure is inspired in the method of uncertainty analysis described in Chapter 1: repeated experimentation under identical conditions. By subjecting the system to random testing many times, over a large class of conditions that characterize the uncertainty of the input, we are observing the uncertain behavior of the system response and learning about its properties. The results may not be as mathematically rigorous as the analytical derivation of the output density and the simulations may be subject to further numerical errors, but they provide an approximate evaluation of the output frequency. In the case of complex systems, Monte Carlo simulations maybe the only practical alternative. Today, computers are standard tools of analysis. They provide an affordable means to test many engineering random systems.

Summarizing, using a Monte Carlo simulation experiment, we are attempting to generate random numbers, X, from a known system-input probability density function, $f_X(x)$, and transform each of them into system-output random numbers, Y, of an unknown probability density function, $f_Y(y)$ (see Figure 4.1). The procedure involves the following steps:

1. Generate a large sample of random numbers that follow the system-input density function. This generally consists of two parts.
   • Generate Uniformly-distributed random numbers, U.
   • Transform Uniform random numbers, U, into those following the system-input probability density function, X (see Figure 4.3).

2. Transform each input realization, X, into an output realization, Y, by using the system operator function (See Figure 4.1).

3. Analyze the output sample by computing its statistical properties and possibly fitting a probability density function.

Generation of Random Numbers from a Specified Density Function
   From above, the first step in a Monte Carlo simulation exercise is to generate random numbers, X, that follow a known probability density function, $f_Y(y)$, given by the input of an engineering system (see Figure 4.1). Each random number represents a synthetic realization, or one simulated data point, of the system input process, X. This first step in Mote Carlo simulations is composed of two, as described above (see Figure 4.3): first generate Uniformly-distributed random numbers, U, in the interval from 0 to 1 (step 1.1), and then transform them into random numbers, X, of the specified probability density function, $f_X(x)$, which models the input to the engineering system (step 1.2). We describe these stages in the following sections.

   To obtain random numbers from a known density function, there are some traditional devices that can help. For instance, telephone numbers picked at random from a phone directory are known to follow a Uniform probability density function. Errors in precision-instrument observations have been known to follow a Gaussian probability density function.

With the advent of inexpensive digital computers, these methods have become impractical. Many computer software packages offer random numbers from various probability distributions as intrinsic functions. The computer uses a mathematical algorithm programmed to calculate numbers that closely resemble the specified probability density function. These numbers are not truly random, since the entire series is predictable and produced by a deterministic mathematical procedure, which contradicts the concept of "random." For this reason, some authors call these numbers pseudo-random. However, if they fit the specified density function they provide a valid means to test a model.

The quality of computer-generated random numbers varies widely. In other words, some packages produce random numbers that truly fit the said probability density function and some do not. It is imperative that the engineer tests the random numbers for quality prior to their application. It is clear that numbers that do not fit the specified density function will lead to erroneous results and incorrect conclusions.

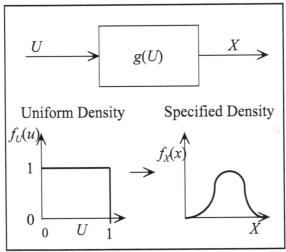

Figure 4.3: Transforming Uniform Numbers, U, into
Random Numbers, X, of a Specified Distribution

As stated, many software packages calculate random numbers of various distributions at the call of a function. Using the concepts described in section 4.1, we will first describe the procedure to generate random numbers, X, of any density function based on random numbers, U, from a Uniform probability distribution, $f_U(u)$. This is particularly useful if one needs random numbers from an unusual distribution or from one not available by the software at hand. Thus, at the heart of most

Monte Carlo simulations there is a procedure that generates random numbers, U, from a Uniform distribution between 0 and 1 and uses a function to transform them into numbers, X, that follow the specified probability density function, $f_X(x)$ (see Figure 4.3). Let us adapt and apply to Figure 4.3 what we learned in section 4.1 concerning input-out systems. In other words, we have a system function $g(U)$ with an input given by Uniformly-distributed random numbers, U, and an output, X, given by random numbers from a specified probability density function, $f_X(x)$.

In this system, the input probability density function is Uniform, $f_U(u)=1$, $0 \le U \le 1$, and the output density function $f_X(x)$ is known (i.e., the density function of the desired random numbers). What we need is the transformation $g(U)$. Applying equation (4.2),

$$f_X(x)=f_U(u)\cdot\left|\frac{du}{dx}\right|_{u=g^{-1}(x)} = \left|\frac{du}{dx}\right| = \left|\frac{dg^{-1}(x)}{dx}\right|$$

$$= \frac{dg^{-1}(x)}{dx}, \quad 0 \le X \le 1, \quad \frac{dg^{-1}(x)}{dx} > 0$$

(4.4)

Multiplying by dx and integrating from $-\infty$ to x,

$$\int_{-\infty}^{x} f_X(\xi)d\xi = \int_{-\infty}^{x} dg^{-1}(\xi)d\xi$$

However, from equation (3.28) the integral of the density function is the cumulative distribution function. Thus,

$$F_X(x)=g^{-1}(x)=U$$

(4.5)

Simply stated, equation (4.5) indicates that sample values of Uniformly-distributed random numbers between 0 and 1, U, are equal to the cumulative distribution function of X, $F_X(x)$. This provides us with the procedure to obtain the corresponding random numbers of X having the pre-specified density function $f_X(x)$. Figure 4.4 is a graphical representation of the procedure stated by equation (4.5): start with a Uniform random number generator, U, from a Uniform distribution, $f_U(u)$ (drawn tilted 90 degrees counter clockwise for clarity); then, according to equation (4.5), each Uniformly distributed random number, u, is equated to the corresponding ordinate value of the cumulative distribution function of the desired numbers X, $F_X(x)$, from which the corresponding x number is found. If we are going to generate thousands of random numbers, a graphical process such as Figure 4.4 is impractical. Instead, we apply equation (4.5) to obtain a suitable transformation of u

values into x values.

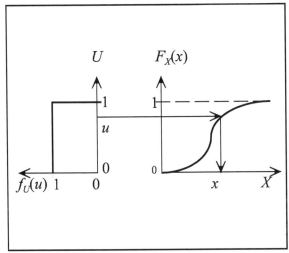

Figure 4.4: Graphical Explanation of Equation (4.5)

Example 4.4: Water Level Subject to Random Demand

Because of variable water demand in a community, the water level, $X(m)$ in a semi-spheric tank at the end of a day is best described as a random variable. Previous historical records and analysis indicate that a good model of the probability density function is given by

$$f_X(x)=a\sin^2(X), \quad 0\leq X\leq \pi$$

(1) Find the value of a such that the laws of probability are satisfied. (2) Derive the cumulative distribution function of water level. (3) Derive a mathematical expression to transform Uniformly-distributed random numbers $0\leq U\leq 1$ into samples of water level X. If $U=0.5$, calculate the value of X.

Solution
(1) According to properties 1 and 2 of probability density functions (see below equation (3.26)), $f_X(x)$ must be positive and its integral must be equal to 1. Thus,

$$\int_{-\infty}^{\infty}f_X(x)dx=\int_{0}^{\pi}a\sin^2(x)dx=a\frac{\pi}{2}=1 \quad \Rightarrow \quad a=\frac{2}{\pi}$$

and $f_X(x)>0$.

(2) The cumulative distribution of X is obtained from equation (3.28):

$$F_X(x) = \int_{-\infty}^{x} f_X(\xi)d\xi = \int_{0}^{x} \frac{2}{\pi} \sin^2(\xi)d\xi$$

After consulting a table of integrals,

$$F_X(x) = \frac{2}{\pi}\left[\frac{\xi}{2} - \frac{\sin(2\xi)}{4}\right]_0^x = \frac{1}{\pi}\left[x - \frac{\sin(2x)}{2}\right], \qquad 0 \le X \le \pi$$

(3) From equation (4.5), we equate the Uniform random variable U to the cumulative distribution function of X, $F_X(x)$:

$$F_X(x) = g^{-1}(x) = U \quad \therefore \quad \frac{1}{\pi}\left[X - \frac{\sin(2X)}{2}\right] = U$$

which could be written as

$$X - \frac{\sin(2X)}{2} - \pi U = 0, \qquad 0 \le X \le \pi$$

For a given Uniform random number U=0.7977, the corresponding value of X=1.571 may be found by solving this nonlinear equation. There are several numerical procedures to arrive at this result (e.g., trial and error, Newton iteration, etc.), which may be consulted in any numeral methods book.

Example 4.5 Transforming a Uniform Random Number into one with a Specified Density (Numerical Method)

Write a Maple program to solve Example 4.4.

Solution

[(1) Equate the integral of the density function to 1 and solve for a.

```
> restart:
  f:=x->a*sin(x)^2:
  int(f(x),x=0..Pi)=1:
  a:=solve(%,a);
```

$$a := \frac{2}{\pi}$$

[(2) The cumulative distribution function of X is obtained by integrating its density function.

```
> int(f(xi),xi=0..x):
  F:=unapply(%,x);
```

$$F := x \rightarrow -\frac{\cos(x)\sin(x) - x}{\pi}$$

[(3) Apply equation (4.5): Equate the cumulative of X to a random number u=0.5; solve for x.

```
> F(x)=0.7977:
  x:=fsolve(%,x);
```

$$x := 2.080270096$$

For question (3), we used the numerical solver in Maple, fsolve(,), which has two arguments: the first specifies the equation to solve, and the second the variable to solve for. One should note that for a general, unspecified, value $U=u$, this equation is difficult to solve analytically (i.e., algebraically) with the analytical solver in Maple, solve(,). In the next section we will introduce the method of decomposition as a new procedure to solve nonlinear equations analytically.

Analytical Decomposition of the Transformation Equation

Examples 4.4 and 4.5 illustrate the common case when the inverse of equation (4.5) leads to a nonlinear expression, whose explicit solution in closed analytical form is difficult. Given a Uniformly-distributed random number $U=u$, one needs to approximate the resulting equation numerically in order to transform the random number into its corresponding one $X=x$ that follows the desired probability density function. If one needs to generate 10,000 random numbers, then we need to solve that many times the nonlinear inverse equation. Notwithstanding the increasing speed of computer processors, this traditional approach does not seem efficient. Under the philosophy that an analytical solution, if available, is usually simpler, faster, and more intuitive than a numerical one, we now introduce a relatively new approach to solve nonlinear (differential) equations, called *the method of decomposition* (Adomian, 1994). Consider the general form of the nonlinear equation given in Example 4.4, question (3):

$$X+h(U)N(X)+g(U)=0 \qquad (4.6)$$

where

X=the dependent (output) variable
$g(U)$, $h(U)$=functions of the independent (input) variable U
$N(X)$=a nonlinear expression of the output variable

In Examples 4.4 and 4.5, $h(U)=-1$, $N(X)=\sin(2X)/2$, and $g(U)=-\pi U$. Let we write equation (4.6) as

$$X=-g(U)-h(U)N(X) \qquad (4.7)$$

and expand the nonlinear term, $N(X)$, as an infinite series. Equation (4.7) becomes

$$X=-g(U)-h(U)\sum_{i=0}^{\infty} A_i \qquad (4.8)$$

We may conceive the solution as a series $X=X_0+X_1+X_2+...+X_n$, where the first term, $X_0=-g(U)$, that is the first term in the right side of equation

(4.8). Each successive term may be calculated based on the previous ones if we express the A_i as an appropriate series expansion. If the series converges, we will have built an analytical series solution. Thus, define the A_i as the Adomian polynomials given by (Adomian, 1994)

$$A_0 = N(X_0),$$

$$A_1 = X_1 \frac{dN(X_0)}{dX_0},$$

$$A_2 = X_2 \frac{dN(X_0)}{dX_0} + \frac{X_1^2}{2!} \frac{d^2N(X_0)}{dX_0^2},$$

$$A_3 = X_3 \frac{dN(X_0)}{dX_0} + X_1 X_2 \frac{d^2N(X_0)}{dX_0^2} + \frac{X_1^3}{3!} \frac{d^3N(X_0)}{dX_0^3}$$

(4.9)

$$A_4 = X_4 \frac{dN(X_0)}{dX_0} + \left[\frac{X_2^2}{2!} + X_1 X_3 \right] \frac{d^2N(X_0)}{dX_0^2} + \frac{X_1^2 X_2}{2!} \frac{d^3N(X_0)}{dX_0^3} + \frac{X_1^4}{4!} \frac{d^4N(X_0)}{dX_0^4},$$

$$A_5 = X_5 \frac{dN(X_0)}{dX_0} + \left[X_2 X_3 + X_1 X_4 \right] \frac{d^2N(X_0)}{dX_0^2} + \left[\frac{X_1 X_2^2}{2!} + \frac{X_1^2 X_3}{2!} \right] \frac{d^3N(X_0)}{dX_0^3}$$

$$+ \frac{X_1^3 X_2}{3!} \frac{d^4N(X_0)}{dX_0^4} + \frac{X_1^5}{5!} \frac{d^5N(X_0)}{dX_0^5}$$

$$\vdots$$

and the solution to equation (4.7) as the series

$$X = X_0 + X_1 + X_2 + \cdots$$

$$X_0 = -g(U)$$

$$X_1 = -h(U)A_0$$

(4.10)

$$X_2 = -h(U)A_1$$

$$\vdots$$

$$X_n = -h(U)A_{n-1}$$

The polynomials A_n are generated for each nonlinearity so that A_0 depends only on X_0, A_1 depends only on X_0 and X_1, A_2 depends only on

X_0, X_1, X_2, etc. All of the X_n components are analytic and calculable. ΣX_n constitutes a generalized Taylor series about the function X_0. The computational strategy reduces to an alternative application of equations (4.10) and (4.9). In other words, begin by calculating X_0 from Equation (4.10). Next, apply equation (4.9) to derive A_0. Then, apply equation (4.10) again to derive X_1. This procedure is repeated until X_i is smaller than an acceptable error. Sometimes the magnitude of the nonlinear equation parameters is such that the series converges rapidly and the n-term partial sum $\Phi_n = \Sigma_{j=0}^{n-1} X_j$, called "the approximant," serves as an accurate enough and practical solution. The surprising feature of decomposition series is that when it converges, only two or four terms are needed to obtain an accurate solution. It is also recommended to calculate an even number of terms in a truncated series.

Example 4.6: Analytical Approximation of the Water-Level Random Numbers Transformation Equation

Using decomposition, solve the water-level random number's transformation equation of Example 4.4. In other words, derive a general analytical expression that transforms Uniform random numbers, U, into water-level random numbers, X.

Solution

Example 4.4, part (3), resulted in an equation which could be written in the form of equation (4.6) with $h(U)=-1$, $N(X)=\sin(2X)/2$, and $g(U)= -\pi U$. Equation (4.7) becomes:

$$X=\pi U+N(X), \quad N(X)=\frac{\sin(2X)}{2}, \quad 0\leq X\leq\pi$$

From equations (4.10) and (4.9), $=X_0+X_1+X_2+\cdots$, where

$$X_0=\pi U$$

$$X_1=A_0=N(X_0)=\frac{\sin(2X_0)}{2}=\frac{\sin(2\pi U)}{2}$$

$$X_2=A_1=X_1\frac{dN(X_0)}{dX_0}=X_1\frac{d}{dX_0}\left(\frac{\sin(2X_0)}{2}\right)=\frac{\sin(2\pi U)}{2}\cos(2\pi U)$$

$$X_3=A_2=X_2\frac{dN(X_0)}{dX_0}+\frac{X_1^2}{2!}\frac{d^2N(X_0)}{dX_0^2}=\frac{\sin(2\pi U)}{2}\cos^2(2\pi U)-\frac{\sin^3(2\pi U)}{4}$$

$\vdots$

Example 4.7: Transforming a Uniform Random Number into one with a Specified Density (Analytical Method)

In Example 4.5, $U=0.7977$. (1) Using the results of Example 4.6, calculate the numerical value of the first four decomposition terms. (2) Approximate the value of X. (3) Estimate the magnitude of the absolute error when using only 4 terms in the series.

Solution

(1) From Example 4.6 substitute $U=0.7977$ into the series equations to obtain the sequence $X_0=2.5061$, $X_1=-0.4777$, $X_2=-0.1410$, and $X_3=0.1764$. The series appears to be convergent. It is interesting to remark that for certain values of U convergence is slow. The user must always check for convergence prior to the use of the series in actual simulations.

(2) Adding the first 4 terms in the series, $\hat{X}=2.5061-0.4777-0.1410+0.1764=2.0637$, where $\hat{X}$ is and estimate of X.

(3) For an exact solution Y must satisfy the nonlinear equation (i.e., upon substituting the calculated value of X it yields 0). For an approximate solution, $\hat{X}$, the difference from zero obtained after substitution is an estimate of the absolute error. Let $\hat{e}$ be an estimate of the absolute error, thus,

$$\hat{X}-\frac{\sin(2\hat{X})}{2}-\pi U=\hat{e}=0.0352$$

Example 4.8

Write a Maple program to solve Example 4.7.

Solution

From Example 4.4, the random number's transformation equation to solve is

$$X-\frac{\sin(2X)}{2}-\pi U=0, \quad 0\leq X\leq\pi$$

where U is a Uniform random number between 0 and 1, and X represent the corresponding random number for water level. This equation is of the form given by equation (4.6) with $h(U)=-1$, $N(X)=\sin(2X)/2$, and $g(U)=-\pi U$. The following Maple worksheet programs equations (4.9) and (4.10) to derive a decomposition solution for X.

```
  Define the A polynomials (equation (4.9)) and the X's decomposition terms
  (equation (4.10)) for the nonlinear operator N
> restart:
   #Nonlinear operator N
   N:=X0->sin(2*X0)/2:
   #
   #First 3 derivatives of N
   diff(N(X0),X0):  dN:=unapply(%,X0):
   diff(N(X0),X0$2):  dN2:=unapply(%,X0):
   diff(N(X0),X0$3):  dN3:=unapply(%,X0):
   #
   #Alternative derivation of A's and X's
   A0:=X0->N(X0):
   X1:=X0->-h*A0(X0):
   A1:=X0->X1(X0)*dN(X0):
   X2:=X0->-h*A1(X0):
   A2:=X0->X2(X0)*dN(X0)+X1(X0)^2/2!*dN2(X0):
   X3:=X0->-h*A2(X0):
   #
   #Add the first 4 terms
   X0+X1(X0)+X2(X0)+X3(X0):
   X:=unapply(%,X0):
  Define the terms of a nonlinear equation (4.6) of the form X+hN(X)+g=0
> h:=-1:
   g:=U->evalf(-Pi*U):
   X0:=U->-g(U):         #Equation (4.10)
  (1) Calculate the first for terms in the series.
> U:=0.79773:  X0(U);  X1(X0(U));  X2(X0(U));  X3(X0(U));
```

$$2.506142708$$

$$-0.4776835447$$

$$-0.1411178776$$

$$0.1763079392$$

```
  (2) Approximate the solution of X.
> X(X0(U));
```

$$2.063649224$$

```
  (3) Calculate the error with 4 terms.
> X(X0(U))+h*N(X0(U))+g(U);
```

$$0.035190061$$

The Heart of Monte Carlo Simulations: Generation of Uniform Random Numbers

We have seen two approaches to solve a nonlinear equation that transforms Uniformly-distributed random numbers $0 \le U \le 1$ into those of a specified density function X (see Figure 4.4): one analytical and one numerical. We are now ready to simulate a random component of a system that follows a particular probability density function, provided we have a good source of Uniform random numbers. The most common computational algorithm is the linear congruential method. Under this approach, a sequence of numbers is recursively produced by the successive application of the equation

$$U_{i+1}' = (cU_i' + d) \bmod m \tag{4.11}$$

where

$U_i' =$ an integer number in the series

$c, d, m=$ large integer numbers (e.g., of the order of 10^6)

mod=modulo operator. For two integers w and z, w mod z is the remainder after dividing of w by z

The sequence is initiated with a seed value U_0', such that $0 \le U_0' \le m$. The generated sequence is of a period *less than* m. Each U_i' value is adjusted to the desired range $a \le U_i \le b$ by the equation

$$U_i = \left(\frac{b-a}{m}\right) U_i' + a \tag{4.12}$$

We remark again that the numbers generated by equations (4.11) and (4.12) are not truly random numbers, since the entire series is predictable. Some authors prefer to call these numbers *pseudo-random numbers*. If the constants c, d, and m are properly selected, the generated numbers closely follow a theoretical Uniform probability density function (see Figure 3.12). Under these circumstances the numbers may be used in simulations. In many cases, however, the resulting numbers depart widely from the required theoretical distributions and should be discarded. This is particularly so when the sample size, N, of numbers is small (e.g., less than 100). It is imperative that the modeler tests the numbers by plotting the resulting frequency histogram (see Chapter 7 for more details), calculating their mean, variance, and serial correlation (to be covered later). Calculated moments of random numbers must conform to those of the theoretical distribution and must pass a goodness-of-fit test to a theoretical distribution. In addition, random numbers should also exhibit a zero serial correlation coefficient, for all lags. Most calculators, mathematics software, spreadsheets, and statistics software provide routines for the generation of Uniformly-distributed random numbers. These numbers must not be used blindly without prior testing since in some cases the software-generated numbers are unsuitable.

Example 4.9: Generation of Uniform Random Numbers

Generate a sequence of $N=1000$ Uniform random numbers between $a=0$ and $b=1$ using equations (4.11) and (4.12). Plot the resulting frequency histogram; calculate their mean and standard deviation; calculate the lag-one serial correlation coefficient; and compare with the theoretical values. Repeat the exercise this time using the numbers produced by the Maple library. State your conclusions.

Solution

The Maple worksheet follows. The manner in which the numbers are calculated and stored in arrays is similar to that of Example 1.1 (see Chapter 1).

[Initialize stats modules. Enter parameter values. Choose large integers for c, d, m, and seed.

```
> restart: with(stats): with(stats[statplots]):
    N:=1000: a:=0.0: b:=1.0: c:=12345: d:=9876:
    m:=1199017: seed:=1243: Up[0]:=seed:
```

[Apply the linear congruential equation using the loop command, for. Note the command syntax. It starts with the word "for", then specifies the the range of values of the sub-index *i* over which the next equation is iteratively executed, and ends with the words "en do".

```
> for i from 1 to N do Up[i]:=(c*Up[i-1]+d) mod m end do:
```

Adjust the numbers to be within the specified limits from a to b, using the sequence, seq(,), command.

```
> U:=[seq(Up[i]*(b-a)/m+a,i=1..N)]:
```

[Plot a frequency histogram of the numbers. Compare the histogram to the theoretical Uniform distribution plotted as a Heaviside Unit step function.

```
> histogram(U,color=gray):
    plot(Heaviside(u)*Heaviside(1-u),u=-0.1..1.1,
        color=black,thickness=2):
    plots[display](%,%%,labels=[` U `,` fU(u) `]);
```

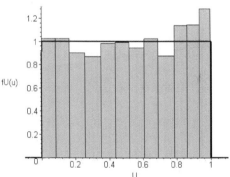

[Calculate mean and standard deviation of the numbers. Compare with the theoretical values mu=0.5 and sigma=sqrt(1/12) (see equation (3.39))

```
> describe[mean](U); describe[standarddeviation](U);
    sqrt(1/12.);
```

$$0.5142472691$$
$$0.2958721617$$
$$0.2886751346$$

[Now check for lag-one serial correlation, which should be close to zero.

```
> U1:=[seq(U[i+1],i=1..N-1),0.5]:
    describe[linearcorrelation](U, U1);
```

$$-0.04660268058$$

Now generate and repeat tests for *N* Uniformly distributed random numbers using the Maple random[ ]( )

```
> Um:=[random[uniform](N)]:
    histogram(Um,color=gray):
    plot(Heaviside(u)*Heaviside(1-u),u=-0.1..1.1,
        color=black,thickness=2):
    plots[display](%,%%,labels=[` U `,`fU(u)`]);
```

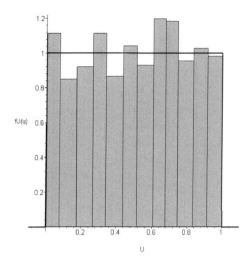

```
> describe[mean](Um); describe[standarddeviation](Um);
                      0.5077277790
                      0.2860012755
> Um1:=[seq(Um[i+1],i=1..N-1),0.5]:
  describe[linearcorrelation](Um, Um1);
                      0.02358306594
```

Note that to plot the theoretical Uniform density we use the Heaviside Unit Step function introduced in Chapter 3. The theoretical values of the mean and standard deviation are 0.5 and 0.2887, respectively (see equation (3.39)). The two series of random numbers appear to fit reasonably well the theoretical Uniform distribution. The last test, the serial lag-one correlation coefficient, will be covered in Chapter 12. The goodness-of-fit test, a measure of the ability of the numbers to fit the theoretical distribution will be introduced in Chapter 7.

Generation of Gaussian Random Numbers

A common approach to the problem of synthesis of Standard Gaussian, random numbers $x_i \sim N(0, 1)$ is to first produce two sequences, u_i and v_i, of Uniform random numbers in the interval from 0 to 1. For each pair of them, a value of x_i is calculated according to the transformation

$$x_i = \sqrt{-2\ln(u_i)}\cos(2\pi v_i) \qquad (4.13)$$

Standard Normal numbers, x_i, may be transformed into Normal numbers $y_i \sim N(\mu_Y, \sigma_Y)$ with mean μ_Y and standard deviation σ_Y by applying equation (3.41):

$$y_i = \mu_Y + \sigma_Y x_i \qquad (4.14)$$

Example 4.10: Synthesis of Lake Concentration Data

In Example 3.22, Chapter 3, lake concentration of TCE in $\mu g/L$, $C \sim N(650, 150)$. (1) generate 1000 data values using equations (4.11)-(4.14), plot their frequency histogram with the theoretical density function, and calculate the sample mean and standard deviation. (2) Repeat the item (1) using the Maple library.

Solution

(1) Simnilar to Example 4.9, generate two sequences of Uniform random numbers, u[ ] and v[ ], in the interval 0 to 1, using equations (4.11) and (4.12).

```
> restart: with(stats): with(stats[statplots]):
  N:=1000: a:=0.0: b:=1.0: c1:=12345: d1:=9876:
  m1:=1199017: seed1:=1243: up1[0]:=seed1: c2:=32567:
  d2:=7862: m2:= 2287653: seed2:=2719: up2[0]:=seed2:
  pii:=evalf(Pi): muY:=650: sigmaY:=150:
  for i from 1 to N do up1[i]:=(c1*up1[i-1]+d1) mod m1 end do:
  for i from 1 to N do up2[i]:=(c2*up2[i-1]+d2) mod m2 end do:
  u:=[seq(up1[i]*(b-a)/m1+a,i=1..N)]:
  v:=[seq(up2[i]*(b-a)/m2+a,i=1..N)]:
```

Apply equation (4.13) to transform the uniform series u[ ] and v[ ] into the Standard Normal series x[ ].

```
> x:=[seq( sqrt(-2*ln(u[i]))*cos(2.*pii*v[i]), i=1..N)]:
```

Transform Standard Normal series x[ ] into Normal series y[ ]~ N(muY, sigmaY) (equation (4.14)).

```
> y:=[seq(sigmaY*x[i]+muY,i=1..N)]:
```

Plot Gaussian random numbers stored in y[ ] along with the theoretical Gaussian curve.

```
> histogram(y,color=gray):
  plot(statevalf[pdf,normald[muY,sigmaY]](C),C=200..1200,
    color=black,thickness=2):
  plots[display](%,%%,labels=[`C`,`fC(c)`]);
```

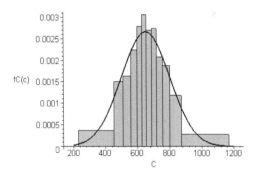

Calculate the sample mean and standard deviation of the Gaussian nombers y[ ]. The theoretical values are 650 and 150, respectively.

```
> describe[mean](y): describe[standarddeviation](y):
```
$$658.0052091$$
$$148.6489914$$

(2) Generate N Normally-distributed random numbers using the Maple random[ ]( ) command. Inside the square brackets specify "normald[mu,sigma]" distribution with mean muY and standard deviation sigmaY. When these limits are not specified, Maple assumes muY~0 and sigmaY~1 (i.e., Standard Normal). Store the numbers in the array ym[ ].

```
> ym:=[random[normald[650,150]](N)]:
```

Plot the results as before.

```
> histogram(ym,color=gray):
  plot(statevalf[pdf,normald[650,150]](C),C=200..1200,
    color=black,thickness=2):
  plots[display](%,%%,labels=[`C`,`fC(c)`]);
```

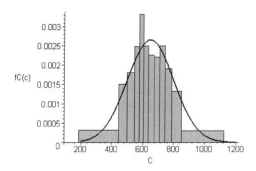

Finally, calculate the mean and standard deviation of the synthetic data.
```
> describe[mean](ym); describe[standarddeviation](ym);
```
643.7496450
149.5863793

Application of a Transformation Equation: Generation of Random Numbers of Other Probability Distributions

To generate random numbers from a particular probability density function that represents the input of an engineering system, one first generates a series of Uniformly-distributed random numbers and then transform them into numbers of the required density by applying the transformation equation (4.5) (see Figure 4.4), as described earlier. This process is programmed as an intrinsic function in most mathematics programs. To generate random numbers of common distributions –such as the Normal, Lognormal, Weibull, Exponential, and Gamma distributions–, the user needs to call an appropriate function in a program (e.g., the statevalf[ ]( ) command in Maple). Sometimes, however, one needs to generate random numbers of unusual probability distributions not available in standard mathematics libraries. In such case, it is necessary to program your own Monte Carlo simulation model and the examples in this chapter are helpful.

Example 4.11: Simulation of Tank Water-Level

In Example 4.4, generate one year of end-of-day water levels in the distribution tank. Use a numerical procedure to solve the transformation equation. Calculate the statistics of the numbers and compare them with the theoretical ones. Calculate the lag-one serial correlation coefficient and see that it is close to zero. Finally, plot the frequency histogram of the numbers along with the theoretical density function.

Solution

```
> restart: with(stats): with(stats[statplots]):
  a:=2/Pi: N:=365: pii:=evalf(Pi):
  #Theoretical density, its mean and std. deviation
  fX:=x->a*sin(x)^2:
  muX:=int(x*fX(x),x=0..pii);
  sigmaX:=sqrt(int((x-muX)^2*fX(x),x=0..pii));
```

$$muX := 1.570796327$$

$$sigmaX := 0.5678618084$$

Generate Uniform random numbers and store them in the array U[ ].
```
> U:=[random[uniform](N)]:
```
Define the transformation equation (see Example 4.4) as a function of a Uniform number, u.
```
> eqn:=u->x-sin(2*x)/2-pii*u=0;
```

$$eqn := u \rightarrow x - \frac{1}{2}\sin(2x) - pii\,u = 0$$

Use the sequence, seq(,), command to evaluate x for each value, u, contained in the vector U[ ].
Note how the arguments in the sequence command were filled. The first argument in the sequence
is actually another command, fsolve(,) which will solve numerically the transformation equation.
The second argument in the sequence specifies that the first argument is to be executed for each
of the values conteined in the vector U[ ].
```
> X:=[ seq( fsolve(eqn(i), x), i=U) ]:
```
Calculate the mean and standard deviation of the numbers, which should be close to muX and
sigmaX.
```
> describe[mean](X); describe[standarddeviation](X);
```

$$1.553188343$$

$$0.5720512328$$

Calculate the lag-one serial correlation coefficient (see Example 4.9).
```
> X1:=[seq(X[i+1],i=1..N-1),muX]:
  describe[linearcorrelation](X, X1);
```

$$0.05975048506$$

Plot the histogram of the random numbers with the theoretical density function.
```
> histogram(X,color=gray):
  plot(2*sin(xi)^2/pii,xi=0..pii,color=black,thickness=2):
  plots[display](%,%%,labels=[` X `,` fX(x )`]);
```

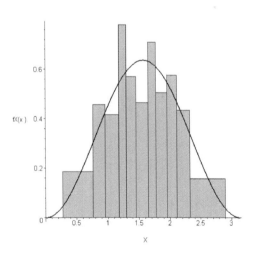

The simulated numbers approximate well the theoretical mean and
standard deviation. The generated numbers have a low serial correlation,
and the shape of the frequency histogram suggests a reasonable fit.
However, a fit test is needed to assess the latter objectively (Chapter 7).

Example 4.12

An engineering system has a random component with a probability density function given by $f_X(x) = a(1+x^2-x^3-x^4)$, $0 \le x \le 1$, $a=1.132076$. (1) Derive an expression to transform Uniform numbers into X-distributed numbers with the above distribution and solve it using decomposition. (2) Test your model by generating 10,000 samples of X; compare the numbers' moments with respect to the theoretical ones; calculate the lag-one serial correlation; and plot the numbers' histogram along with the theoretical density.

Solution

[(1) Define the probability density function, f(x).
[> restart: a:=1.132076: f:=x->a*(1+x^2-x^3-x^4):
[Define the cumulative distribution function, F(x).
[> int(f(xi),xi=0..x):
[F:=unapply(%,x):
[Transformation equation (4.6), $X+hN(x)+g(U)=0$, is obtained by setting F(x)=U,
[where U is a Uniform number (see equation 4.5).
[> eqn:=F(x)=U;

$$eqn := 1.132076000\,x + 0.3773586667\,x^3 - 0.2830190000\,x^4 - 0.2264152000\,x^5 = U$$

[From above, identify the first coefficient on the left hand side, b.
[> b:=op(1,lhs(eqn))/x;

$$b := 1.132076000$$

[Next, to obtain the nonlinear function, N(x), divide the second, third,
[and fourth terms on the left hand side by b.
[> (op(2,lhs(eqn))+op(3,lhs(eqn))+op(4,lhs(eqn)))/b:
[N:=unapply(%,x):

$$N := x \rightarrow 0.3333333334\,x^3 - 0.2500000000\,x^4 - 0.2000000000\,x^5$$

[Similarly, h=1, and g(U) is derived by dividing U by b and changing its sign
[(move to the left hand side of equality).
[> h:=1:
[g:=U->-U/b:
[Similar to Example 4.8, define the A polynomials (equation (4.9)) and the X's
[decomposition terms (equation (4.10)) for the nonlinear operator N
[> #First 3 derivatives of N
    diff(N(X0),X0): dN:=unapply(%,X0):
    diff(N(X0),X0$2): dN2:=unapply(%,X0):
    diff(N(X0),X0$3): dN3:=unapply(%,X0):
    #
    #Alternative derivation of A's and X's
    A0:=X0->N(X0):
    X1:=X0->-h*A0(X0):
    A1:=X0->X1(X0)*dN(X0):
    X2:=X0->-h*A1(X0):
    A2:=X0->X2(X0)*dN(X0)+X1(X0)^2/2!*dN2(X0):
    X3:=X0->-h*A2(X0):
    #
    #Add the first 4 terms
    X0+X1(X0)+X2(X0)+X3(X0):
    X:=unapply(%,X0):
[Define the first term in the decomposition series.
[> X0:=U->-g(U):   #Equation (4.10)

(2) Theoretical density, its mean and std. deviation

```
>  muX:=int(x*f(x),x=0..1);
   sigmaX:=sqrt(int((x-muX)^2*f(x),x=0..1));
```

$$muX := 0.4339624667$$

$$sigmaX := 0.2550412827$$

Generate Uniform random numbers and store them in the array Unum[ ].

```
>  N:=10000: with(stats): with(stats[statplots]):
   Unum:=[random[uniform](N)]:
```

Apply the decoimposition solution X( ) to each Uniform random number and store in the array Xnum[ ].

```
>  Xnum:=[ seq( X(X0(i)), i=Unum ) ]:
```

Calculate the mean and standard deviation of the numbers, which should be close to muX and sigmaX.

```
>  describe[mean](Xnum); describe[standarddeviation[1]](Xnum);
```

$$0.4376637669$$

$$0.2550266066$$

Calculate the lag-one serial correlation coefficient (see Example 4.11).

```
>  X1num:=[seq(Xnum[i+1],i=1..N-1),muX]:
   describe[linearcorrelation](Xnum, X1num);
```

$$-0.0009542724006$$

Plot the histogram of the random numbers with the theoretical density function.

```
>  histogram(Xnum,color=gray):
   plot(f(x),x=0..1,color=black,thickness=2):
   plots[display](%,%%,labels=[` X` ,` f(x)`]);
```

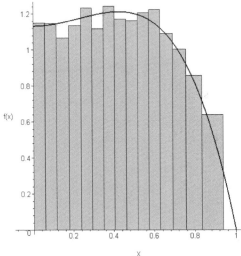

Example 4.12 illustrates the process of generating samples from an unusual distribution for which a random number generator is not available in software libraries. It further shows the use of analytical decomposition for the solution of the equation that transforms Uniform numbers into X numbers following the specified $f_X(x)$ distribution. We will take advantage of what we have learned when we discuss the solution of nonlinear stochastic differential equations in Chapter 13.

Summary: Generation of Random Numbers for a Monte Carlo Simulation Model

On page 108 we introduced the three basic steps in a Monte Carlo simulation model. *Step* 1 involves the generation of random samples for a system component, X, described by a known probability density function, $f_X(x)$. *Step* 2 consists in passing each sample of X through the engineering system by evaluating the system transfer function, $Y=g(X)$, at each sample of X (see Figure 4.1). $Y=g(X)$ is, in general, a solution to a differential equation governing the system. This repetitive evaluation of g for all samples of X produces samples of the system response or system output, Y. We will return to this problem in Chapter 13 when we discuss random differential equations. *Step* 3 of a Monte Carlo simulation consists in evaluating the statistical properties of the output, Y, and perhaps the fitting of a probability density function, $f_Y(y)$ if desired. Fitting probability distributions to data is discussed in Chapter 7.

Much of the effort in this chapter has been geared toward the generation of random numbers from a specified distribution, that is, *Step* 1 in Monte Carlo simulations. Let us now summarize the options we have:

☐ The simplest case involves the generation of random numbers, X, from a common distribution, $f_X(x)$, such as the Normal, Lognormal, Exponential, or Weibull distribution. Samples from these are easily obtainable from one of the intrinsic functions in most computer programs (see Examples 1.1 and 4.11).

☐ For the general problem of random numbers, X, from a probability distribution, $f_X(x)$, for which standard computer libraries are not available, the modeler must build his own generator in two stages as follows (see Figure 4.5):

  • Start by generating Uniform numbers, U, in the interval from 0 to 1, which could be accomplished by calling a standard intrinsic function in most computer programs, or from the application of equations (4.11)-(4.12) (see Example 4.9).

  • Derive an expression that turns Uniform numbers, U, into numbers, X, of the desired density, $f_X(x)$. This transformation is built by equating the cumulative distribution function of X to the number U, that is, set $F_X(x)=U$ and solve for U (see equation (4.5) and Figure 4.4).

- The solution of the transformation equation may be obtained numerically by invoking an equation solver function in a computer program (see Examples 4.5, and 4.11), or by applying analytical decomposition equations (4.9) and (4.10). With decomposition, a solution may be manually derived (see Examples 4.6 and 4.7), or programmed in a computer (see Examples 4.8 and 4.12).

☐ All random numbers, regardless of their source, must be tested for quality prior to their use. This includes ascertaining that the moments (e.g., the mean and the variance) of the random numbers are close to those of the theoretical distribution from whence they came; the serial correlation coefficient of the numbers is close to zero (i.e., uncorrelated numbers); and that the empirical frequency histogram of the numbers fits the theoretical distribution (see Examples 4.9 through 4.12).

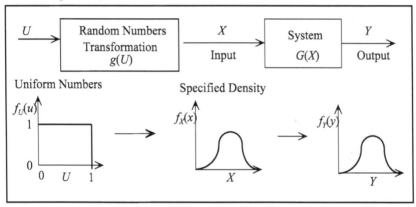

Figure 4.5: Monte Carlo Simulation Model Starting with Uniform Numbers

4.3 SIMULATION OF SYSTEMS WITH SEVERAL RANDOM VARIABLES

System Sensitivity to Uncertainty in One or More Variables

Most engineering systems have a combination of several input variables, a system operator function with one or more uncertain variables, and a combination of several output variables. Not all system variables are random. Most systems have one or more deterministic components and one or more uncertain components which are best described as random variables with a given probability density function. There is also the very common situation when a system variable is partially deterministic and partially random. In other words, a variable may have a large portion of its magnitude governed by a well-known process or physical law, in addition to a smaller uncertain part regarded as an error in measurement, environmental fluctuation, or any other uncontrolled variability. The latter is usually characterized as a random variable. Even if a fundamental law is not known, it is common to conceive a variable, say w, as the summation of its mean, $\overline{w}$, plus or minus a smaller random fluctuation, w', characterized as a random variable. $\overline{w}$ is the deterministic component estimated as the mean of a sufficiently large sample, and w' is a random variable with samples sometimes greater and sometimes less than $\overline{w}$. With this scheme, $E\{w\}=\overline{w}$, and $E\{w'\}=0$.

The complexity in the uncertainty analysis of a system may be substantially reduced by a judicious study of the effect of each system component on overall system response. The sensitivity of system output to variability in one input variable may be observed by calculating the system output variable when the input variable takes on its maximum and minimum values, respectively, while keeping all other variables constant. By observing output sensitivity to the maximum range of variability in the input one may assess the quantitative importance of the input variable. Depending on the system operator, $G(\ )$, a wide variability in the input may be translated into an even wider variability in the output. Thus, the input uncertainty may be amplified. Conversely, input variability might actually be reduced after it passes through the system. This is the case of many energy dissipation systems. By repeating the above sensitivity analysis with each uncertain input variable, the engineer may objectively quantify the relative importance of each of the uncertain variables. By this process she can decide which variables may be regarded as deterministic (i.e., output magnitude is not very sensitive to input uncertain variability), and which may be regarded as truly random. The latter are the most difficult to control and the ones that have the greatest impact on the predictability or forecasting of the system output. Measurement effort and uncertainty analysis of these variables will be the major focus of system modeling.

Systems with Several Independent Random Variables

Examples 1.1 and 1.2 (Chapter 1) introduced the tools of Monte Carlo simulations of a system with several random variables. Sensitivity of system response to various levels of input uncertainty was studied. Armed with the theory covered in the last chapters, the reader would benefit by reexamining those examples. It was noted that, depending on the input level of uncertainty (i.e., the variance), output uncertainty may be substantially amplified to the point that the output becomes completely chaotic and unpredictable. It was also observed that the system was more sensitive to certain input variables than to others. In Examples 1.1 and 1.2 output velocities were substantially more sensitive to input head variances than to hydraulic conductivity variances. We found the reason in the particular mathematical form of the system operator: certain operations such as differentiation are inherently more unstable than others (e.g., integration). Monte Carlo simulations are valuable exercises for planning an appropriate measurement strategy that would minimize an output variance by minimizing that of the most sensitive intervening variables.

An important simplifying assumption of the systems studied thus far, is that of *independence* among the various random variables (see equation (2.24), Chapter 2, for the concept of independent events). Two system random variables are independent if sample values of one do not affect statistically those of the other. More precisely, two random variables are *independent* if their joint probability density function equals the product of the individual, marginal, density functions as we will see in Chapter 5. Often, the independence of two or more system random variables is *assumed* as a convenient, a preliminary, or an approximate model. Modeling of random systems with dependent random variables requires information on their joint density function, which substantially increases the complexity and data requirements. It is important to observe whether independence is a modeling assumption or the result of conclusive correlation tests performed on samples (see Chapter 5); that way one can assess the accuracy of our modeling conclusions.

A common decision in the analysis of engineering systems is to suppose that the random variables in the system input, while dependent on one another, are independent of the random variables in the system operator. The operator, $G(X)$, is usually composed of intrinsic system parameters, material, or mechanical properties. It is reasonable to assume that these system parameters are statistically independent of the system input, which are usually composed of external environmental excitations or system forcing functions. In Examples 1.1 and 1.2, Chapter 1, it was assumed that the errors in hydraulic head measurements (i.e., the system input) was statistically independent of the system parameter, the hydraulic conductivity, which is a coefficient that depends on the fluid and the porous media properties. This is a reasonable assumption. By assuming independence between these variables, it was possible to generate random numbers from the two distributions as a set of two independent series. With the above concepts, let us give a preliminary classification of random systems:

1. *Random input systems*: systems with one or more input components

characterized as random variables.

2. *Random parameter systems*: systems with one or more of its operator's parameters, physical, chemical, or mechanical properties defined as random variables.

3. *Composite systems*: a combination of the above two classes.

We will extend this preliminary list when we cover systems governed by differential equations. Most examples covered in this chapter refer to random input systems. Examples 1.1 and 1.2 in Chapter 1 refer to a composite system.

Example 4.13: Forecasting Groundwater Velocity Statistics

Modify the program in Example 1.1, Chapter 1, to generate 2000 samples of groundwater flow velocity; calculate its mean, standard deviation, and coefficient of variability; and plot the frequency histogram.

Solution

```
[ Start the statistics module. An define variables.
[ > restart: with(stats): with(stats[statplots]):
    deltax:=100: h1_mean:=125: h2_mean:=122: K_mean:=50:
    N:=2000: Sh1:=125*0.01: Sh2:=122*0.01: Sk:=50*0.1:
[ Generate Gaussian random numbers representing head and conductivity measurements.
[ > h1_data:=[ random[normald[h1_mean,Sh1]](N) ]:
    h2_data:=[ random[normald[h2_mean,Sh2]](N) ]:
    K_data:= [ random[normald[K_mean,Sk]](N) ]:
[ Calculate the hydraulic gradient data by subtracting the array h1_data
[ from the array h2_data and dividing by deltax.
[ > dh_dx:=(h2_data-h1_data)/deltax:
[ Sequentially apply equation (1.1) to calculate the output velocity samples.
[ > u_data:=[ seq(-K_data[i]*dh_dx[i],i=1..N) ]:
[ Calculate the mean and the standard deviation of the output velocity.
[ > u_bar:=describe[mean](u_data);
    sigma_u:=describe[standarddeviation](u_data);
    Cv:=sigma_u/u_bar;
```

$$u\_bar := 1.478405238$$
$$sigma\_u := 0.9049436720$$
$$Cv := 0.6121079990$$

```
[ > histogram(u_data,color=gray,
```

Velocity, U(m/month)

With N=2000, we now have more reliable values of the statistics: with the current level of uncertainty in heads and conductivity, the velocity has a sample mean $\bar{u}$=1.478$m/month$, a sample standard deviation σ_u=0.905$m/month$, and a coefficient of variability of C_v=0.612. From Example 1.2, the system is more sensitive to uncertainty in the heads. To reduce the output variance (i.e., to increase output predictability), increase the accuracy in head's measurements or modify the way the hydraulic gradient is calculated. This could be accomplished by increasing the number of measurements while reducing the spacing between measurements, Δx. One may also use a more accurate finite difference formula for the spatial derivative. It is clear that random system simulation is useful in planning and improving engineering models. We will later use the above information to fit probability distributions.

Example 4.14

A system operator is given by $U=v+We^{-Kx}$, where $0 \leq x < \infty$ is the system input, K is a decay parameter such that $K \sim N(0.1, 0.01)$, W is a scale parameter such that $W \sim N(1, 0.1)$, v=1 is a constant, and U is the system output. (1) Given a fixed value of x=10, and assuming the random variables are independent, construct a model to predict the statistical properties of the output (i.e., a sample mean, standard deviation, skewness coefficient, and frequency histogram). (2) Now allow x to vary. For each sample of K and W, U is an exponentially-decaying curve of x. Using 10 samples of K and 10 samples of W, generate and plot 10 corresponding samples of $U(x)$.

Solution

```
[ (1) Start the statistics module and enter data.
[ > restart: with(stats): with(stats[statplots]):
    mu_k:=0.1: mu_w:=1: sigma_k:=0.01: sigma_w:=0.1:
    X:=10: v:=1: N:=1000:
[ Generate u and k random numbers.
[ > k_data:=[ random[normald[mu_k,sigma_k]](N) ]:
    w_data:=[ random[normald[mu_w,sigma_w]](N) ]:
[ Calculate sample output values.
[ > u_data:=[ seq( v+w_data[i]*exp(-k_data[i]*X), i=1..N) ]:
[ Calculate the statistics and plot the output frequency histogram.
[ > u_bar:=describe[mean](u_data);
    sigma_u:=describe[standarddeviation](u_data);
    Cv:=sigma_u/u_bar;
    gama:=describe[skewness](u_data);
    histogram(u_data,color=gray,labels=[`U`,`fU(u)`]);
```

$$u\_bar := 1.370276477$$
$$sigma\_u := 0.05172819743$$
$$Cv := 0.03775019005$$
$$gama := -0.2264597490$$

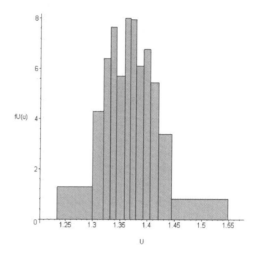

[(2) Using the for command, generate 10 curves and store them in the vector g[ ].

```
> for i from 1 to 10 do
    g[i]:=plot(v+w_data[i]*exp(-k_data[i]*x),x=0..50,
        color=black,labels=[` X `,` U `]) end do:
  plots[display]( seq( g[i], i=1..10) );
```

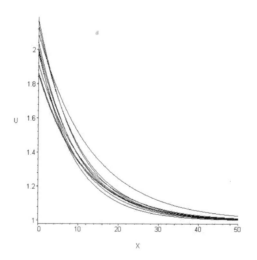

We emphasize that the histogram above corresponds to that of the random variable U at $x=10$. At another value of x, U is another random variable. For each x there is a different random variable with different statistics. Thus, U in reality is *a family* of random variables, or a *random process*. For a fixed sample value of the random parameters (i.e., for a fixed sample of K and W), U is a smooth output sample curve (see the last graph in (2)). For a fixed value of x, U is a random variable (see the graph in (1)). We will return to this concept in the chapter on random processes.

It is interesting to note that, in this case, the variance of U decreases as x increases. In other words the exponential curves get closer to one another indicating a decrease in system uncertainty.

4.4 ANALYTICAL DERIVATION OF SECOND-ORDER STATISTICS

Sometimes the modeler seeks to obtain the mean and the variance of the system output and does not desire to spend time in lengthy numerical simulations of random variables. Furthermore, in many practical circumstances Monte Carlo simulation is not possible. This occurs when the probability density function of the system input or the system parameter is not available. The engineer often faces the real-life scenario of having to study, design, or forecast a system with insufficient data. He knows that at least one of the components is a random variable, but there is not sufficient data to warrant a good fit with a theoretical probability distribution. Most of the time, the reason behind insufficient data is economical: it is expensive to collect data; it requires qualified personnel, sophisticated equipment, and effort. Sometimes the inherent nature of the system makes it difficult to collect data. This is the case of environmental systems that require long historical records, or structural systems that imply destructive tests.

In such cases of insufficient data, the engineer must perform his analysis with a limited set of samples from which the first and second-order sample statistics (i.e., the mean and the variance) of the input are estimated. With the mean and the variance of the input quantities, or of the parameters, it is only possible to obtain at best the corresponding mean and variance of the output. These statistics give preliminary information on system uncertainty. The method to obtain the mean and the variance of the output is indeed simple. The procedure consists in applying the rules of expectation, as discussed in Chapter 3, to the system operators' equation. Depending on the mathematical complexity of the operator, sometimes it is possible to arrive at simple results. In principle, if the input density function is $f_X(x)$ and the system operator is $G(X)$ (see Figure 4.1), the mean of the output, μ_Y, is given by

$$\mu_Y = E\{Y\} = E\{G(X)\} = \int_{-\infty}^{\infty} G(x)f_X(x)dx \qquad (4.15)$$

which may be inferred from the definition of the expectation operator, equation (3.30). If the input random variable is discrete, then equation

(4.15) becomes

$$\mu_Y = E\{Y\} = E\{G(X)\} = \sum_{i=1}^{m} G(x_i)p_X(x_i) \qquad (4.16)$$

where

$p_X(x_i)$=probability mass function of X

m=number of class intervals

Equation (4.16) may be inferred from the definition of the expectation operator for discrete random variables, equation (3.5). If the input density function is not known, but the first two moments of the input are, the corresponding moments of the output may be deduced by applying the properties of the expectation operator (see properties 1 through 4, below equation (3.5)).

Example 4.15

A system operator equation is given by $Y=G(X)=aX^2+bX+c$, where X is a random variable with an unknown distribution; a, b, and c are constants, and Y is the system output. If the mean and the variance of X are known and denoted as μ_X and σ_X^2, respectively, derive expressions for the mean and variance of the output, μ_Y and σ_Y^2 respectively.

Solution

Taking expectation on both sides of the operator's equation, and using the properties of expectation (see properties 1, 2, and 3 below equation (3.5), Chapter 3),

$$\mu_Y = E\{Y\} = E\{aX^2+bX+c\} = E\{aX^2\}+E\{bX\}+E\{c\}$$

$$= aE\{X^2\}+bE\{X\}+c = a(\mu_X^2+\sigma_X^2)+b\mu_X+c$$

where equation (3.8) was used in the last line. Hence, the mean of the output is given as a function of the mean and the variance of the input, and the system parameters a, b, and c. Now from equation (3.8),

$$\sigma_Y^2 = E\{Y^2\}-\mu_Y^2 = E\{[aX^2+bX+c]^2\}-\mu_Y^2$$

However, after solving, the right-hand side contains $\overline{X^4}$ and $\overline{X^3}$, that is the fourth and the third moments of X with respect to the origin. These moments are not provided with the data. Thus, in this system, the output variance may not be found with only the input mean and variance.

Example 4.16

In Example 4.14, (1) derive an analytical expression to estimate the mean of the output. (2) Using the expression in (1), calculate the numerical value of the mean at $x=1$.

Solution

(1) Taking expectation on both sides of the system operator's equation, using the properties of expectation, and noting that the random variables are independent (see property 4 of expectation below equation (3.5)), then

$$\mu_U = E\{U\} = v + E\{W\}E\{e^{-Kx}\}$$

Expanding the exponential as a Taylor series,

$$\mu_U = v + \mu_W\left[1 - \mu_K x + E\{K^2\}\frac{x^2}{2!} - E\{K^3\}\frac{x^3}{3!} + E\{K^4\}\frac{x^4}{4!} - \cdots\right]$$

Now, since $K \sim N(\mu_k, \sigma_k)$, we can transform it into a standard Normal variable $Y \sim N(0, 1)$ by applying equation (3.41). Thus, substituting $K = Y\sigma_K + \mu_K$ into the last equation,

$$\mu_U = v + \mu_W[1 - \mu_K x + E\{(Y\sigma_K + \mu_K)^2\}\frac{x^2}{2!} - E\{(Y\sigma_K + \mu_K)^3\}\frac{x^3}{3!}$$

$$+ E\{(Y\sigma_K + \mu_K)^4\}\frac{x^4}{4!} - \cdots]$$

Since Y is a Standard Normal variable, all of the odd moments in the series are zero and all the even moments may be computed by the formula $E\{Y^n\} = 1 \times 3 \times \cdots (n-1)\sigma_Y^n$, for n even (see Papoulis, 1984 for proof), where $\sigma_Y^4 = E^2\{Y^2\}$ and $\sigma_Y = 1$ for a standard Normal. Let us truncate the series and compute the first few terms only:

$$\mu_U \approx v + \mu_W\left[1 - \mu_K x + E\{(Y\sigma_K + \mu_K)^2\}\frac{x^2}{2!}\right] = v + \mu_W\left[1 - \mu_K x + (\sigma_K^2 + \mu_K^2)\frac{x^2}{2}\right]$$

(2) At $x=1$ and substituting the values of the moments into the last equation,

$$\mu_U \approx 1 + 1 \times \left[1 - 0.1 \times 1 + (0.01^2 + 0.1^2)\frac{1}{2}\right] = 1.90505$$

which is very close to the Monte Carlo-calculated mean in Example 4.14. Notice that as x increases, the series diverges and the results are incorrect. For small values of x, the analytical method is simple and accurate. Specifically, it is easy to see that in the derived equation for the mean the following restriction applies:

$$\mu_W \mu_K x < 1 \quad \Rightarrow \quad x < 10$$

When the system operator's equation results from a solution to a differential equation, the above results apply directly. However, if the system differential equation is nonlinear, and a simple analytical solution is sought, the method of decomposition described earlier in this chapter constitutes a valuable tool to construct a series that may converge fast to the true nonlinear solution. From the truncated series, one may estimate the first two moments of the output as described in this section. We will return to this important aspect of uncertainty analysis when we study random differential equations.

PROBLEMS

4.1 Solve Example 4.1 if the input density function is a Laplace random variable with parameter $\alpha = 1$.

4.2 Solve Example 4.2 if $Y = X^2 + 2X + 1$.

4.3 In Example 4.3, assume $a = c = 1$. Plot the input density, the system function, and the output density. State your conclusions about the effect of the system on the shape of the output density function.

4.4 A system component is modeled as a Rayleigh random variable, X. In order to conduct a Monte Carlo simulation, an engineer needs random numbers that follow this probability distribution. (1) Derive a mathematical expression that would transform Uniformly-distributed random numbers, U, into Rayleigh-distributed ones. (2) If the Rayleigh parameter $\alpha = 1$ and a Uniformly-distributed random number $U = 0.435$, calculate the corresponding value of the Rayleigh-distributed random number.

4.5 A system component has a probability density function given by

$$f_X(x) = \frac{4}{3}(1 - X^3), \quad 0 \le X \le 1$$

(1) Derive an expression that transforms Uniform random numbers, U, in the interval from 0 to 1, into random numbers, X, that follow the above density function. (2) Using decomposition derive the analytical form of the first four terms of the solution to the transformation expression in (1).

4.6 In Problem 4.5, $U=0.500$. (1) calculate the numerical value of the first four decomposition terms. (2) Approximate the value of $X(3)$. Estimate the magnitude of the absolute error when using only 4 terms in the series.

4.7 Write a Maple program to solve Problem 4.5. *Hint*: modify and adapt the program in Example 4.8.

4.8 (1) Modify the program in Problem 4.7 when $U=0.5$ specifically, and calculate the numerical value of the first four decomposition terms. (2) Approximate the numerical value of X. (3) Calculate the error of using 4 decomposition terms.

4.9 In Example 4.9, (1) rerun the program with $N=100$. Do the numbers fit the theoretical Uniform distribution well? Is $N=100$ a representative sample? (2)Repeat with $N=3000$. What do you conclude about the effect of the sample size in the quality of the numbers as a whole?

4.10 In Problem 3.18, Chapter 3, annual sales in millions of dollars $S\sim N(350,\ 125)$. (1) generate 50 years of sales data using equations (4.11) through (4.14), plot their frequency histogram with the theoretical density function, and calculate the sample mean and standard deviation. (2) Repeat the item (1) using the Maple library. State how good the model is. Justify your conclusion. *Hint*: Modify the program in Example 4.10 to run with the current data.

4.11 A system component follows a Rayleigh density function with $\alpha=1$. Modify Example 4.11 to generate 5000 samples of this distribution. Use a numerical procedure to solve the transformation equation. Calculate the statistics of the numbers and compare them with the theoretical ones. Calculate the lag-one serial correlation coefficient and see that it is close to zero. Finally, plot the frequency histogram of the numbers along with the theoretical density function.

4.12 In problem 4.5, generate 4000 realizations of the system component X using the derived decomposition solution of the transformation equation. As before, compare the resulting random numbers with the theoretical model (i.e., moments, correlation, and histogram). *Hint*: Modify and adapt the programs in Examples 4.8 and 4.12, and Problem 4.7.

4.13 A system operator is given as $Y=aX+b$, where X is an input such that $\leq X \leq$, a and b are the system parameters such that $a\sim N(0.01, 0.001)$ and $b\sim N(0.1, 0.01)$, and Y is the system output. (1) Given $X=0.5$, and assuming the random variables are independent, construct a model to predict the statistical properties of the output (i.e., the sample mean, standard deviation, skewness, and histogram). (2) For each sample of a and b, Y is a straight line. Using 10 samples of a and 10 samples of b, generate and plot 10 corresponding samples of Y.

4.14 (1) Using the rules of expectation, derive analytical expressions for the mean and the variance of the output of the system in Problem 4.13. (2) Calculate the numerical values of the output's mean and standard deviation when $X=0.5$ and compare them with those obtained with the Monte Carlo simulations in Problem 4.13.

4.15 A system operator is given by $h=z+v\cos(\omega t)$, where z is a constant, $v\sim N(1, 0.1)$ the random amplitude, $\sim N(0,1$ the random phase, and t is time. Assuming independence between v and ω, (1) derive an expression for the mean of the output h; and (2) use an appropriate approximation to derive an expression for the variance of the output.

4.16 A system operator is given by the nonlinear equation $Y=aX+bZe^{-Y}$, where X is a deterministic function, Z is a random variable with mean μ_Z and variance σ_Z^2, and a and b are constants. Use the method of decomposition to derive an expression for the mean of the output Y.

5 SYSTEMS WITH JOINTLY-DISTRIBUTED RANDOM VARIABLES

Most engineering systems of practical importance are affected by more than one random variable jointly interacting to produce the system response. For instance, air temperature is related to relative humidity in the generation of conditions leading to precipitation. Soil porosity and soil texture are related to other variables, such as water content and soil-water diffusivity. Chemical concentration in a solution is related to surface tension and viscosity. Although we may construct a theoretical deterministic functional relationship among these variables, which may be valid under certain laboratory-controlled conditions, the fact remains that under general field conditions most engineering variables jointly interact in an uncertain way. Under general conditions, the complexity of the system and our inability to accurately measure many of its components necessitates the use of uncertainty analysis to objectively evaluate its properties. As another example, the life of a particular structural element and the frequency of loading are two random variables by themselves, each having its own (*marginal*) probability density function. Yet, the two variables may be correlated in a way that uncertainty analysis alone may quantify. We then define *the joint probability mass function*, if the two variables are discrete, or *the joint probability density function*, if the two random variables are continuous. We may also define *the joint cumulative distribution function*, and the joint moments. The aim of the exercise is to quantify the uncertain characteristics of the two variables together.

5.1 TWO DISCRETE RANDOM VARIABLES

The Joint Probability Mass Function

Consider a discrete random variable X that can take on at most a countable number of values x, and a discrete random variable Y that can take on at most a countable number of values y. Then the *joint (or bivariate) probability mass function* of X and Y is defined as

$$p_{XY}(x, y) = P(X=x, Y=y) \qquad (5.1)$$

where

$p_{XY}(x, y)$=probability that X takes on values x and Y takes on values y

The usual rules of probability apply to the joint probability mass function:

1. $0 \leq p_{XY}(x_i, y_j) \leq 1$, $i=1, 2,..., j=1, 2,...,$ and $p_{XY}(x, y)=0$ for all other x, y.

2. $\sum_{i=1}^{\infty}\sum_{j=1}^{\infty} p_{XY}(x_i, y_j)=1$.

A graphical representation of the joint probability mass function is a three-dimensional figure with the possible values of X and Y in the horizontal plane and the corresponding values of $p_{XY}(x, y)$ in the vertical axis. The shape of this graph is a representation of the uncertainty behavior of two random variables. In other words, it indicates the values of X and Y with the greatest chances of occurrence and those with the lowest.

Example 5.1: Home Construction with Several Plans
A residential home builder has 3 model plans available to his customers with prices $160,000, $200,000, and $240,000 respectively. He offers the plans to be built on 2 types of lots in a large subdivision with prices $40,000 and $80,000, respectively. Each home model may be built in either lot type. The number of homes built over the last 5 years is summarized in Table 5.1:

Table 5.1: Homes Sold by Model Price and Lot Type

Lot Price / Home Price	$40,000	$80,000
160,000	36	30
200,000	24	12
240,000	12	6

If the builder has constructed 120 homes in the same period, define and plot the joint probability mass function of home price and lot price.

Solution
Let us define the random variables X=home price, and Y=lot price. Since the sample size is N=120, then from the data table, equation (5.1), and the frequency definition of probability (see equation (2.4), Chapter 2), we build Table 5.2.

Table 5.2: Joint Probability Mass Function of Home Price and Lot Type

Lot Price / Home Price	$40,000	$80,000
160,000	0.30	0.25
200,000	0.20	0.10
240,000	0.10	0.05

Thus, $P(X=160,000, y=40,000)=36/20=0.30$, etc. Figure 5.1 shows a three-dimensional bar graph of the joint probability density function of home price and lot type. The highest probability of occurrence, as represented by past system performance, is the combination $X=160,000$, $Y=40,000$. Thus, this builder has more success with the lower-income sector. Another frequent occurrence is the combination $X=160,000$, $Y=80,000$. The least frequent occurrence is $X=240,000$, $Y=80,000$. This is valuable information for the planning of future resources and company objectives.

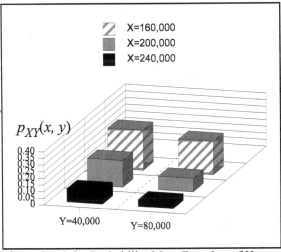

Figure 5.1: Joint Probability Mass Function of Home
Price and Lot Type in Example 5.1

The Joint Cumulative Distribution Function

Similar to the univariate case, the *joint (or bivariate) cumulative distribution function,* is defined as

$$F_{XY}(x, y)=P(X\leq x, Y\leq y) \tag{5.2}$$

where

$F_{XY}(x, y$ =probability that X is less than or equal to x, and Y is less than of equal to y

For discrete random variables X and Y, the joint distribution function may be expressed in terms of the joint probability mass function as

$$F_{XY}(x, y)=\sum_{i=1}^{i=x} \sum_{j=1}^{j=y} p_{XY}(i, j) \tag{5.3}$$

The joint distribution function represents the probability that the random variable X does not exceed the value x, and the random variable Y does not exceed the value y. The joint distribution function has the following properties (compare them with those of univariate functions in Chapter 3):

1. $F_{XY}(x, y)$ is a non-decreasing function. If $a<b$, then $F_{XY}(a, y)<F_{XY}(b, y)$, and $F_{XY}(x, a)<F_{XY}(x, b)$.

2. $F_{XY}(-\infty, y)=F_{XY}(x,-\infty)=F_{XY}(-\infty,-\infty)=0$.

3. $F_X(\infty, \infty)=1$.

4. The distribution function is continuous from the right on either independent variable. That is, $\lim\limits_{x \to a^+} F_{XY}(x, y)=F_{XY}(a, y)$, $\lim\limits_{y \to a^+} F_{XY}(x, y)=F_{XY}(x, a)$.

5. The probability that the random variables lie in the interval $a<X\le b,\ c<Y\le d$ is

$$P(a<X\le b,\ c<Y\le d)=F_{XY}(b, d)-F_{XY}(a, d)-F_{XY}(b, c)+F_{XY}(a, c) \quad (5.4)$$

which can be found by applying the axioms of probability after suitably defining rectangular regions in the X-Y plane.

Example 5.2
   In Example 5.1, (1) calculate the probability that a customer chooses (1) a home of $200,000 or less to be built in a lot of $40,000 or less; and (2) a home plan greater than $200,00 in a lot greater than $40,000.

Solution
(1) From Equation (5.3) and Table 5.1,

$$F_{XY}(200,000,\ 40,000)=P(X\le200,000,\ Y\le40,000)$$

$$=p_{XY}(160,000,\ 40,000)+p_{XY}(200,000,\ 40,000)$$

$$=0.30+0.20=0.50$$

(2) From Table 5.2, there is only one value that satisfies the condition $X>200,000,\ Y>40,000$:

$$P(X>200,000,\ Y>40,000)=p_{XY}(240,000,\ 80,000)=0.05$$

Example 5.3

In Example 5.1, (1) derive the probability mass function of the total sale price (i.e., home plan plus lot price). (2) Calculate the expected value of the total sale price.

Solution

(1) Total sale price is the summation of home price plus lot price. Let $T=X+Y$. Values of T are obtained after an accounting of all the possibilities of adding each value of X with each other of Y. The probability of occurrence associated with a particular T is calculated by adding each value of the contributing probabilities. For example

$$P(T=240,000)=p_{XY}(200,000, \ 40,000)+p_{XY}(160,000, \ 80,000)$$

$$=0.20+0.25=0.45$$

Repeating this process for all possible realizations of T we obtain Table 5.3 which summarizes the probability mas function:

Table 5.3: Probability Mass Function of Total Sale Price T in Example 5.3

t	200,000	240,000	280,000	320,000
$p_T(t)$	0.30	0.45	0.20	0.05

Finally, we verify our calculations by making sure that $\Sigma p_T(t)=1$.

(2) From equation (3.5),

$$E\{T\}=\sum_{i=1}^{4} t_i \ p_T(t_i)$$

$$=200,000\times0.30+240,000\times0.45+280,000\times0.20+320,000\times0.05$$

$$=240,000$$

Marginal Distributions

The marginal (or univariate) distribution of the random variable X alone is obtained after a summation of the joint probability mass function over all values of Y:

$$p_X(x)=\sum_{y} p_{XY}(x, \ y) \tag{5.5}$$

Similarly, the marginal distribution of the random variable Y alone is obtained after a summation of the joint probability mass function over all values of X:

$$p_Y(y) = \sum_x p_{XY}(x, y) \tag{5.6}$$

Example 5.4

In Example 5.1, calculate the values of the marginal distributions of home price, $p_X(x)$, and lot price, $p_Y(y)$.

Solution

From Table 5.2, we create a new column for the values of $p_X(x)$. Application of equation (5.5) translates into adding the cell values of the joint probability mass function in each row (see Table 5.4). Now create a new row for the values of $p_Y(y)$. Application of equation (5.6) translates into adding the cell values of the probability mass function in each column.

Table 5.4: Derivation of Marginal Probability Mass Functions of Home Price and Lot Type in Example 5.4

X \ Y	40,000	80,000	$p_X(x)$
160,000	0.30	0.25	0.55
200,000	0.20	0.10	0.30
240,000	0.10	0.05	0.15
$p_Y(y)$	0.60	0.40	1.00

We check our calculations by verifying that the rules of probability are satisfied by the marginal mass functions, that is $\sum_x p_X(x) = \sum_y p_Y(y) = 1$.

Conditional Joint Functions

In Chapter 2 we studied the situation when a probabilistic event is conditioned by the fact that another has already occurred. A generalization of this situation is the case of a system of two random variables X and Y, such that one of them is restricted to operate, or has already occurred, over a subset of its possible values. Thus, from equation (2.19) we define the conditional joint probability mass function as

$$p_{X|Y}(x|y) = \frac{p_{XY}(x, y)}{p_Y(y)} \tag{5.7}$$

where

$p_{X|Y}(x, y)$=conditional joint probability mass function of random variable X, given that random variable Y has occurred

The analogy between equations (2.19) and (5.7) is clear: given that $Y=y$, the set of possible values is limited to that of Y only. Thus, the probability of X given that $Y=y$ (i.e., $p_{X|Y}(x, y)$) is simply the probability of the values common to X and Y (i.e., $p_{XY}(x, y)$) over the probability of Y only. It is possible to generalize equation (5.7) to situations when, rather than restricting one of the variables to a fixed value (e.g., $Y=y$), we set it to be within a specified range (e.g., $a<Y<b$). In such a case, the denominator in equation (5.7) is replaced by the probability that the said variable is within that range (e.g., $P(a<Y<b)=F_Y(b)-F_Y(a)$). Clearly the use of the cumulative distribution function facilitates the corresponding calculation. Mathematically,

$$P(X|a<Y<b)=p_{X|Y}(x|a<Y<b)=\frac{F_{XY}(x, b)-F_{XY}(x, a)}{F_Y(b)-F_Y(a)}, \quad a<b \quad (5.8)$$

Note that equation (5.8) is a function of X only (i.e., Y is already restricted to be between a and b). This resulting function could be further used to calculate the probability that $c<X<d$ by using the usual rules of probability. In this case the resulting number is the probability that $c<X<d$ given that $a<Y<b$. If we let a be the minimum value of the random variable Y (recall that $F_Y(-\infty)=0$), then equation (5.8) becomes the *conditional joint cumulative distribution function*:

$$P(X|Y\le b)=F_{X|Y}(x|Y\le b)=\frac{F_{XY}(x, b)}{F_Y(b)} \quad (5.9)$$

We will further clarify these concepts when we cover the conditional theory for continuous variables (see section 5.2).

Example 5.5
    In Example 5.1, a customer buys a $40,000 lot. What is the probability that she also orders a $160,000 model to be built?

Solution
    From equation (5.7),

$$P(X=160,000|Y=40,000)=p_{X|Y}(160,000, 40,000)$$

$$P(X=160,000 \mid Y=40,000) = \frac{p_{XY}(160,000,\ 40,000)}{p_Y(40,000)} = \frac{0.30}{0.60} = 0.50$$

where the values in Table 5.4 have been used. The contractor should calculate several conditional probabilities that would ultimately help determine a policy associated with the maximization of profits (e.g., targeting certain promotions associated with the highest probabilities of success).

5.2 TWO CONTINUOUS RANDOM VARIABLES

The Joint Probability Density Function

For systems affected by two continuous random variables, X and Y, we define the *joint (or bivariate) probability density function* $f_{XY}(x,\ y)$ of the variables X and Y jointly taking on values x and y, respectively. As before, the rules of probability must apply to the joint probability density function:

1. $0 \le f_{XY}(x,\ y) \le 1,\ \forall\ x,\ y$.

2. $\int\limits_{-\infty}^{\infty}\int\limits_{-\infty}^{\infty} f_{XY}(x,\ y)dxdy = 1$.

3. For any two-dimensional region in the $x,\ y$ plane,

$$P(a<X<b,\ c<Y<d) = \int\limits_a^b\int\limits_c^d f_{XY}(x,\ y)dxdy \qquad (5.10)$$

Rule 1 requires that the joint probability density function be between 0 and 1 for all values of X and Y. A graph of this function is a three-dimensional solid with the horizontal plane representing the possible values the random variables X and Y may take, and the vertical axis representing the values of $f_{XY}(x,\ y)$. Rule 2 above requires that the volume of this solid (instead of the area for univariate density functions, see section 3.3, Chapter 3) be equal to 1. Rule 3 states those partial subset volumes under $f_{XY}(x,\ y)$ are numerically equal to the probability that X and Y are within specified limits. The reader should review the probability rules of single random variables (Chapter 3), since those of the bivariate functions are straightforward extensions of the basic case.

Example 5.6: Life Time of a Two-Component Drill

A drill tip used in groundwater exploration has two mobile components whose life time is best described by their joint probability density function. Define X=life time of the first component (*hours*), and Y=life time of the second component (*hours*). According to historical records of component failure under identical conditions, the manufacturer has determined that the joint probability density function of component life time is given by

$$f_{XY}(x, y)=2\times10^{-6}e^{-0.001x-0.002y}, \quad x, y>0$$

which indicates that the joint density function is a double exponential function that decays with time (i.e., as time increases, the probability that the component operates is lower). (1) Verify that $f_{XY}(x, y)$ is a true density function. (2) Plot $f_{XY}(x, y)$.

Solution

(1) From rule 1 above we verify that $f_{XY}(x, y)\geq0$. From rule 2,

$$\int_{-\infty}^{\infty}\int_{-\infty}^{\infty}f_{XY}(x, y)dxdy=\int_{0}^{\infty}\int_{0}^{\infty}2\times10^{-6}e^{-0.001x-0.002y}dxdy$$

$$=2\times10^{-6}\int_{0}^{\infty}e^{-0.001x}dx\int_{0}^{\infty}e^{-0.002y}dy$$

$$=2\times10^{-6}\times\frac{1}{0.001}\times\frac{1}{0.002}=1$$

(2) The following Maple worksheet shows the procedure. The reader is encouraged to explore the Maple help system for three-dimensional plotting features and options available.

```
[ Define the joint probability density function as the mapping from (x,y) into f(x,y).
[ > f:=(x,y)->0.000002*exp(-0.001*x)*exp(-0.002*y);
```
$$f:=(x, y)\rightarrow 0.2\ 10^{-5}\ e^{(-0.001\ x)}\ e^{(-0.002\ y)}$$
```
[ Plot f(x,y) as a three-dimensional graph between appropriate limits of x and y.
[ > plot3d(f(x,y),x=0..1000,y=0..1000,axes=BOXED,
       labeldirections=[HORIZONTAL,HORIZONTAL,VERTICAL],
       labelfont=[TIMES,ROMAN,16],
       labels=[`X(hours)`, `Y(hours)`, `f(x,y)`],
       tickmarks=[3,3,4],color=gray);
```

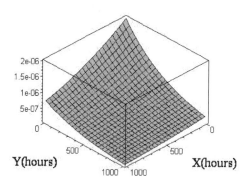

The Joint Cumulative Distribution Function

Similar to the case of two discrete random variables (see section 5.1), the *joint (or bivariate) cumulative distribution function* of two continuous random variables X and Y is defined as

$$F_{XY}(x, y)=P(X \leq x, \ Y \leq y) \tag{5.11}$$

where

$F_{XY}(x, y)$ = probability that X is less than or equal to x, and Y is less than of equal to y

For the continuous random variables X and Y, the joint distribution function may be expressed in terms of the joint probability density function as

$$F_{XY}(x, y)= \int_{-\infty}^{x} \int_{-\infty}^{y} f_{XY}(u, v)dudv \tag{5.12}$$

See the analogy between equation (5.3) (discrete random variables) and equation (5.12) (continuous random variables).

Example 5.7

In Example 5.6, what is the probability that both components fail in under 1000 hours of use?

Solution

From equation (5.12),

$$P(X \leq 1000, \ Y \leq 1000)=F_{XY}(1000, \ 1000)= \int_{0}^{1000} \int_{0}^{1000} f_{XY}(x, y)dxdy$$

$$P(X \leq 1000, \ Y \leq 1000) = 2 \times 10^{-6} \int_0^{1000} e^{-0.001x} dx \int_0^{1000} e^{-0.002y} dy$$

$$= 2 \times 10^{-6} \left(\frac{1-e^{-1}}{0.001} \right) \left(\frac{1-e^{-2}}{0.002} \right) = 0.547$$

The Marginal Probability Density Functions

The same principle outlined in equations (5.5) and (5.6) for the derivation of marginal probability mass functions of discrete random variables applies to continuos random variables, except that integration replaces summation. Thus, if $f_{XY}(x, y)$ is the joint probability density function of the continuous random variables X and Y, then the marginal probability density function of X alone is obtained by integrating the joint density over the entire range of Y:

$$f_X(x) = \int_{-\infty}^{\infty} f_{XY}(x, y) dy \tag{5.13}$$

Similarly, the marginal probability density function of Y alone is obtained by integrating the joint density over the entire range of X:

$$f_Y(y) = \int_{-\infty}^{\infty} f_{XY}(x, y) dx \tag{5.14}$$

Example 5.8

In Example 5.6, find the probability that the first component lasts longer than 2000 hours.

Solution

We first derive the marginal distribution of the first component. From equation (5.13),

$$f_X(x) = \int_0^{\infty} 2 \times 10^{-6} e^{-0.001x - 0.002y} dy = 2 \times 10^{-6} e^{-0.001x} \times \left[-\frac{e^{-0.002y}}{0.002} \right]_0^{\infty}$$

$$= 0.001 e^{-0.001x}, \quad x \geq 0$$

Similarly, from equation (5.14), one easily finds that

$f_Y(y) = 0.002e^{-0.002y}$, $y \geq 0$. From equations (2.18) and (3.28),

$$P(X > 2000) = 1 - F_X(2000) = 1 - \int_{-\infty}^{2000} f_X(x)dx$$

$$= 1 - \int_0^{2000} 0.001e^{-0.001x} dx = 1 - 0.001 \times \left[-\frac{e^{-0.001x}}{0.001} \right]_0^{2000}$$

$$= 1 - 0.865 = 0.135$$

Conditional Joint Functions

By analogy with jointly discrete variables (see equation (5.6)), we define the *conditional joint probability density function* of the continuous random variable X given the continuous random variable Y by replacing the joint probability mass functions by the joint probability density functions:

$$f_{X|Y}(x|y) = \frac{f_{XY}(x, y)}{f_Y(y)} \tag{5.15}$$

where

$f_{X|Y}(x, y)$ = conditional joint probability density function of random variable X, given that random variable Y has occurred

In other words, the conditional density of $X|Y$ is simply the joint density of X and Y divided by the marginal density of Y. Equations (5.8) and (5.9) are also valid for jointly continuous random variables. The graph of the conditional joint density is a subset region of the three-dimensional joint density.

Example 5.9

From Examples 5.6 and 5.8, (1) define the probability of life of the second component if the first one lasts longer than 2000 hours. (2) Use the result in (1) to calculate the probability that the second component lasts longer than 1000 hours given that the first lasts longer than 2000.

Solution
(1) From equations (5.15), (5.8), and (5.9),

$$f_{Y|X}(y|x \le 2000) = \frac{F_{XY}(2000, y)}{F_X(2000)}$$

Conversely,

$$f_{Y|X}(y|x > 2000) = \frac{1 - F_{XY}(2000, y)}{1 - F_X(2000)} = \frac{\int_{2000}^{\infty} f_{XY}(x, y)dx}{\int_{2000}^{\infty} f_X(x)dx}$$

$$= \frac{\int_{2000}^{\infty} 2 \times 10^{-6} e^{-0.001x - 0.002y} dx}{\int_{2000}^{\infty} 0.001 e^{-0.001x} dx}$$

$$= \frac{2 \times 10^{-6} e^{-0.002y} \times \left[\frac{e^{-0.001x}}{-0.001} \right]_{2000}^{\infty}}{\left[-e^{-0.001x} \right]_{2000}^{\infty}} = 0.002 e^{-0.002y}$$

which defines the probability of life of the second component once the first has lasted longer than 2000 hours. Note it is a function of Y only (i.e., X is already restricted to be greater than 2000.

(2) Since $f_{Y|X}(y|x > 2000)$ is a density function, we calculate the required chance by applying the usual rules of probability:

$$P(X > 2000 | Y > 1000) = \int_{1000}^{\infty} f_{Y|X}(y|x > 2000)dy = \int_{1000}^{\infty} 0.002 e^{-0.002y} dy$$

$$= \left[-e^{-0.002y} \right]_{1000}^{\infty} = 0.135$$

Statistically Independent Random Variables

The concept of statistical independence of events (see equations (2.24) and (2.25)) may be generalized to jointly distributed random variables. Thus, we say that two random variables X and Y are *statistically independent* if their joint probability density function equals the product of their marginal density functions. In other words,

$$f_{XY}(x, y) = f_X(x)f_Y(y) \tag{5.16}$$

Similarly, X and Y are statistically independent if the conditional density of X given Y equals the marginal density function of X:

$$f_{X|Y}(x|y) = f_X(x), \quad f_{Y|X}(y|x) = f_Y(y) \tag{5.17}$$

Equations (5.16) and (5.17) imply that if two random variables are statistically independent, the specific values taken by one variable do not effect the frequency behavior of the other. We remark that the fact that two random variables are statistically independent does not necessarily mean they are not functionally related in any other form. Their independence refers to our limited definition of independence as stated by equations (5.16) or (5.17) only. The variables may or may not be functionally, or physically, related in a way that our test is unable to identify. Conversely, if two random variables are not statistically independent, it only means they are related in their frequency behavior. In other words, we may not conclude that two variables are physically, functionally, or nonlinearly related simply because according to our test they are classified as statistically dependent. The statistical literature is plagued by erroneous conclusions that imply that there might be a cause-effect relationship between two variables when in fact the only conclusion should be that they are statistically dependent.

5.3 SPECIAL MOMENTS OF TWO RANDOM VARIABLES

In Chapter 3 we applied the properties of the expectation operator to define certain statistical measures, called moments, of single random variables (e.g., see equations (3.30) through (3.32)). Some of these moments had special names, such as the mean, the variance, the standard deviation, the skewness, and the coefficient of variability. These moments serve to characterize particular features of a density function, such as the location of its center of mass, the average spread of the data, the degree of symmetry, etc. Similarly, we may also construct certain meaningful statistical measures of two random variables operating in a system. These measures help us infer the degree to which the two variables may be interrelated.

The Covariance of Two Random Variables

The *covariance* of the random variables X and Y is defined as the expectation of the product of the variables minus their respective means:

$$C_{XY}=E\{(X-\mu_X)(Y-\mu_Y)\} \qquad (5.18)$$

where

C_{XY}=covariance of X and Y
$E\{\}$=expectation operator (equation (3.5) for discrete variables and
  equation (3.30) for continuous variables)
μ_X=the mean of X
μ_Y=the mean of Y

Recall the properties of expectation (see properties 1 through 4 below equation (3.5), Chapter 3). Thus, equation (5.18) reduces to,

$$C_{XY}=R_{XY}-\mu_X\mu_Y \qquad (5.19)$$

where

$R_{XY}=E\{XY\}$=the *correlation function* of X and Y

If $X=Y$, equations (5.18) and (5.19) become the variance of X, see equations (3.7), (3.8), and (3.31).

The Correlation Coefficient of Two Random Variables
  The *correlation coefficient* of random variables X and Y is defined as their covariance divided by the product of their standard deviations. In other words, the correlation coefficient is the covariance normalized to yield a numerical value of less than one:

$$\rho_{XY}=\frac{C_{XY}}{\sigma_X\sigma_Y} \qquad (5.20)$$

where

ρ_{XY}=correlation coefficient of X and Y
σ_X=standard deviation of X
σ_Y=standard deviation of Y

The correlation coefficient has the following properties:

1. $-1\le\rho_{XY}\le1$.

2. If X and Y are statistically independent (see equation (5.16) for definition), then $\rho_{XY}=0$.

3. If $\rho_{XY}=0$, they are called *uncorrelated*.

4. If X and Y are uncorrelated (i.e., $\rho_{XY}=0$), they *may or may not* be statistically independent. In other words, statistical independence is a stronger condition which implies the random variables are also uncorrelated. However, the opposite is not necessarily true.

5. If Y is a linear combination of X (e.g., $Y=aX+b$, with a, b constants), then $\rho_{XY}=\pm1$.

The closer ρ_{XY} is to 0 the lesser the degree of statistical relationship between X and Y. As ρ_{XY} approaches $|1|$ in magnitude, the greater the degree of statistical relationship between the random variables. For example a value of $\rho_{XY}=0.8$ implies that the magnitude of one of the variables explains 80% of that of the other. When $\rho_{XY}=\pm1$, there is a perfect correlation between the variables, to the point that by knowing the value of one of them, the other is also known. It is important to remark that the correlation coefficient is a measure of *linear* statistical relationship between the said random variables. That is to say, a high correlation coefficient means that the two variables linearly correlate. It does not necessarily mean that there exists a cause-effect, or physical, relationship between them as some statistical studies claim. On the other hand, finding a nearly zero correlation coefficient does not necessarily mean the two variables are not related in any other unknown nonlinear fashion.

Example 5.10

In Examples 5.6 and 5.8, calculate (1) the mean life of the first component; (2) the mean life of the second component; (3) the variance of X; (4) the variance of Y; (5) the correlation of X and Y; (6) the covariance of X and Y; and (7) the correlation coefficient of X and Y. Are X and Y statistically independent? Are they uncorrelated?

Solution
(1) From equation (3.30),

$$\mu_X=E\{X\}=\int_{-\infty}^{\infty}xf_X(x)dx=10^{-3}\int_{0}^{\infty}xe^{-0.001x}dx$$

Applying integration by parts with $u=x$ and $dv=e^{-0.001x}dx$,

$$\mu_X=10^{-3}\left(\left[-10^3xe^{-0.001x}\right]_0^{\infty}+\int_{0}^{\infty}10^3e^{-0.001x}dx\right)=1000\,hours$$

(2) By a similar procedure,

$$\mu_Y = E\{Y\} = \int_{-\infty}^{\infty} y f_Y(y) dy = 2 \times 10^{-3} \int_0^{\infty} y e^{-0.002y} dy = 500 hours$$

(3) From equation (3.31),

$$\sigma_X^2(x) = E\{(X-\mu_X)^2\} = \int_{-\infty}^{\infty} (x-\mu_X)^2 f_X(x) dx = \int_0^{\infty} (x-1000)^2 \times 10^{-3} e^{-0.001x} dx$$

Breaking the parentheses results in three integrals easily solvable with integration by parts, consulting a table of integrals, or using Maple. Thus, $\sigma_X^2 = 10^6$ *hours*^{2}, and the standard deviation $\sigma_X = 10^3$ *hours*.

(4) By a similar procedure, $\sigma_Y^2 = 250000 hours^2$, and $\sigma_Y = 500 hours$.

(5) The correlation between X and Y is defined as (see below equation (5.19))

$$R_{XY} = E\{XY\} = \int_{-\infty}^{\infty} \int_{-\infty}^{\infty} xy f_{XY}(x,y) dx dy = 2 \times 10^{-6} \int_0^{\infty} x e^{-0.001x} dx \int_0^{\infty} y e^{-0.002y} dy$$

$$= 500000 hour^2$$

(6) From equation (5.19)

$$C_{XY} = R_{XY} - \mu_X \mu_Y = 500000 - 1000 \times 500 = 0.0 hour^2$$

(7) From equation (5.20)

$$\rho_{XY} = \frac{C_{XY}}{\sigma_X \sigma_Y} = \frac{0.0}{1000 \times 500} = 0.0$$

Since in this particular instance $f_{XY}(x, y) = f_X(x) f_Y(y)$ (as it can be easily proved by performing the product), then X and Y are statistically independent. This implies that they are also uncorrelated, a fact corroborated in the item (7).

5.4 THE BIVARIATE GAUSSIAN DENSITY FUNCTION

One of the most important joint probability distributions is the *bivariate Gaussian* (or Normal) probability density function. It is used extensively in engineering models of a wide class of phenomena. Let $X \sim N(\mu_X, \sigma_X)$, and $Y \sim N(\mu_Y, \sigma_Y)$. Denoting $r = \rho_{XY}$, the correlation coefficient of X and Y, and defining the Standard Normal variables (see equation (3.41), Chapter 3)

$$x_n = \frac{X - \mu_X}{\sigma_X}, \quad y_n = \frac{Y - \mu_Y}{\sigma_Y}, \tag{5.21}$$

then the bivariate Gaussian probability density function id given by

$$f_{XY}(x, y) = \frac{e^{-\frac{x_n^2 - 2rx_n y_n + y_n^2}{2(1 - r^2)}}}{2\pi\sqrt{1 - r^2}} \tag{5.22}$$

When $r=0$, equation (5.22) reduces to the product of two univariate Gaussian random variables, which according to equation (5.16) characterizes two statistically independent Gaussian variables.

Example 5.11
Two jointly-distributed Gaussian random variables X and Y have the following parameter values: $\mu_X = 1$, $\mu_Y = 1$, $\sigma_X = 1$, $\sigma_Y = 1$, and $r = 0.5$. Plot its joint probability density function.

Solution
The following Maple worksheet illustrates the procedure.

```
[ Define parameter values.
[ > mux:=1: muy:=1: sigmax:=1: sigmay:=1: r:=0.5:
[ Define the Standard Gaussian variables as functions of x and y, respectively.
[ > xn:=x->(x-mux)/sigmax: yn:=y->(y-muy)/sigmay:
[ Define the exponent as a function of x and y.
[ > expo:=(x,y)->(xn(x)^2-2*r*xn(x)*yn(y)+yn(y)^2)
               /(2*(1-r^2)):
[ Define the joint density function fXY(x,y).
[ > fXY:=(x,y)->exp(-expo(x,y))/(2*Pi*sqrt(1-r^2)):
[ Plot the density function as a three-dimensional figure.
[ > plot3d( fXY(x,y),x=-2..4,y=-2..4,style=patch,
        axes=box,color=gray,
        labels=[`X`,`Y`,`fXY(x,y)`],
        labeldirections=[HORIZONTAL,HORIZONTAL,VERTICAL]
        labelfont=[TIMES,ROMAN,16]);
```

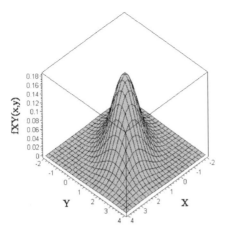

Click on the graph and then on the graph rotation keys in the tool bar to observe the graph move around its horizontal or vertical axis. This is a nice way to familiarize oneself with the three-dimensional form of the density function.

5.5 SYSTEMS FORCED BY JOINTLY-DISTRIBUTED RANDOM VARIABLES

Consider the situation of an engineering system with an input composed of two jointly-distributed random variables, X and Y. The system operator function, $G(X, Y)$, is now a function of two random variables. It transforms the input variables into the system output, Z (see Figure 5.2). If the input is characterized by the joint probability density function of X and Y, $f_{XY}(x, y)$, then the system output, Z, is best characterized by its probability density function $f_Z(z)$. In practice, it is difficult to obtain the output density function, depending on the mathematical form of $f_{XY}(x, y)$ and that of $G(X, Y)$. The procedure involves a generalization of the principles described in section 4.1, Chapter 4. As before, the statistics of the output may be obtained by applying the definition of expectation. Thus, for two jointly-distributed continuous variables in the input, the mean of the output μ_Z is given as a generalization of equation (4.15), Chapter 4:

$$\mu_Z = E\{Z\} = E\{G(X, Y)\} = \int\limits_{-\infty}^{\infty}\int\limits_{-\infty}^{\infty} G(x, y)f_{XY}(x, y)dxdy \tag{5.23}$$

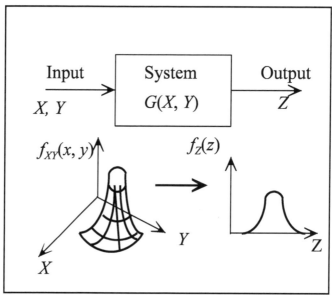

Figure 5.2: A System Forced by Jointly-Distributed Random
Variables

Similarly, if the input has two jointly-distributed discrete random variables, then equation (4.16) becomes

$$\mu_Z = E\{Z\} \; = \; E\{G(X, \; Y)\} = \sum_{i=1}^{m} \sum_{j=1}^{n} G(x_i, \; y_j) p_{XY}(x_i, \; y_j) \qquad (3.24)$$

Output of Sums of Random Variables: The Central Limit Theorem

For the special case when the input random variables are statistically independent and the operator is $Z=G(X, \; Y)=X+Y$, then the probability density function of the output is given by

$$f_Z(z) = \int_{-\infty}^{\infty} f_X(z-y) f_Y(y) dy \qquad (5.25)$$

which is called the *convolution integral* of the density functions $f_X(x)$ and $f_Y(y)$. Furthermore, it can be shown that if $Z=G(X, \; Y)=X+Y$, and X and Y are Gaussian, whether or not they are independent, then the output density function, $f_Z(z)$, is also Gaussian with a mean $\mu_Z=\mu_X+\mu_Y$, and a variance $\sigma_Z^2=\sigma_X^2+2r\sigma_X\sigma_Y+\sigma_Y^2$, where r is the correlation coefficient between X and Y. If in addition X and Y are independent, then $r=0$, and we have that the summation of two independent Gaussian variables results in another Gaussian variable with mean equal to the summation of the input means, and with variance equal to the summation of the input variances.

These results suggest that the summation of independent Gaussian variables produce another Gaussian variable. In general, however, as the number of component random variables increases, their summation tends to be Gaussian, regardless of the density function of the component variables. More formally, if X_1, X_2,···, X_N are *independent* random variables, with means μ_1, μ_2, · · · , μ_N, and standard deviations σ_1, σ_2,···, σ_N, respectively, then the summation of their standardized variables follows a Standard Gaussian variable, provided that their standard deviations are of the same order (i.e., not one σ_i is dominating). In other words,

$$Z = \frac{1}{\sqrt{N}} \sum_{i=1}^{N} \left(\frac{X_i - \mu_i}{\sigma_i} \right) \sim N(0, 1) \tag{5.26}$$

as N becomes large. This is called *the central limit theorem* and constitutes one of the most important concepts in probability with applications in modeling, statistical inference, and sampling theory. The idea that a summation of independent random variables approaches a Gaussian random variable has inspired many in the past. The central limit theorem has been suggested as evidence that *all* random phenomena tend to be Gaussian in the limit. Many uncertain variables in fields as diverse as thermodynamics, sediment transport, and hydrodynamic dispersion, to name a few, have been assumed to be Gaussian in the name of the central limit theorem. A usual argument in engineering relates to the situation of a particular uncertain variable that results from many other unknown and complex environmental fluctuations, but whose combined output is allegedly Gaussian. Thus, a useful theorem has been overused and invoked out of convenience when an engineering model encounters difficulty (e.g., when the density function is unknown). One needs to carefully observe the restrictions of the central limit theorem to discard many of the justifications for its use: the component random variables must be independent, and not a single one should dominate the *summation*. Many uncertain variables do not result from a summation of other variables, are not independent of each other, and do not have similar variances. A more appropriate approach is to *assume* that a particular unknown distribution is Gaussian as a first approximation without invoking the central limit theorem.

With these thoughts on the limitations of the central limit theorem, let us generalize the values of the mean and the variance of summations of

N random variables:

$$Z=\sum_{i=1}^{N} X_i$$

$$\mu_Z=\sum_{i=1}^{N} \mu_i \qquad\qquad (5.27)$$

$$\sigma_Z^2=\sum_{i=1}^{N} \sigma_i^2 + \sum_{i=1}^{N}\sum_{j=1}^{N} \sigma_i\sigma_j\rho_{ij}, \quad \forall \; i \neq j$$

where
ρ_{ij}=correlation coefficient between X_i and X_j

Thus, the mean of a summation of N random variables, has a mean equal to the summation of the component means. The variance of a summation of N random variables equals the summation of the component variances plus the product of the component standard deviations times their correlation coefficients. If the component random variables are independent, then the correlation coefficients are zero and the variance of the output is simply the summation of the variances of the input. For an in depth treatment on the problem of probability distributions of several random variables and functions of several random variables the reader is referred to Papoulis (1984).

Example 5.12: Construction of a Steel Frame
After years of experience in many projects, a company has found that the setup and construction of a large steel frame require 20 independent labor teams to be assembled in series (i.e., the critical path in PERT). Each team spends an average time of $T_i=4days$ with a variance of $1day$. According to the data, a Uniform distribution exists in the time spent per team. What are the probability density function, the mean, and the variance of the project completion time, T?

Solution
The T_i times are independent and their variances are equal. Since the total project time is the summation of all the tasks, the conditions of the central limit theorem are satisfied and from equation (5.26),

$$T=\sum_{i=1}^{20} T_i \sim N(\mu_T, \; \sigma_T^2)$$

$$\mu_T = \sum_{i=1}^{20} \mu_i = \sum_{i=1}^{20} 4 = 80 days$$

$$\sigma_T^2 = \sum_{i=1}^{20} \sigma_i^2 = \sum_{i=1}^{20} 1 = 20 \ days^2 \ \Rightarrow \ \sigma_T = \sqrt{20} = 4.47 days$$

It is important to question the validity of the assumption of independence between   the different tasks: is each team managed independently of the others? Do they share equipment and human resources? Does the delay of completion in one team affect the timing of the next? These practical questions should be answered before deciding on independence. On the other hand, if one does not know whether or not the teams are independent, or does not have correlation data to include in the calculations, then the engineer should simply state in the consulting report to the customer that she *assumes* independence as a first approximation to the real process.

Example 5.13: Cumulative Error in Highway Polygonal

A geometrical highway design requires the laying of an open polygonal line. A surveying company has found from past experience that a one-*km* line may be measured to an accuracy of $1000m \pm 0.1\%$ of error given by a Uniform probability density function. (1) Calculate the mean and the variance of the error in one *km* (i.e., derive expressions that show equation (3.39) is correct). (2) Identify the form of the probability density function of the error after 10 1-*km* markers are placed in the field, calculate its mean, and its standard deviation. (3) Estimate the probability that the error is within $\pm 1m$ after 10 one-*km* markers.

Solution

(1) Let E be the error in one *km* with a Uniform probability density function between -0.1% of 1000 *m* and 0.1% of 1000 *m*. From equation (3.37)

$$f_E(e) = \frac{1}{2a}, \quad -a \le E \le a, \quad a = 0.001 \times 1000 = 1 \ m$$

$$\mu_E = E\{E\} = \int_{-a}^{a} e \cdot f_E(e) de = \int_{-a}^{a} \frac{e.de}{2a} = \frac{1}{2a} \left[\frac{e^2}{2} \right]_{-a}^{a} = 0$$

$$\sigma_E^2 = E\{(E-\mu_E)^2\} = \int_{-a}^{a} e^2 f_E(e)de = \int_{-a}^{a} \frac{e^2 de}{2a} = \left[\frac{e^3}{6a}\right]_{-a}^{a} = \frac{a^2}{3}$$

$$\sigma_E = \sqrt{\sigma_E^2} = \frac{a}{\sqrt{3}} = \frac{1}{1.732} = 0.57735m$$

which yields the same result as equation (3.39). Figure 5.3 shows the density function of the error in one *km*.

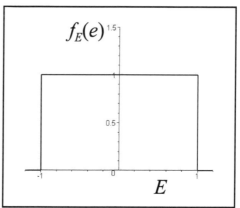

Figure 5.3: Density Function of Error in 1*km*

(2) Let E_T be the error after 10 one-*km* markers. *Assuming* that each one-*km* error is independent of the others (i.e., the magnitude of the error in one reach does not affect that of the others), then by the central limit theorem, and equation (5.27),

$$E_T = \sum_{i=1}^{10} E \sim N(\mu_{E_T}, \sigma_{E_T}^2)$$

$$\mu_{E_T} = \sum_{i=1}^{10} \mu_E = 10 \times 0 = 0$$

$$\sigma_{E_T}^2 = \sum_{i=1}^{10} \sigma_E^2 = 10 \times \frac{a^2}{3} = 3.3333m^2 \quad \Rightarrow \quad \sigma_{E_T} = \sqrt{3.3333} = \pm 1.8257m$$

Figure 5.4 shows the probability density function of the error after 10 *km*.

(3) We follow the procedure outlined in Example 3.22, Chapter 3, to calculate areas under a Normal curve, by first transforming the Normal

density followed by E_T into a Standard Normal density $X \sim N(0, 1)$ (see equation (3.41)), and then reading the corresponding area in Table A.1, Appendix A:

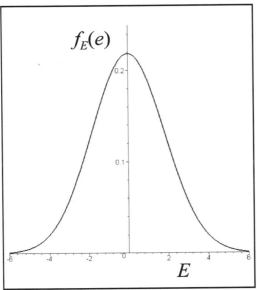

$f_E(e)$

Figure 5.4: Density Function of Error in 10km

$$x_1 = \frac{-1 - \mu_{E_T}}{\sigma_{E_T}} = \frac{-1 - 0}{1.8257} = -0.5477, \quad x_2 = 0.5477$$

$$P(-1 \le E_T \le 1) = P(x_1 \le X \le x_2) = P(X \le x_2) - [1 - P((X \le x_2)] = 2P(X \le x_2) - 1$$

$$= 2 \times 0.708 - 1 = 0.416$$

PROBLEMS

5.1 A residential home builder has 5 types of models and 4 lot sizes. The complete sales data is given in Table 5.5. Build a table of the joint probability mass function and plot it.

5.2 In Problem 5.1, what is the probability that a customer will order a home with a price greater than $240,000 in a lot with a price greater than $80,000?

5.3 In Problem 5.1, (1) build the table of the probability mass function of the total sale price (i.e., home plan plus lot price). (2) Plot the probability mass function of total sale price. (3) Calculate the expected value of the total sale price.

Table 5.5: Homes Sold by Price and Lot Type

Home Price \ Lot Price	$40,000	$80,000	$120,000	$150,000
160,000	36	30	10	1
200,000	24	12	15	10
240,000	12	6	10	18
300,000	3	8	15	12
350,000	0	2	9	16

5.4 In Problem 5.1, (1) calculate the values of the marginal probability mass function of home price and that of lot price; and (2) plot the marginal functions.

5.5 In Problem 5.1, if a customer buys an $80,000 lot, what is the probability that he orders a $240,000 home?

5.6 Customers of a busy supermarket may choose to pay using the self-service automated facility or the traditional cashier-assisted system. Let $X=$ proportion of time the automated unit is in use, and $Y=$proportion of time the cashier system is in use in a day. From historical data, the management has determined that the joint probability density function of X and Y is given by

$$f_{XY}(x, y)=\frac{6}{5}(x+y^2), \quad 0 \le X \le 1, \ 0 \le Y \le 1$$

(1) Verify that it is a true joint probability density function.(2) Plot it.

5.7 In Problem 5.6, (1) derive the marginal probability density functions of X and Y; and (2) calculate the probability that the traditional cashier system is in use between 25% and 75% of the time.

5.8 In Problem 5.6, define the probability density function of the automated system if the traditional system operates at greater than 80% capacity?

5.9 In Problem 5.6, use the result in Problem 5.8 to calculate the probability that the automated system operates between 40% and 60% of the time given that the traditional system operates at greater than 80% capacity.

5.10 In problem 5.6, calculate (1) the mean usage of the first component; (2) the mean usage of the second component; (3) the variance of X; (4) the variance of Y; (5) the correlation of X and Y; (6) the covariance of X and Y; and (7) the correlation coefficient of X and Y. Are X and Y statistically independent? Are they uncorrelated?

5.11 Solve Example 5.12 if each task is serially correlated to the next one with a coefficient of 0.2, but uncorrelated with any other task.

5.12 In Example 5.13, calculate the probability that the error after the 10 markers is within $\pm 1m$ if the error in 1 km has a serial correlation coefficient of 0.1 with that of the next one, but it is uncorrelated with that of any other. Assume that independence still holds.

"We do not know the number of opportunities [events] *that will come our way before our life comes to a close. This inescapable end should induce us not to put off till the morrow what may be achieved this very day . . . From a mystical point of view, it is possible to establish a correlation between problems calling on probability theory and the choices we encounter in daily life."*

Anonymous.

6 ESTIMATION THEORY IN ENGINEERING

Until now we have studied the fundamental concepts of probability and its applications to modeling of engineering systems. For most problems we already had a probabilistic model available. We had a probability distribution somehow built from observational data. In this chapter we will begin to establish a connection between experimental data and the parameters of a probabilistic model. We will study the fundamentals of statistical inference theory. As its name implies, this theory attempts to estimate the possible values of the mean and the variance of a theoretical probability density function underlying an uncertain process, based on limited sample data. This is useful in many engineering system applications. Sometimes an engineer is interested in finding the general, true, values based on an experiment as a first step in identifying the governing density function. In other applications, the engineer is not interested in the true parameter values, but rather in the assessment of quality control of products and services. For instance, the engineer needs a means to assess whether or not a manufacturer that guarantees a certain mean in a serial product is indeed delivering the promised quality.

6.1 STATISTICS AND UNCERTAINTY ANALYSIS

In section 1.2, Chapter 1, we described a general procedure of uncertainty analysis in engineering. We conceived the process as composed of four main steps. Step 1 is the gathering of experimental data of an uncertain phenomenon under identical conditions. Step 2 is the estimation of the *statistics* of the experimental data. We emphasized this is mainly a deductive process whereby the engineer calculates the properties of a sample data. Step 3 is the inductive process of construction of a theoretical probabilistic model (i.e., a probability distribution) that describes the uncertain behavior of a phenomenon in question. Step 4 is the application of the model in the context of an engineering project. Until now we have dwelt on the types of probabilistic models available and their applications in simulation. We now move closer to the real world and begin to focus on Step 2 and the basis of statistics.

Population Parameters versus Sample Statistics

When we studied probability distributions in Chapter 3, we considered the mean, μ_X, the variance, σ_X^2, the coefficient of variability, C_v, and the skewness coefficient, γ_X of a random variable X, as calculated from equations (3.30), (3.31), (3.32), and (3.33), respectively. These and other *parameters* arise from moment measures taken on the *complete set* of values a random variable may take. For instance, if X is continuous, the mean is obtained after integration of the first-order moments with respect to the origin over the entire set of possible samples of X (see equation (3.30)). If the random variable is discrete, integration is replaced by a summation over the entire set of possible outcomes of the random variable (see equation (3.6)). When we obtain these measures based on the entire set of possible outcomes of a random variable, they are called *population parameters*. The population parameters of a probability distribution are *constant* values that reflect the characteristics of a random variable when all possible values are taken into consideration. Thus, the *population mean, population variance,* and *population skewness* are constant deterministic parameters based on the whole ensemble of values the random variable may take.

In the real world, one rarely has the population parameters. The best one can do, is to conduct an experiment to obtain a limited set of values. From this set of values, called a *sample*, we *estimate* or *infer* the possible values of the population parameters. From a sample of N values, we calculate the *sample mean* from the following equaiton:

$$\bar{X} = \hat{\mu}_X = \frac{1}{N}\sum_{i=1}^{N} x_i \tag{6.1}$$

where
$\bar{X}$=the sample mean of X
$\hat{\mu}_X$=estimate of population mean (the "hat" denotes "estimate")
x_i=i-th sample value of X
N=sample size (i.e., number of observations of X)

Note that equation (3.6), although mathematically identical to equation (6.1), is conceptually different. Equation (3.6) yields the population mean, μ_X, of a discrete random variable when all possible outcomes, N, are used in the calculation. In contrast, equation (6.1) yields the sample mean, $\bar{X}$, which is an *estimate* of the population mean, μ_X, when a limited sample of size N of a discrete or a continuous random variable X is used in the calculation.

The sample variance may be computed from

$$S_X^2 = \hat{\sigma}_X^2 = \frac{1}{N-1}\sum_{i=1}^{N}\left(x_i - \bar{X}\right)^2 \tag{6.2}$$

where
 S_X^2=the sample variance of X
 $\hat{\sigma}_X^2$=estimate of the population variance

Note again the similarity between equation (3.10), which is the population variance, σ_X^2, of a discrete variable X, and equation (6.2), which is a sample variance or an *estimate* of the population variance when a sample of size N is taken from a discrete or a continuous random variable X. The latter has N-1 in the denominator indicating a loss in confidence in the estimate by the fact that we are using a limited sample instead of all population values. In addition, equation (6.2) has the sample mean, $\bar{X}$, that is an uncertain estimate of μ_X rather than μ_X itself. The sample standard deviation is calculated as before as the square root of the sample variance, that is $S_X=\hat{\sigma}_X=\sqrt{S_x^2}$.

Sample higher-order moments are calculated with formulae similar to those of corresponding population parameters, except that sample statistics are used instead of population parameters, and the coefficient N is the sample size instead of the population size (if countable). The value of N is also affected by the loss in degrees of freedom incurred upon by the calculation of sample statistics instead of population parameters. For example, in Chapter 3 we learned that the population skewness coefficient was calculated as a ratio of the population third moment about the mean to the population standard deviation to the power of 3. In contrast, the sample skewness coefficient is calculated as

$$\hat{\gamma}_X=\frac{N\sum_{i=1}^{N}\left(x_i-\bar{X}\right)^3}{(N-1)(N-2)S_X^3} \qquad (6.3)$$

where
 $\hat{\gamma}_X$=sample skewness coefficient (the "hat" denotes "estimate")

The important concept to remember is that the population parameters of a random variable are theoretical constant values obtained when all possible realizations of a random variable are available and used in calculations, whereas the sample statistics are estimates of corresponding values based on a limited sample taken from the process under observation. The sample size, N, depends on technical and budgetary constraints of the project.

Figure 6.1 illustrates the concept of a natural phenomenon, or an engineering process classified as a random variable X. The prototype, or true, phenomenon is represented as a gray rectangle containing all possible outcomes of X, each point being one outcome. From the ensemble of these outcomes, X has as population parameters: the mean, μ_X, the variance, σ_X^2, the coefficient of skewness, γ_X, etc. Except for simple, countable, discrete variables, these parameters are usually unknown. Instead, we design Experiment 1 that consists in taking N_1 identical and independent measurements of X, from which we estimate the possible values of the population parameters. These estimates are called the sample statistics: the sample mean $\bar{X}_1$ (the sub-index 1 indicating it is the sample mean of Experiment 1), the sample variance, S_1^2, the sample skewness coefficient, $\hat{\gamma}_1$, etc. The N_1 identical and independent measurements constitute a *random sample* of X. We may repeat the observations and conduct another test, Experiment 2, that consist in taking N_2 measurements of X, from which we estimate another set of possible values of the population parameters (i.e., another set of statistics): $\bar{X}_2$, S_2^2, $\hat{\gamma}_2$, etc. $\bar{X}_1$ and $\bar{X}_2$ are two *different* estimates, two different measures, of the same population mean, μ_X. We may repeat the experiments again, each time computing a different set of statistics of the same population parameters. The fact that each experiment yields a different value of the same statistic is evident from the very process of error in measurement, instrument resolution, and especially the limited sample size.

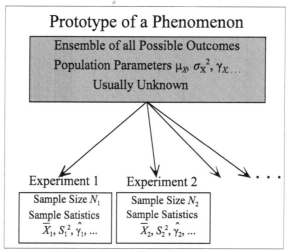

Figure 6.1: Statistics and Parameters

Probability Distribution of the Sample Mean

From above, we derive the concept that sample statistics are themselves random variables. In other words, a random variable X has a probability density function, $f_X(x)$, which we call the *parent distribution*, with a population mean μ_X. On the other hand, the sample mean of X, $\bar{X}$, has a probability density function, $f_{\bar{X}}(\bar{x})$, which is approximately Normal if the sample size N is large. This follows from the central limit theorem (see Rosenkrantz, 1997, Chapter 6 for details). The mean of the sample mean is $E\{\bar{X}\}=E\{\hat{\mu}_X\}=\mu_X$ (i.e., the mean of the sample mean is the population mean). The variance of the sample mean is σ_X^2/N, or the standard deviation of the sample mean is $\sigma_X/\sqrt{N}$. In other words, X follows a parent probability density function $f_X(x)$ with a population mean μ_X and a population standard deviation σ_X, whereas the sample mean of X, $\bar{X} \sim N(\mu_X, \sigma_X/\sqrt{N})$. Note that the variance of $\bar{X} \to 0$ as $N \to \infty$.

Now, the population variance of X is σ_X^2. On the other hand, the sample variance of X, S_X^2, has a probability density function of its own with a mean equal to the population variance of X, σ_X^2. Mathematically, $E\{S_X^2\}=E\{\hat{\sigma}_X^2\}=\sigma_X^2$.

Example 6.1: Distribution of Sample Mean of a Company Stock
    The price of the stock of a company at the New York Stock Exchange fluctuated rapidly during a volatile market day. From long historical records, a seasoned investor knows that, for that time of the year, the stock has a mean value of $59.125 with a standard deviation $2.250. If the investor signs on an electronic trading site through the Internet 6 times in the day to consult the real-time price, determine the probability density function of the stock price, the probability density function of the sample mean, the expected value of the sample mean, and the standard deviation of the sample mean.

Solution
    The probability density function of the stock price X, $f_X(x)$, is unknown. The probability density function of the sample mean $\bar{X}$ is approximately Normal with a mean equal to the population mean of X, and a standard deviation of $\sigma_X/\sqrt{N}=2.250/\sqrt{6}=0.919$. Thus, in summary, $f_X(x)$ is unknown with $\mu_X=59.125$ and $\sigma_X=2.250$, whereas $\bar{X} \sim N(59.125, 0.919)$.

6.2 POINT ESTIMATORS

    In the previous section we introduced the distinction between the population parameters and the sample statistics. While the population parameters are deterministic constants, the sample statistics estimate the values of the parameters and are themselves random variables with their own density functions. In practice, when we run an experiment, we need a procedure to estimate the possible value of the population parameters. The mathematical procedure that we use to achieve this is called a *point estimator*. As the name implies, a point estimator is an operation performed on the sample values of a random variable, the result being a numerical value (i.e., a sample, a point, a statistic) of the parameter being sought. For example a point estimator of the population mean yields a sample mean.

A good point estimator produces a value of the sample mean which is close in magnitude to the population, true, value. There are other properties point estimators should have. A good point estimator is *unbiased*. In other words, the expectation of the estimate of the parameter is equal to the parameter itself. The bias, if any, and the variance of the estimator should approach zero as the sample size approaches infinity. This makes a *consistent* estimator. An unbiased and consistent estimator is also an *accurate* estimator. See Ziemer (1997, Chapter 6) for exercises and examples on the theoretical properties of point estimators.

Estimation with the Method of Moments

We have already discussed the measure of moments of density functions in Chapter 3. In section 6.1 of this chapter we presented the method of moments as a practical tool to estimate the statistics of random variables. Equations (6.1), (6.2), and (6.3) are the moments' estimators of the mean, the variance, and the coefficient of skewness of a random variable, respectively. Several examples in Chapters 3, 4, and 5 specifically applied the moments' estimates of the mean, variance, etc., to a given sample. Moments are by far the most commonly used procedure of point estimation of population parameters.

Estimation with the Method of Maximum Likelihood

Another common procedure of point estimation of population parameters is the method of *maximum likelihood*. It consists in finding the value of a parameter such that a function called the likelihood function is maximized. As any other estimation procedure, we start by running an experiment to obtain N independent samples of a random variable X that possesses a probability density function $f_X(x)$. Let θ be a parameter of X to be estimated. If X_1 is the first sample of X, the conditional density function of X_1 is $f_{X_1}(x_1|\theta)$ if it came from X, that is the density function of X evaluated at X_1. The conditional density function of X_2 is $f_{X_2}(x_2|\theta)$, etc. Since the samples X_i, $i=1,...,N$ are independent, then the joint conditional probability density function of the samples, called the *likelihood function,* is simply the product of the individual densities:

$$f_{X_N}(x_1, x_2,..., x_N|\theta)=L(\theta)=\prod_{i=1}^{N}f_{X_i}(x_i|\theta)$$

(6.4)

where

$L(\theta)$=the likelihood function
$f_{X_N}(x_1, x_2, \ldots, x_N|\theta)$=joint conditional density of the samples
$\prod$=the product operator

The final step is to find an estimate of θ, $\hat{\theta}$ (the "hat" denotes "estimate"), such that the likelihood function is maximum. To this end we differentiate the likelihood function with respect to the parameter θ and equate it to zero:

$$\frac{dL(\theta)}{d\theta}\Big|_{\theta=\hat{\theta}}=0$$

(6.5)

From equation (6.5) the $\hat{\theta}$ is found.

Example 6.2
For the Exponential random variable, equation (3.35) states that the density function is $f_T(t)=\omega e^{-\omega t}$, t, $\omega>0$, where ω is a parameter. Find the maximum likelihood estimate of ω.

Solution
From equation (6.4) the likelihood function is

$$L(\omega)=\prod_{i=1}^{N}\omega e^{-\omega t_i}$$

where t_i is the i-th independent sample of T. Solving,

$$L(\omega)=\omega^N e^{-\omega\sum_{i=1}^{N}t_i}$$

To eliminate the exponential, let us take logarithms:

$$\ln(L(\omega))=\ln(\omega^N)+\ln\left(e^{-\omega\sum_{i=1}^{N}t_i}\right)=N\ln(\omega)-\omega\sum_{i=1}^{N}t_i$$

Now differentiate the logarithm of the likelihood function with respect to ω and equate it to zero:

$$\frac{\partial}{\partial \omega}\left[\ln(L(\omega))\right] = \frac{N}{\omega} - \sum_{i=1}^{N} t_i \quad \Rightarrow \quad \hat{\omega} = \frac{N}{\displaystyle\sum_{i=1}^{N} t_i} = \frac{1}{\overline{T}}$$

where $\overline{T}$ is the sample mean of T. Thus, $\hat{\omega}$ is the maximum likelihood estimate of ω.

Example 6.3

Derive the maximum likelihood estimators of the parameters μ and σ^2 of the Normal distribution.

Solution

Taking independent samples x_1, x_2, ..., x_N from a Normal distribution, the likelihood function is the product of the Normal density evaluated at each of the x_i:

$$L(\mu,\ \sigma^2) = \frac{1}{(2\pi)^{N/2}\sigma^N} e^{-\sum_{i=1}^{N}(x_i-\mu)^2/2\sigma^2}$$

Taking logarithms,

$$\ln(L(\mu,\ \sigma^2)) = \ln\left(\frac{1}{(2\pi)^{N/2}\sigma^N}\right) - \frac{\displaystyle\sum_{i=1}^{N}(x_i-\mu)^2}{2\sigma^2}$$

$$= -N\ln(\sigma) - \frac{N}{2}\ln(2\pi) - \frac{\displaystyle\sum_{i=1}^{N}(x_i-\mu)^2}{2\sigma^2}$$

The latter contains two unknowns, μ and σ^2. Thus, we differentiate the log-likelihood function with respect to each of these variables, set them to zero, and solve the resulting two equations simultaneously.

$$\frac{\partial \ln(L)}{\partial \mu} = \frac{\displaystyle\sum_{i=1}^{N}(x_i-\mu)}{\sigma^2} = 0, \qquad \frac{\partial \ln(L)}{\partial \sigma} = -\frac{N}{\sigma} + \frac{\displaystyle\sum_{i=1}^{N}(x_i-\mu)^2}{\sigma^3} = 0$$

Solving these equations simultaneously,

$$\hat{\mu} = \overline{X}, \qquad \hat{\sigma}^2 = \frac{1}{N}\sum_{i=1}^{N}(x_i-\overline{X})^2$$

It is interesting to note that $\hat{\sigma}^2=(N-1)S^2/N$, where S^2 is the moments' estimator of σ^2. Since $E\{\hat{\sigma}^2\}=(N-1)\sigma^2/N \neq \sigma^2$, then the maximum likelihood estimator is biased.

6.3 INTERVAL ESTIMATORS

Confidence Intervals

As an alternative to point estimators of population parameters, which provide an evaluation of sample statistics, there are the *interval estimators*, which give a range of values where the population parameter is possibly contained. For example, using the method of moments, equation (6.1), we use a set of sample values to calculate that an estimate of the mean temperature in a boiler at a given time is $\hat{\mu}=\bar{X}=120^0C$. This is a point estimate of the population mean temperature. On the other hand, we may state that the mean temperature is somewhere between 115^oC and 125^oC. This is an interval estimate. Since we are not completely sure about the true population value of the mean, obviously we need to characterize our uncertainty associated with the given interval. We could simply say that we are completely sure that the mean temperature is in between 0^oC and 500^oC, but this statement is of little practical value. Instead, we would like to give a small interval and yet retain a large enough degree of certainty associated with it. Clearly, the smaller the interval, the more accurate our estimate (i.e., the closer to the true value), but the less certainty. For instance, we may say that we are 98% sure that the mean temperature is between 115^oC and 125^oC. Alternatively, we may say that we are only 95% sure that the true mean temperature is between 110^oC and 130^oC.

More formally, for some small α, called the *level of significance*, we want to be $(1-\alpha)\%$ sure that a population parameter θ is contained in the interval $L<\theta<U$. Let us express the uncertainty in probabilistic terms:

$$P(L<\theta<U)=1-\alpha, \quad \alpha>0 \qquad (6.6)$$

where
    α=the level of significance
    L=the interval lower limit
    U=the interval upper limit

In practice, the level of significance, or confidence level, is chosen according to technical or budgetary constraints. It is an acceptable design criterion. Decreasing the level of significance (i.e., increasing the level of certainty) usually requires more samples, more investment in experimentation. Increasing the confidence level results in fewer costs, but possibly unacceptable security. Given the confidence level, the direct problem consists in finding the confidence limits, L and U. Another practical problem consists in finding the number of samples to take, given an acceptable confidence interval U-L. We may explore confidence intervals for any population parameter θ. In this section we study confidence intervals for the mean and the variance of a Normal random variable.

Case 1: Confidence Interval for the Mean, Variance is Known

We are seeking to define a confidence interval for the population mean, μ_X, of a random variable X. We run an experiment and take N independent samples of X, from which we calculate the sample mean $\bar{X}$. From section 6.1, the sample mean is another random variable whose mean is the population mean of X, μ_X, that is $E\{\bar{X}\}=\mu_X$. We also stated that the variance of the sample mean is the population variance of X divided by the sample size, that is $Var(\bar{X})=\sigma_X^2/N$. Figure 6.2 illustrates the probability density function of the sample mean of a random variable X, that is $\bar{X}\sim f_{\bar{X}}(\bar{x})$. Notice that the mean of the sample mean is the population mean. If we define a confidence level α, then from equation (6.6),

$$P(L<\mu_X<U)=1-\alpha \qquad (6.7)$$

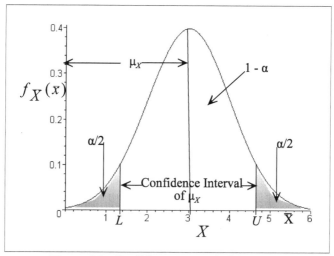

Figure 6.2.: Confidence Interval of the Mean

This is represented in Figure 6.2 as the clear, unshaded, area. Each tail of the density function of the sample mean, $\bar{X}$, has an area of $\alpha/2$ (i.e., the total area under the density function must equal 1). Thus, given α, we seek to calculate the abscissa values L and U, which define the confidence intervals.

In section 6.1, we saw that $\bar{X} \sim N(\mu, \ \sigma/\sqrt{N})$ if N is large. Let us define the statistic

$$Z = \frac{\bar{X} - \mu_X}{\sigma_X/\sqrt{N}} \tag{6.8}$$

Since Z is a linear combination of a Gaussian random variable, then it follows another Gaussian variable, more precisely a Standard Gaussian variable. Thus $Z \sim N(0, 1)$. Figure 6.3 shows the Standard Gaussian density function of Z, $f_Z(z)$. For a given value of α, the unshaded area under the Gaussian density has an area equal to $1 - \alpha$. Due to the symmetry of the Gaussian density with respect to $Z = 0$, this area is limited by the abscissa values $-Z_{1-\alpha/2}$ and $Z_{1-\alpha/2}$, where $Z_{1-\alpha/2}$ is the abscissa (i.e., the value of Z) corresponding to a cumulative area of $1 - \alpha/2$. This value may be read in tables of Standard Gaussian areas, such as Table A.1, Appendix A, or obtained with a computer program, such as Maple (see Example 3.21, Chapter 3). Thus, from Figure 6.3,

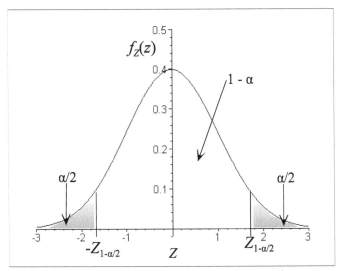

Figure 6.3: Standard Gaussian Density Function

$$P\left(-Z_{1-\frac{\alpha}{2}} < Z < Z_{1-\frac{\alpha}{2}}\right) = 1 - \alpha \tag{6.9}$$

Substituting equation (6.8) into (6.9),

$$P\left(-Z_{1-\frac{\alpha}{2}} < \frac{\bar{X}-\mu_X}{\sigma_X/\sqrt{N}} < Z_{1-\frac{\alpha}{2}}\right) = 1-\alpha \qquad (6.10)$$

which may be written as

$$P\left(-\frac{\sigma_X Z_{1-\frac{\alpha}{2}}}{\sqrt{N}} < \bar{X}-\mu_X < \frac{\sigma_X Z_{1-\frac{\alpha}{2}}}{\sqrt{N}}\right) = 1-\alpha \qquad (6.11)$$

Now, isolating μ_X in the middle of the inequalities (i.e., sending $\bar{X}$ to the other side of the inequalities, multiplying by -1 and rearranging),

$$P\left(\bar{X}-\frac{\sigma_X Z_{1-\frac{\alpha}{2}}}{\sqrt{N}} < \mu_X < \bar{X}+\frac{\sigma_X Z_{1-\frac{\alpha}{2}}}{\sqrt{N}}\right) = 1-\alpha \qquad (6.12)$$

Comparing equation (6.12) with equation (6.7), we conclude that the limits of the confidence band for the population mean are:

$$L=\bar{X}-\frac{\sigma_X Z_{1-\frac{\alpha}{2}}}{\sqrt{N}}, \qquad U=\bar{X}+\frac{\sigma_X Z_{1-\frac{\alpha}{2}}}{\sqrt{N}} \qquad (6.13)$$

Given the significance level, α, the population standard deviation, σ_X, and a sample of size N, one may compute the confidence limits from equation (6.13).

Example 6.4

A machine produces precision pins. From past records it is known that the standard deviation of the diameter is $0.001cm$. However, the mean diameter varies with daily adjustment. One day, 120 pins were sampled and the sample mean was $0.2119cm$. Find the 99% confidence interval for the mean diameter.

Solution

Let X be the pin diameter in cm. We need the confidence interval for the mean diameter, μ_X, when the standard deviation, $\sigma_X=0.001cm$, is known. This is Case 1 with $N=120$, $\bar{X}=0.2119cm$, and $1-\alpha=0.99$, or $\alpha=0.01$. Figure 6.4 shows the Standard Gaussian density function for the Z statistic. We need the abscissa value, Z, corresponding to a cumulative area under the density function of 0.995. From Table A.1, $Z_{1-\alpha/2}=Z_{0.995}=2.575$. Now from equation (6.13),

$$L = \bar{X} - \frac{\sigma_X Z_{1-\frac{\alpha}{2}}}{\sqrt{N}} = 0.2119 - \frac{0.001 \times 2.575}{\sqrt{120}} = 0.2117 cm$$

$$U = \bar{X} + \frac{\sigma_X Z_{1-\frac{\alpha}{2}}}{\sqrt{N}} = 0.2119 + \frac{0.001 \times 2.575}{\sqrt{120}} = 0.2121 cm$$

We are 99% confident that $0.2117 < \mu_X < 0.2121 cm$.

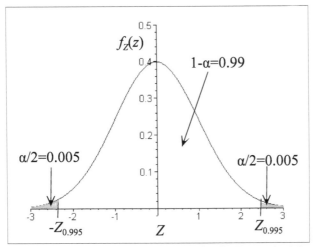

Figure 6.4: Standard Gaussian Density for Example 6.4

Case 2: Confidence Interval for the Mean, Variance is not Known

As in Case 1, we want to find the confidence interval for the mean of a random variable X, μ_X. However, in Case 2 the population variance of X, σ_X^2, is unknown. If $X \sim N(\mu_X, \sigma_X)$, then the statistic

$$T = \frac{\bar{X} - \mu_X}{S_X/\sqrt{N}} \qquad (6.13)$$

follows a *Student's t* probability density function with N-1 degrees of freedom. Note that the statistic T, as defined in equation (6.13), is similar to the statistic Z in equation (6.8), except that the population standard deviation, σ_X, is replaced by the sample one, S_X. For a given sample size N, the Student's t probability density function with N degrees of freedom is given by

$$f_T(t) = \frac{\Gamma\left(\dfrac{N+1}{2}\right)}{\sqrt{\pi N} \cdot \Gamma\left(\dfrac{N}{2}\right) \cdot \left(1 + \dfrac{t^2}{N}\right)^{(N+1)/2}}, \quad -\infty < T < \infty, \; N = integer \quad (6.14)$$

where

$\Gamma(\;)$ = gamma function as defined by equation (3.47)

The Student's t is symmetric about $T=0$. It resembles a Normal curve, and the Standard Normal curve serves as an approximation for large values of T (e.g., $T>30$). Figure 6.5 illustrates the Student's t density function with m degrees of freedom. For a given confidence level a, the cumulative area under the density function, $F_T(t_{m,1-a}) = P(T \le t_{m,1-a})$, is shown as the shaded area. Notice that $t_{m,1-a}$ is the *abscissa value* (or the inverse) of the t density function with m degrees of freedom and a cumulative area of $1-a$. Because of symmetry, the abscissa value $t_{m,1-a}$ is equal to $-t_{m,1-a}$. Table A.2, Appendix A, has abscissa values of t for some selected areas, $1-a$, and degrees of freedom, m. These can be called in computer programs too.

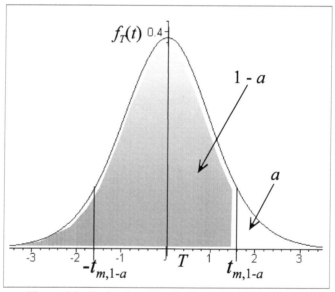

Figure 6.5: Student's t Distribution with N Degrees of Freedom and $1-a$ Cumulative Area

Now, imagine that we have a given confidence level, α, split as small equal areas on the left and the right tails of the Student's t

distribution with say N-1 degrees of freedom. This situation is depicted in Figure 6.6. The abscissa values corresponding to the limits of the shaded areas are $t_{N-1,1-\alpha/2}$ and $-t_{N-1,1-\alpha/2}$, respectively. Thus, returning to equation (6.13), if T follows a Student's t distribution with N-1 degrees of freedom, N is the sample size, and α is a given a confidence level, we can write a statement similar to equation (6.9) as

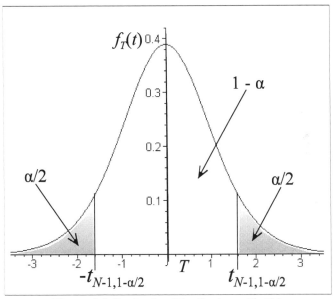

Figure 6.6: Student's t Distribution with N-1
Degrees of Freedom and $\alpha/2$ Tail Areas

$$P(-t_{N-1,\,1-\frac{\alpha}{2}} < T < t_{N-1,\,1-\frac{\alpha}{2}}) = 1-\alpha \qquad (6.15)$$

Substituting T for its value in equation (6.13) and solving for μ_X as in Case 1, we obtain

$$P\left(\overline{X} - \frac{t_{N-1,\,1-\frac{\alpha}{2}}\,S_X}{\sqrt{N}} < \mu_X < \overline{X} + \frac{t_{N-1,\,1-\frac{\alpha}{2}}\,S_X}{\sqrt{N}}\right) = 1-\alpha \qquad (6.16)$$

which gives the lower and upper limits of the confidence interval for the population mean:

$$L = \overline{X} - \frac{t_{N-1,\,1-\frac{\alpha}{2}}\,S_X}{\sqrt{N}}, \quad U = \overline{X} + \frac{t_{N-1,\,1-\frac{\alpha}{2}}\,S_X}{\sqrt{N}} \qquad (6.17)$$

We emphasize that strictly speaking equation (6.17) is valid when the parent distribution of X is Normal. For other distributions, this result is only an approximation.

Example 6.5

Four determinations of pH, p, of a certain solution were 7.90, 7.94, 7.91, and 7.93. Assuming that $p \sim N(\mu_p, \sigma_p)$, (1) find the 99% confidence interval for the population mean of p, and (2) repeat the problem using Maple.

Solution

(1) We are asked to calculate the confidence interval for the population mean of the random variable p. Since the population variance, σ_p^2, is unknown we are in Case 2 with $N=4$, and $\alpha=0.01$. Let us first calculate the sample mean and sample standard deviation from equations (6.1) and (6.2), respectively:

$$\bar{p}=\frac{\sum\limits_{i=1}^{4} p_i}{4}=7.92 \qquad S_p=\sqrt{\frac{\sum\limits_{i=1}^{4} (p_i-\bar{p})^2}{3}}=0.0183$$

From equation (6.17),

$$L=\bar{p}-\frac{t_{4-1,\,1-\frac{0.01}{2}}\,S_p}{\sqrt{4}}=7.92-\frac{5.8409\times0.0183}{2}=7.867$$

$$U=\bar{p}+\frac{t_{4-1,\,1-\frac{0.01}{2}}\,S_p}{\sqrt{4}}=7.92+\frac{5.8409\times0.0183}{2}=7.973$$

where the value $t_{N-1,1-\alpha/2}=t_{3,0.995}=5.8409$ was read in Table A.2. We are 99% confident that $7.867 < \mu_p < 7.973$.

(2)
```
> with(stats): alpha:=0.01: N:=4: p:=[7.9,7.94,7.91,7.93]:
```
Compute sample mean and sample standard deviation. For the latter, notice the constraint of 1, "standarddeviation[1]," indicating *sample* standard deviation (see equation (6.2)). The default value is 0, indicating *population* deviation.
```
pbar:=describe[mean](p);
Sp:=describe[standarddeviation[1]](p);
```
$$pbar := 7.920000000$$
$$Sp := 0.01825741858$$

Find the abscissa (inverse) value of the student's t distribution with N-1 degrees of freedom and cumulative area of 1-alpha/2.
```
> t:=statevalf[icdf,studentst[N-1]](1-alpha/2);
```
$$t := 5.840909310$$

Compute the confidence limits.
```
> L:=pbar-t*Sp/sqrt(N); U:=pbar+t*Sp/sqrt(N);
```
$$L := 7.866680037$$
$$U := 7.973319963$$

Case 3: Confidence Interval for the Variance, Mean is Known

We now turn our attention to the calculation of the confidence interval of the population variance of the random variable X, σ_X^2, when the population mean, μ_X, is known. If $X \sim N(\mu_X, \sigma_X)$, then the statistic

$$V = \frac{NS_X^2}{\sigma_X^2} \tag{6.18}$$

follows a *Chi Squared*, χ^2, probability density function with N degrees of freedom. Again $S_X^2 = \hat{\sigma}_X^2$ is the sample variance of X (the "hat" denoting "estimate" of a variance). The Chi-Squared density function with N degrees of freedom is given by

$$f_V(v) = \frac{v^{\frac{N}{2}-1} e^{-\frac{v}{2}}}{2^{\frac{N}{2}} \Gamma\left(\frac{N}{2}\right)}, \qquad V>0, \ N=integer \tag{6.19}$$

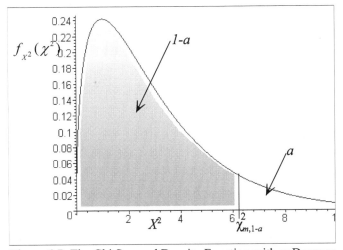

Figure 6.7: The Chi Squared Density Function with m Degrees of Freedom and $1-a$ Cumulative Area

The Chi Squared probability density function results from the sum of N squares of independent Standard Gaussian random variables. Figure 6.7 illustrates the Chi-Squared density function with m degrees of freedom. For a given confidence level a, the cumulative area under the density function, $F_X^2(\chi_{m,1-a}^2) = P(X^2 \le \chi_{m,1-a}^2)$, is shown as the shaded area. Notice that $\chi_{m,1-a}^2$ is the *abscissa value* (or the inverse) of the Chi-Squared density function with m degrees of freedom and a cumulative area of 1-a. Table A.3, Appendix A, has abscissa values of Chi Squared for some selected areas 1-a, and degrees of freedom m. These can be called in computer programs too.

Now, imagine that we have a given confidence level, α, split as small equal areas on the left and the right tails of the Chi-Squared distribution with N degrees of freedom. This situation is depicted in Figure 6.8. The abscissa values corresponding to the limits of the shaded areas are $\chi^2_{N,\alpha/2}$ and $\chi^2_{N,1-\alpha/2}$, respectively. Notice the Chi-Squared distribution is not symmetric. Returning to equation (6.18), if V follows a Chi-Squared distribution with N degrees of freedom, N is the sample size, and α is a given confidence level, we can write a statement similar to equation (6.9) as

$$P(V_{N,\frac{\alpha}{2}} < V < V_{N,1-\frac{\alpha}{2}}) = 1-\alpha \tag{6.20}$$

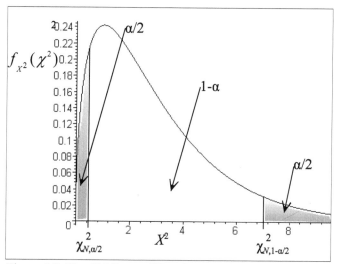

Figure 6.8: The Chi Squared Density Function with N Degrees of Freedom and $\alpha/2$ Tail Areas

Substituting V for its value in equation (6.18) and solving the inequalities for σ^2_X, as in previous cases, we obtain

$$P\left(\frac{NS^2_X}{V_{N,1-\frac{\alpha}{2}}} < \sigma^2_X < \frac{NS^2_X}{V_{N,\frac{\alpha}{2}}} \right) = 1-\alpha \tag{6.21}$$

which gives the lower and upper limits of the confidence interval for the population variance:

$$L = \frac{NS_X^2}{V_{N,1-\frac{\alpha}{2}}}, \qquad U = \frac{NS_X^2}{V_{N,\frac{\alpha}{2}}} \qquad (6.22)$$

Example 6.6

A structure is designed to rest on 100 piles. Nine test piles were driven at random locations into the supporting soil stratum and loaded until failure occurred according to Table 6.1. From past experience in nearby sites, it is known that the mean pile capacity is $85 Ton$. Assuming that pile capacity, Q, follows a Normal probability density function, (1) calculate the 98% confidence interval for the variance of pile capacity, and (2) repeat the problem using Maple.

Table 6.1: Test Pile Capacity for Example 6.6

Test Pile	1	2	3	4	5	6	7	8	9
Pile Capacity Q (*Ton*)	82	75	95	90	88	92	78	85	90

Solution

(1) We are required to calculate the confidence interval for the variance of Q, σ_Q^2, when the mean $\mu_Q = 85 Ton$ is known. This is Case 3 with $N=9$, and $\alpha = 0.02$. Let us first calculate the sample standard deviation from equation (6.2), except that we use the population mean $\mu_Q = 85 Ton$, since it is known:

$$S_Q^2 = \frac{\sum_{i=1}^{9} (Q_i - 85)^2}{8} = 45.75 Ton^2$$

From equation (6.22),

$$L = \frac{NS_Q^2}{V_{9,\,0.99}} = \frac{9 \times 45.75}{21.6660} = 19.00, \qquad U = \frac{NS_Q^2}{V_{9,\,0.01}} = \frac{9 \times 45.75}{2.0879} = 197.21$$

where the values of V were read in Table A.3. We are 98% confident that the variance of pile capacity is between 19.00 and 197.21 Ton^2, or that the standard deviation is between 4.36 and 114.04 Ton.

(2)

```
[ Initialize statistics module and enter data.
  [ > restart: with(stats): alpha:=0.02: N:=9: muQ:=85.:
    Q:=[82.,75.,95.,90.,88.,92.,78.,85.,90.]:
    N:=describe[count](Q):
  [ Compute sample standard deviation. Since the population mean is known, we use
    the formula instead of the describe command, which uses thesample mean.
  [ > SQ2:=sum((Q[i]-muQ)^2,i=1..N)/(N-1);
```

$$SQ2 := 45.75000000$$

```
  [ Find the abscissa (inverse) value of the Chi Squared distribution with N degrees
    of freedom and cumulative area of alpha/2. Repeat for a cumulative area of 1-alpha/2.
  [ > Vl:=statevalf[icdf,chisquare[N]](alpha/2);
    Vu:=statevalf[icdf,chisquare[N]](1-alpha/2);
```

$$Vl := 2.087900736$$
$$Vu := 21.66599433$$

```
[ Compute the confidence limits for the variance.
  [ > L:=N*SQ2/Vu; U:=N*SQ2/Vl;
```

$$L := 19.00443588$$
$$U := 197.2076512$$

```
  [ > l:=sqrt(L); u:=sqrt(U);
```

$$l := 4.359407744$$
$$u := 14.04306417$$

```
[ We are 98% confident that the standard deviation is between 4.36 and 14.04Ton.
```

Case 4: Confidence Interval for the Variance, Mean is not Known

We want to find the confidence interval of the population variance of the random variable X, σ_X^2, when the population mean, μ_X, is unknown. Case 4 is similar to Case 3, except that the population mean is not known. If $X \sim N(\mu_X, \sigma_X)$, then the statistic

$$W = \frac{(N-1)S_X^2}{\sigma_X^2} \tag{6.23}$$

follows a *Chi-Squared*, χ^2, probability density function with $N-1$ degrees of freedom. Equation (6.20) now becomes

$$P(W_{N-1,\frac{\alpha}{2}} < W < W_{N-1,1-\frac{\alpha}{2}}) = 1-\alpha \tag{6.24}$$

Again, $W_{N-1,\alpha/2}$ is the abscissa W of the Chi Squared probability density function with $N-1$ degrees of freedom corresponding to a cumulative area of $\alpha/2$. Substituting equation (6.23) into equation (6.24) and solving for the population variance as before, we obtain

$$P\left(\frac{(N-1)S_X^2}{W_{N-1,1-\frac{\alpha}{2}}} < \sigma_X^2 < \frac{(N-1)S_X^2}{W_{N-1,\frac{\alpha}{2}}}\right) = 1-\alpha \tag{6.25}$$

which gives the lower and upper limits of the confidence interval for the population variance:

$$L = \frac{(N-1)S_X^2}{W_{N-1,1-\frac{\alpha}{2}}}, \qquad U = \frac{(N-1)S_X^2}{W_{N-1,\frac{\alpha}{2}}} \qquad (6.26)$$

Example 6.7

In Example 6.5, calculate the 95% confidence interval for the population variance of p.

Solution

We are asked to calculate the confidence interval of σ_p^2 when the population mean μ_p is not known. Thus, we are under Case 4 with $N=4$, $\alpha=0.05$, and $S_p^2 = 0.0183^2 = 0.000335$. From equation (6.26),

$$L = \frac{(4-1)S_p^2}{W_{3,0.975}} = \frac{3 \times 0.00033489}{9.3484} = 0.000107$$

$$U = \frac{(4-1)S_p^2}{W_{3,0.025}} = \frac{3 \times 0.00033489}{0.2158} = 0.004655$$

Taking the square root, we are 95% confident that the population standard deviation of p, σ_p, is between 0.010 and 0.068.

6.4 STATISTICAL TESTS

We have seen point estimators (sections 6.1 and 6.2) and interval estimators (section 6.3) to conjecture the possible value of a population parameter, θ, of a random variable, X. A third approach consists in testing the validity of statistical proposals, called *hypotheses*, concerning the population parameters. This approach is particularly useful when we are interested in inferring if a particular sample belongs to a population. Examples of this appear every day in engineering practice. For instance, an engineer maybe interested in deciding if a certain volume of ready-mixed concrete she ordered actually has the specified failure strength, X. To assess this, she takes a series of N samples, pours them into testing cylinder shapes, cures them, and fails them in the laboratory according to the standard ASTM protocol. From the N samples, she computes the sample mean strength, X. Due to uncertainty in the testing process, the limited sample size, and natural heterogeneity inherent in the material manufacturing, the sample mean will usually be different from the specified population mean. In other words, it is possible that the sample mean is different from that of the population, even though the population mean has the right strength. That is, if the engineer took a large enough sample, she will obtain a sample mean close to the population mean. The question is: how different should the sample mean be before she concludes that the true population mean strength, μ_X, is not the specified one (i.e., the one she ordered) and rejects the batch? To achieve this, one needs an objective means, a statistical test, to calculate a region of acceptance, and that of rejection, of a sample mean. More precisely, we state a hypothesis that the population mean equals the specified one, and accept or reject the hypothesis according to the results of the test.

There are several ways to approach the problem of delineating the region of acceptance, called the *critical region*. We will opt for the simplest avenue that would make use of concepts already built. Imagine that we choose a small significance level α. When we studied confidence levels for the mean, we found that the probability that the statistic Z (equation 6.8) is between the two limits $-Z_{1-\alpha/2}$ and $Z_{1-\alpha/2}$ is $1-\alpha$ (see equation (6.9) and Figure 6.3). We used this concept to derive the confidence limits associated with a chosen α (see equations (6.10) through (6.13)). Alternatively, we can calculate the sample mean, and thus the corresponding Z statistic, and accept the hypothesis if Z falls within the confidence limits. In other words, we assume that if Z falls beyond the confidence limits, the value of the sample mean is not plausible and we reject the hypothesis. This is the basis of the following tests. Some conceptual problems arise: if we choose a small enough α, we run into the risk of rejecting a hypothesis that is really correct, called the *Type* I *error*. To make this error small, we increase the value of α. Then, we run into the risk, β, of accepting a hypothesis that is really incorrect, called the *Type* II *error*.

In the following, we focus on tests for the mean, μ_X, and the variance, σ_X^2, parameters of a random variable X that follows a Normal distribution (i.e., $X \sim N(\mu_X, \sigma_X^2)$). For random variables that follow other probability distributions, these results constitute only an approximation. As we did for confidence intervals, we subdivide the tests into four cases, depending on whether the test is for the mean or for the variance, and whether the other parameter (i.e., the variance in a test for the mean, or the mean in a test for the variance) is known or not known. Each case has its own test statistic. Within each case, there are three possible scenarios or questions one may ask. Each question has a specific way to formulate the hypothesis and a criterion to accept it or reject it. In practice, we simply select the applicable case that satisfies the conditions and then the specific question within such case according our objectives and data available.

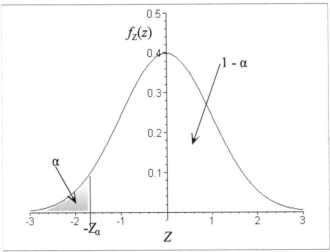

Figure 6.9: Critical Region for Case 1, Question 1

Case 1: Test for the Population Mean, Variance is Known

This is a test about the population mean μ_X, of a Normal random $X \sim N(\mu_X, \sigma_X^2)$ variable. The test statistic, R, is given again by

$$R = \frac{\bar{X} - C}{\sigma_X / \sqrt{N}} \tag{6.27}$$

R follows a Standard Normal random variable, with C a constant. We hypothesize μ_X to be greater, equal, or less than C. This test answers the following 3 questions about μ_X:

Question 1: Is $\mu_X = C$ or is $\mu_X < C$?

The format of this question calls for a left tail test with an area of α. We would reject the hypothesis that $\mu_X = C$, and thus accept the alternative hypothesis that $\mu_X < C$, if the calculated R statistic is less than $-Z_\alpha$, which is equal to $Z_{1-\alpha}$ because of symmetry (see Figure 6.9). The procedure is as follows:

(1) Calculate R from equation (6.27).

(2) State the null hypothesis H_0: $\mu_X = C$, and the alternative hypothesis H_1: $\mu_X < C$.

(3) If $-Z_\alpha < R$ accept H_0. If $R < -Z_\alpha$ reject H_0 and implicitly accept H_1.

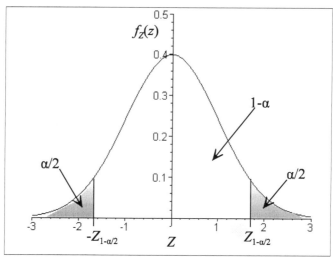

Figure 6.10: Critical Region for Case 1, Question 2

Question 2: Is $\mu_X = C$ or is $\mu_X \neq C$?

Here we have a two-tail test, each tail with an area value of $\alpha/2$. The region of acceptance is bounded by the abscissas $Z_{1-\alpha/2}$ and $-Z_{1-\alpha/2}$. We would reject the hypothesis that $\mu_X = C$ if the statistic R falls in the shaded area of Figure 6.10. The procedure is as follows:

(1) Calculate R from equation (6.27).

(2) State the null hypothesis H_0: $\mu_X = C$, and the alternative hypothesis H_1: μ_X.

(3) If $-Z_{1-\alpha/2} < R < Z_{1-\alpha/2}$ accept H_0. If $R < -Z_{1-\alpha/2}$ or $Z_{1-\alpha/2} < R$ reject H_0 and implicitly accept H_1.

Question 3: Is $\mu_X = C$ or is $C < \mu_X$?

Here we have a right-tail test with an area value of α. The region of acceptance is bounded by the abscissa $Z_{1-\alpha}$. We would reject the hypothesis that $\mu_X = C$ if the statistic R falls in the shaded area of Figure 6.11. The procedure is as follows:

(1) Calculate R from equation (6.27).

(2) State the null hypothesis H_0: $\mu_X = C$, and the alternative hypothesis H_1: $C < \mu_X$.

(3) If $R < Z_{1-\alpha}$ accept H_0. If $Z_{1-\alpha} < R$ reject H_0 and implicitly accept H_1.

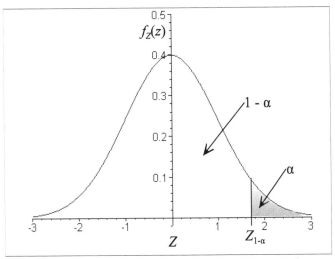

Figure 6.11: Critical Region for Case 1, Question 3

Case 2: Test for the Population Mean, Variance is not Known, $N \geq 60$

This is a test about the population mean, μ_X, of a Normal random variable $X \sim N(\mu_X, \sigma_X^2)$. The difference with Case 1 is that the population variance is not known. The test statistic, R, is now given by

$$R = \frac{\overline{X} - C}{S_X / \sqrt{N}} \tag{6.28}$$

where the sample standard deviation, instead of the population one, is now used. This test is identical to that in Case 1. In other words, there are three possible questions with the same procedure as in Case 1, except that equation (6.28) is used in the calculation of the R statistic, instead of equation (6.27).

Case 3: Test for the Population Mean, Variance is not Known, $N < 60$

This test is identical to Case 2, except that the available sample size N is of less than 60 data values. The test statistic now follows a Student's t probability density function with N-1 degrees of freedom. Thus, the same procedure is to be followed, except that in the decisions in Questions 1, 2, and 3 we replace the Z abscissa values of a Standard Normal distribution with the corresponding ones from a Student's t distribution.

Example 6.8

Four determinations of copper concentration, X, in a certain solution yielded a mean concentration of 8.3% and a standard deviation of 0.03%. Test at the 5% level of significance the hypothesis that the population mean concentration is equal to 8.32% versus the one that the mean is less than that value. Assume that the concentration of copper follows a Normal distribution.

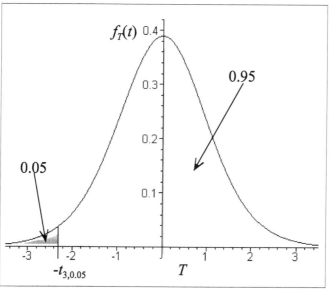

Figure 6.12: Student's t Distribution of Example 6.8

Solution

We are testing for μ_X when σ_X^2 is not known, $\overline{X}=8.30$, $S_X=0.03$, $N=4<60$, $C=8.32$, $\alpha=0.05$, and the question is whether μ_X is equal or less than C. Thus, use Case 3, Question 1. From equation (6.28) and following steps (1), (2), and (3),

$$R=\frac{\overline{X}-C}{S_X/\sqrt{N}}=\frac{8.30-8.32}{0.03/\sqrt{4}}=-1.333$$

The null hypothesis is H_0: $\mu_X=8.32\%$. The alternative hypothesis is

H_1: μ_X<8.32%. We have a left-tail test with $-t_{N-1,\alpha} = -t_{3,0.05} = -t_{3,0.95} =$ -2.3534 from Table A.2 (see Figure 6.12). Since $-t_{3,0.05}<R$ then accept H_0. We say, at the 5% level of significance, there is no evidence against the statement μ_X=8.32%

Example 6.9

The time required to reset a computer-controlled manufacturing robot is being studied. Over the last few weeks, the time (*min*) has been recorded on 10 occasions after the system crashed as follows: 5.84, 8.59, 8.48, 5.44, 10.91, 10.09, 10.32, 10.058, 3.39, and 6.68. Assume that the reset time $T_r \sim N(\mu_T, \sigma_T^2)$. The management is interested in knowing whether the reset time is significantly different from 8.0*min* at the 5% level of significance.

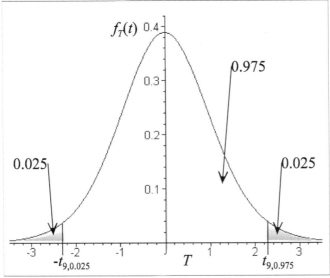

Figure 6.13: Student's t Distribution of Example 6.9

Solution

We are testing for μ_T when σ_T^2 is not known. From equations (6.1), and (6.2), $\overline{T}_r$=7.987 *min*, and S_T=2.529 *min*, N=10<60, C=8.0 *min*, α=0.05, and the question is whether μ_T is equal or different from C (either greater than or less than C). Thus, use Case 3, Question 2. From equation (6.28) and following steps (1), (2), and (3),

$$R = \frac{\overline{X}-C}{S_X/\sqrt{N}} = \frac{7.987-8.0}{2.529/\sqrt{10}} = -0.016$$

The null hypothesis is H_0: $\mu_T=8.0$. The alternative hypothesis is H_1: $\mu_T \neq 8.0$. We have a two-tail test. From Table A.2 the upper bound of the critical region is given by $t_{N-1,1-\alpha/2}=t_{9,0.975}=2.2622$; the lower bound of the critical region is given by $-t_{N-1,\alpha/2}=-t_{9,0.025}=-t_{9,0.975}=-2.2622$ (see Figure 6.13). Since $-t_{9,0.025}<R<t_{9,0.975}$ then accept H_0. We say, at the 5% level of significance, there is no evidence against the statement $\mu_T=8.0\,min$.

Case 4: Test for the Population Variance, $N \leq 100$

This is a test about the population variance, σ_X^2, of a Normal random variable $X\sim N(\mu_X, \sigma_X^2)$. The test statistic, R, is given again by

$$R=\frac{(N-1)S_X^2}{C}$$

(6.29)

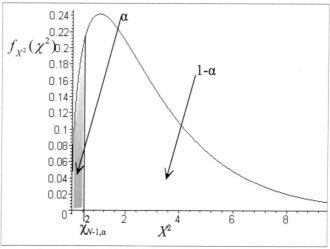

Figure 6.14: The Chi Squared Density Function
for Case 4. Question 1

This test answers the following 3 questions about σ_X^2:

Question 1: Is $\sigma_X^2=C$ or is $\sigma_X^2<C$?

The format of this question calls for a left-tail test with an area of α. We would reject the hypothesis that $\sigma_X^2=C$, and thus accept the alternative hypothesis that $\sigma_X^2<C$, if the calculated R statistic is less than $\chi_{N-1,\alpha}^2$, which is the abscissa value of the Chi-Squared distribution with $N-1$ degrees of freedom and cumulative area of α (see Figure 6.14). The procedure is as follows:

(1) Calculate R from equation from equation (6.29).

(2) State the null hypothesis H_0: $\sigma_X^2 = C$, and the alternative hypothesis H_1: $\sigma_X^2 < C$.

(3) If $\chi_{N-1,\alpha}^2 < R$ accept H_0. If $R < \chi_{N-1,\alpha}^2$ reject H_0 and accept H_1.

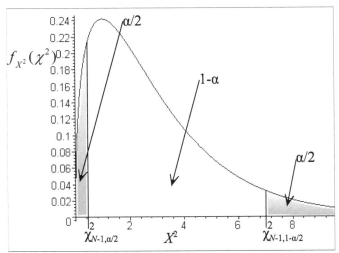

Figure 6.15: The Chi Squared Density Function
for Case 4, Question 2

Question 2: Is $\sigma_X^2 = C$ or is $\sigma_X^2 \neq C$?
   Here we have a two-tail test, each tail with an area of $\alpha/2$. The region of acceptance is bounded by the abscissas $\chi_{N-1,\alpha/2}^2$ and $\chi_{N-1,1-\alpha/2}^2$. We would reject the hypothesis that $\sigma_X^2 = C$ if the statistic R falls in the shaded area of Figure 6.15. The procedure is as follows:

(1) Calculate R from equation (6.29).

(2) State the null hypothesis H_0: $\sigma_X^2 = C$, and the alternative hypothesis H_1: $\sigma_X^2 \neq C$.

(3) If $\chi_{N-1,\alpha/2}^2 < R < \chi_{N-1,1-\alpha/2}^2$ accept H_0. If $R < \chi_{N-1,\alpha/2}^2$ or $\chi_{N-1,1-\alpha/2}^2 < R$, reject H_0 and implicitly accept H_1.

Question 3: Is $\sigma_X^2 = C$ or is $C < \sigma_X^2$?
   Here we have a right-tail test with an area of α. The region of acceptance is bounded by the abscissa $\chi_{N-1,1-\alpha}^2$. We would reject the

hypothesis that $\sigma_X^2 = C$ if the statistic R falls in the shaded area of Figure 6.16. The procedure is as follows:

(1) Calculate R from equation (6.29).

(2) State the null hypothesis H_0: $\sigma_X^2 = C$, and the alternative hypothesis H_1: $C < \sigma_X^2$.

(3) If $R < \chi_{N-1,1-\alpha}^2$ accept H_0. If $\chi_{N-1,1-\alpha}^2 < R$, reject H_0 and implicitly accept H_1.

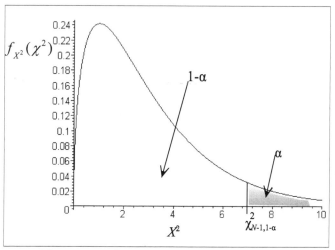

Figure 6.16: The Chi Squared Density Function
for Case 4, Question 3

Case 5: Test for the Population Variance, $N > 100$

This is a test about the population variance, σ_X^2, of a Normal random variable $X \sim N(\mu_X, \sigma_X^2)$. The test statistic, R, is now given by

$$R = \sqrt{\frac{N-1}{2}}\left(\frac{S_X^2}{C} - 1\right) \tag{6.30}$$

This test is identical to Case 4, except that the available sample size, N, is greater than 100 data values. The test statistic now follows a Standard Normal probability density function. Thus, the same procedure of Case 4 is to be followed, except that in the decisions in Questions 1, 2, and 3 we replace the χ^2 abscissa values with corresponding ones from a Standard Normal distribution.

Example 6.10

A thermoelectric plant consumes cooling water depending on electricity demand. During the summer months in a year, 11 samples of water consumption were taken at random. Assume that water consumption $X \sim N(\mu_X, \sigma_X^2)$. From the 11 samples, $S_X^2 = 154.6 (m^3 \times 10^3/day)^2$. The management is interested in knowing whether or not the population variance is greater than $140 (m^3 \times 10^3/day)^2$. Test at the 5% level of significance.

Solution

This is a test for the variance with $\alpha = 0.05$, $N = 11 < 100$, and $C = 140 (m^3 \times 10^3/day)^2$. We use Case 4, Question 3. We follow steps (1), (2), and (3) of the procedure. From equation (6.29),

$$R = \frac{(N-1)S_X^2}{C} = \frac{(11-1) \times 154.6}{140} = 11.043$$

The null hypothesis is H_0: $\sigma_X^2 = 140 (m^3 \times 10^3/day)^2$. The alternative hypothesis is H_1: $\sigma_X^2 > 140 (m^3 \times 10^3/day)^2$. This is a right-tail test with $\chi_{10,0.95}^2 = 18.3070$ read from Table A.3 (see Figure 6.17). Since $R < \chi_{10,0.95}^2$ then accept H_0. We say, at the 5% level of significance, there is no evidence against the statement that $\sigma_X^2 = 140 (m^3 \times 10^3/day)^2$.

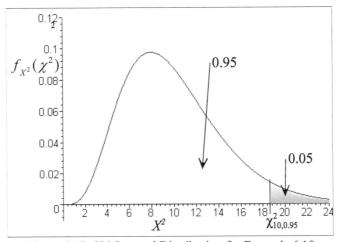

Figure 6.17: Chi Squared Distribution for Example 6.10

Example 6.11

A machine prints components of a microprocessor. Quality control management has decided that if the variance in the amplitude of a particular nerve connector is greater than $0.001\ mm^2$, the machine will be replaced. Amplitudes were measured in 1000 chips yielding a sample variance of $0.00102\ mm^2$. After testing at the 5% level of significance, recommend the appropriate course of action.

Solution

Assume the amplitude $A \sim N(\mu_A, \sigma_A^2)$. This is a test for the variance with $\alpha = 0.05$, $N = 1000 > 100$, $S_A^2 = 0.00102\ mm^2$, and $C = 0.001\ mm^2$. We use Case 5, Question 3. Following steps (1), (2), and (3) in the procedure, then from equation (6.30),

$$R = \sqrt{\frac{N-1}{2}}\left(\frac{S_A^2}{C}-1\right) = \sqrt{\frac{1000-1}{2}}\left(\frac{0.00102}{0.001}-1\right) = 0.447$$

The null hypothesis is H_0: $\sigma_A^2 = 0.001mm^2$. The alternative hypothesis is H_1: $\sigma_A^2 > 0.001mm^2$. This is a right-tail test with $Z_{0.95} = 1.645$ from Table A.1 (see Figure 6.18). Since $R < Z_{0.95}$ then accept H_0. We say, at the 5% level of significance, there is no evidence to support the claim that $\sigma_A^2 > 0.001mm^2$ and the machine should not be replaced at this time.

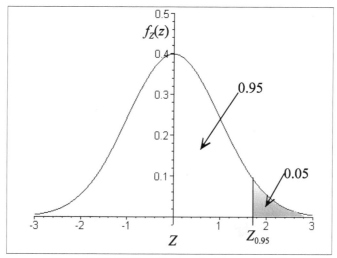

Figure 6.18: Critical Region for Example 6.11

PROBLEMS

6.1 In Example 6.1, the investor signs on 10 times. (1) Determine the density function of the sample mean along with its mean and standard deviation. (2) How many times does he need to check the stock price so that the standard deviation of the sample mean price has a standard deviation of 15 cents or less?

6.2 Take samples k_1, k_2, ..., k_N from a Poisson-distributed random variable, write the likelihood function, take its logarithm and find the maximum likelihood estimator for the parameter λ in the Poisson distribution.

6.3 Ten samples were taken of the mash temperature ($^\circ C$) in a large reflux distiller at a bourbon factory with the following results: 80.3, 80.2, 80.8, 78.3, 81.1, 79.8, 78.5, 79.2, 79.5, 79.7. Assuming that the mash temperature follows a Normal distribution, compute the moments' estimates and the maximum likelihood estimates of the mean and the standard deviation.

6.4 Solve Example 6.4 using Maple.

6.5 A random sample of tensile failure strength of 1-*inch* iron bars produced by the same furnace yielded the following results: 3148, 3033, 2971, 2996, 3102, 3009, 2877, 2966, 2888, and 3022 *psi*. Assuming the failure strength follows a Normal distribution, calculate the 95% confidence interval for the mean strength.

6.6 After many measurements, the mean hydraulic conductivity of an aquifer has been found to be *75m/month*. At a particular location, random soil samples yielded the following conductivity values: 33, 88, 70, 78, 61, 54, 68, 69, 59, 77, 57, 91, 70, 72, 56, and 40 *m/month*. Assuming that the hydraulic conductivity follows a Normal distribution (a strong assumption), find the 95% confidence interval for the variance of the conductivity.

6.7 Solve Example 6.7 using Maple.

6.8 A certain process yields pieces of steel strand with population standard deviation of their breaking strength equal to 500*psi*. A random sample of 9 pieces of strands from the process yields $\overline{X}_B = 12,260\,psi$, where $B \sim N(\mu_b, \sigma_B^2)$ and B is the breaking strength. Can we say at the 5% level of significance that $\mu_B = 13,500\,psi$?

6.9 Hourly wages of employees in a certain company have become a source of contention. An impartial arbitrator finds that industry-wide hourly wages are approximately normally distributed with a mean of $11.65 per hour. An arbitrator examined the earning record of 40 workers selected at random from the payroll list of the company. He found that the sample mean is $11.53 with a sample standard deviation of $0.30. Can we say at the 5% level of significance that the company pays less than the industry average?

6.10 A contractor claims that the joints spacing in a highway project have a standard deviation of 1.5cm. An inspection of 12 joints at random yielded a sample standard deviation of 1.7cm. Can it be said at the 5% level of significance that he delivered a project with a standard deviation greater than 1.5cm?

7 FITTING PROBABILITY MODELS TO DATA

In Chapter 6 we studied methods to estimate the magnitude of population parameters of a probability distribution. Engineering design often requires the calculation of probabilities or risks associated with the occurrence of an event. If a theoretical probabilistic model is available, the procedure to use was described in Chapter 3. We might even be required to synthesize realizations of an uncertain event using Monte Carlo simulation methods described in Chapter 4. In order to do this, the engineer needs not only the population parameters of the underlying probability distribution, but also the distribution itself. In a large class of engineering applications a theoretical model (i.e., the governing probability density function) is not available. Instead, the engineer proceeds with the design of an experiment intended to gather N independent samples from the process. In this chapter we explore methods to fit theoretical probability distributions to data obtained from limited testing.

7.1 EMPIRICAL DISTRIBUTIONS

The Frequency Histogram from Observed Data

The simplest approach to the problem of finding a possible probability distribution to describe an uncertain phenomenon is the construction of a *frequency histogram*. This is a bar graph that approximately shows the variability of absolute frequency, or of relative frequency, with respect to the value of a random variable. As the reader remembers, we already used this concept in connection with the simulation of random systems in Chapter 4. Examples 4.9 through 4.14 used frequency histograms of computer-generated random numbers to approximately assess whether or not the numbers approached the theoretical distributions from where they came. In Chapter 4 we used Maple to generate the histogram automatically. Let us study its basic principle more in detail. To construct a histogram from a random sample of N data points of a continuous random variable, we subdivide the possible range of values of the variable, X, into cells $i=1, 2, \ldots, M$. Each cell or class interval has a mid point abscissa value, X_i, a lower limit, $X_i - \Delta X/2$, and an upper limit, $X_i + \Delta X/2$, where ΔX is the class interval width. The *relative frequency* value assigned to the i-th cell is $\hat{f}_i = n_i/N$, where the "hat" denotes "estimate" or empirical frequency, n_i is the *number* of data points, or *absolute frequency*, with values of X such that $X_i - \Delta X/2 < X < X_i + \Delta X/2$ out of the total number of points, N. The last step consists in plotting i versus $\hat{f}_i$, $i=1$, 2, ..., M. Note: cell limits must not be located at actual data values. For a discrete random variable, the X_i are the fixed values a random variable may take and the class interval width is zero.

Example 7.1: Flood Frequency at a River Section

Table 7.1: Flood Frequency Analysis of Example 7.1

Year	$Q(m^3/s)$	Ordered $Q(m^3/s)$	$Ln(Q)$	Rank m	$T=(N+1)/m$ (year)
1984	16200	28300	10.25	1	28.0
1985	6500	24300	10.10	2	14.0
1986	20300	20500	9.93	3	9.3
1987	14700	20300	9.92	4	7.0
1988	20500	20100	9.91	5	5.6
1989	10100	19900	9.90	6	4.7
1990	7200	19200	9.86	7	4.0
1991	28300	18300	9.81	8	3.5
1992	18300	17900	9.79	9	3.1
1993	24300	17000	9.74	10	2.8
1994	17900	16200	9.69	11	2.6
1995	12100	14700	9.60	12	2.3
1996	10900	14700	9.60	13	2.2
1997	17000	14400	9.57	14	2.0
1998	10300	13600	9.52	15	1.9
1999	19900	12100	9.40	16	1.8
2000	13600	10900	9.30	17	1.7
2001	5700	10400	9.25	18	1.6
2002	8800	10300	9.24	19	1.5
2003	9400	10100	9.22	20	1.4
2004	19200	9400	9.15	21	1.3
2005	6800	8800	9.08	22	1.3
2006	14700	8400	9.04	23	1.2
2007	8400	7200	8.88	24	1.2
2008	20100	6800	8.82	25	1.1
2009	10400	6500	8.78	26	1.1
2010	14400	5700	8.27	27	1.0

The first two columns in Table 7.1 show the year and the maximum flow rate at a cross section in a large river from 1984 to 2010. In this case, the "experiment" is the collection of historical data. Note that a variable such as the daily flow rate has strong deterministic components, which should be studied with the known hydrologic deterministic methods. Only a small component of it would be random (e.g., the errors in measurement). On the other hand, by selecting the maximum annual flow (i.e., one value per year), we build a new variable with strong random components. Construct a frequency histogram for the random variable annual maximum flow rate, Q.

Solution

To facilitate the calculations, let us create a third column in Table 7.1 that contains the rearrange floods in decreasing order of magnitude. The maximum flow is less than $29,000 \, m^3/s$ and the minimum is greater than $5,000 \, m^3/s$ with a total range of $24,000 \, m^3/s$. Normally between 6 and 10 cells should provide a desirable display, that is one with cells not too low in its number of samples. Let us choose a graph with 8 cells with an equal interval length $\Delta X = 3,000 \, m^3/s$. Observing the third column in Table 7.1 count the number of flows, n_i, in each class interval (see Table 7.2) and calculate the relative frequency, $\hat{f}_i$, in each cell. Finally plot $\hat{f}_i$ versus Q_i in a bar chart (see Figure 7.1).

Table 7.2: Preparation of Data for Example 7.1

Cell, i	Range (m^3/s)	Center Value $Q_i(m^3/s)$	Absolute Frequency, n_i	Relative Frequency, $\hat{f}_i = n_i/N$
1	5,000 - 7,999	6,500	4	0.148
2	8,000 - 10,999	9,500	7	0.259
3	11,000 - 13,999	12,500	2	0.074
4	14,000 - 16,999	15,500	4	0.148
5	17,000 - 19,999	18,500	5	0.185
6	20,000 - 22,999	21,500	3	0.112
7	23,000 - 25,999	24,500	1	0.037
8	26,000 - 29,000	27,500	1	0.037
$M=8$			$N = \sum_{1}^{8} n_i = 27$	$\sum_{1}^{8} \hat{f}_i = 1.000$

The histogram illustrates the ranges of values in Q that are more likely to occur, and those with fewer chances in any year. Generally, flows in the range from 8,000 to $11,000 \, m^3/s$ are the most frequent. Extremely high flows greater than $20,000 \, m^3/s$, and extremely low flows of less than $5,000 \, m^3/s$, are less likely to occur. However, drought (low flow) is more frequent than flood (high flow), relatively speaking. There is an anomalous range from 11,000 to $14,000 \, m^3/s$ with low occurrences.

This probably reflects our particular choice of class intervals, or the lack of data, rather than actual behavior of the random variable

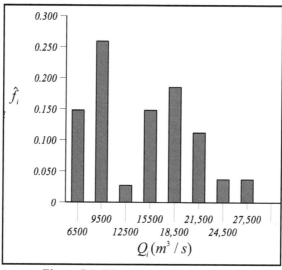

Figure 7.1: Histogram of Example 7.1

The Expected Frequency Histogram from a Theoretical Distribution
    The frequency histogram should be interpreted as a sample, or preliminary, probability density function of a random variable. It is interesting that the histogram satisfies the fundamental rules of probability (see properties 1 through 5, below equation 3.3). In particular, the area under the histogram is equal to 1 and the frequency approaches zero as the random variable tends to plus or minus infinity. While the histogram offers a useful first insight into the uncertain features of a random variable, it is a crude approximation to the underlying, true, probability density function. Evidently, the shape of the histogram depends on the choice of the number of cells. In Example 7.1, we selected the length of the class interval ΔX such that the number of data points, n_i, in most cells would not be too low. While this approach offers a more attractive, uniform, graph, it shows the subjectivity in the histogram shape. In fact, it is difficult to identify the possible probability density function based on the shape of the histogram, especially when there is insufficient data. In Example 7.1, we may speculate that the random variable Q is strictly positive and positively skewed. These features could point toward candidate distributions to study as possible models. Pursuing this avenue, one needs to compare visually or numerically the frequency histogram obtained from the data with the theoretical density function to

be fitted. However, this comparison is not as direct as one wished. The frequency histogram is a *discretized* version of a theoretical probability density function. As we have seen, its shape changes according to the magnitude of the class interval, ΔX. Thus, to compare a frequency histogram obtained from data, $\hat{f}_i$ (the "hat" denotes "estimate" of relative frequency), with a candidate density function, $f_X(x)$, we first need to plot the theoretical density as a frequency histogram, f_i, with the same class interval, ΔX. A frequency interpretation of a density function assumes that

$$f_i = P\left(X_i - \frac{\Delta X}{2} \leq X < X_i + \frac{\Delta X}{2}\right) = f_X(X_i)\Delta X, \quad X\ Continuous$$

$$(7.1)$$

$$f_i = P(X = X_i) = p_X(X_i), \quad X\ Discrete$$

where

    f_i = *expected relative frequency* of $f_X(x)$ in the i-th cell of size ΔX
    $f_X(x)$ = theoretical probability density function of X (continuous)
    $p_X(X_i)$ = probability mass function of X (discrete) evaluated at $X = X_i$

Equation (7.1) simply indicates that the probability that the random variable is in the i-th class interval is the area under the density function bounded by the class interval limits. This is a good approximation for M large. As ΔX decreases in size, the number of points in a cell decreases, and in the limit as ΔX approaches zero, $f_i = f_X(x)$. A more precise calculation of the area is obtained by the difference in the cumulative distribution function at the lower and upper cell limits. As an approximation, Equation (7.1) facilitates a comparison between *observed relative frequencies*, $\hat{f}_i$, with respect to *expected relative frequencies* arising from theoretical distributions, f_i. If one wishes to express this comparison in terms of absolute, rather than relative, frequencies, then multiply the relative frequencies by the sample size N. In other words, the *observed absolute frequency* is $n_i = \hat{f}_i \times N$, and the *expected absolute frequency* $e_i = f_i \times N$ (see section 7.2 for application).

Example 7.2: Comparison of Observed and Expected Frequency Histograms

The first 4 columns of Table 7.3 contain data from 1000 samples of a measurement process, X, with a mean of 10 and a variance of 1. It is suspected that a Normal distribution is an appropriate model. Compare the data versus the theoretical frequency histograms when the class interval $\Delta X = 0.5$.

Table 7.3: Frequency Data for Example 7.3

Cell, i	X Cell Range	Center Value X_i	Observed Absolute Frequency, n_i	Observed Relative Frequency, $\hat{f}_i = n_i/N$	Expected Relative Frequency, f_i
1	6.25 - 6.74	6.5	1	0.001	0.000
2	6.75 - 7.24	7.0	2	0.002	0.002
3	7.25 - 7.74	7.5	9	0.009	0.009
4	7.75 - 8.24	8.0	25	0.025	0.027
5	8.25 - 8.74	8.5	68	0.068	0.065
6	8.75 - 9.24	9.0	125	0.125	0.121
7	9.25 - 9.74	9.5	171	0.171	0.176
8	9.75 - 10.24	10.0	205	0.205	0.200
9	10.25 - 10.74	10.5	159	0.159	0.176
10	10.75 - 11.24	11.0	130	0.130	0.121
11	11.25 - 11.74	11.5	60	0.060	0.065
12	11.75 - 12.24	12.0	31	0.031	0.027
13	12.25 - 12.74	12.5	11	0.011	0.009
14	12.75 - 13.24	13.0	3	0.003	0.002
15	13.25 - 13.74	13.5	0	0.000	0.00

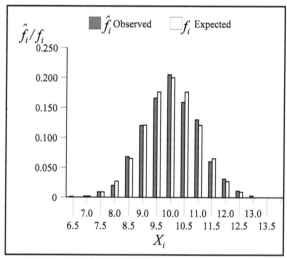

Figure 7.2: Histograms in Example 7.2

Solution

To complete the table, calculate the observed relative frequency as $\hat{f}_i = n_i/1000$. From equation (7.1), the expected or theoretical Normal relative frequency, f_i, is the Normal density function with mean 10 and variance 1 evaluated at the X_i, the mid cell abscissa value, times the class interval width $\Delta X = 0.5$:

$$f_i = f_X(X_i)\Delta X = \frac{e^{-\frac{(X_i - 10)^2}{2 \times 1}}}{\sqrt{2\pi \times 1}} \times 0.5$$

The histograms are shown in Figure 7.2. The Normal distribution appears to be a good model for the data. However, this observation is intuitive and preliminary. We have not evaluated how good the fit is yet.

Example 7.3: A Computer-Generated Histogram

In Example 7.1, we would like to assess the possibility that the random variable Q is Lognormally distributed. Design a computer worksheet to produce an observed frequency histogram and compare it to a corresponding theoretical one.

Solution

The following Maple worksheet starts with the flow data, Q_i, and transforms it into a log-series $X_i = Log(Q_i)$. The next, commands build a frequency data table (such as Table 7.3) in memory. Note that Maple automatically applies Equation (7.1) to each histogram and permits a direct comparison with a theoretical density function. A comparison is made between a histogram of X and a corresponding Normal distribution. Remember, if Q follows a Lognormal distribution, then $X=log(Q)$ follows a Normal distribution (see Chapter 3).

```
[ Initialize statistics module, enter flow data, and store in array Q[ ]. M – number of class intervals.
[ > restart: with(stats): with(stats[statplots]):
    Q:=[28300,24300,20500,20300,20100,19900,19200,18300,
    17900,17000,16200,14700,14700,14400,13600,12100,
       10900,10400,10300,10100,9400,8800,8400,7200,
       6800,6500,5700]:
   N:=describe[count](Q): M:=8:
[ Calculate the logarithm of flow series and store into array X[ ].
[ > [seq(log(Q[i]),i=1..N)]:
[ Calculate sample mean and sample standard deviation of logrithmic data.
[ > Xbar:=describe[mean](X):
    S_X:=describe[standarddeviation[1]](X):
[ Define the Normal density as a function, enter desired class interval width, and minimum cell limit.
  Maple plots histograms with equation (7.1) already imbedded
  (i.e., scales histogram by a factor of 1/deltax).
[ > pp:=evalf(Pi):
    f:=x->exp(-(x-Xbar)^2/(2*S_X^2))/sqrt(2.*pp*S_X^2):
    deltax:=0.25: Xmin:=8.5:
[ Create a vector with the values of the cell limits.
[ > limits:=[ seq( Xmin+i*deltax..Xmin+(i+1)*deltax,
                   i=0..(M-1) ) ]:
limits := [ 8.5 .. 8.75, 8.75 .. 9.00, 9.00 .. 9.25, 9.25 .. 9.50, 9.50 .. 9.75, 9.75 .. 10.00,
   10.00 .. 10.25, 10.25 .. 10.50]
[ Calculate the observed absolute frequency in each cell according to the specified limits.
  Use the transform[tallyinto]( , ) command. In the output, Weight(9.50..9.75,6) means that
  in the interval from 9.50 to9.75 there are 6 data points, etc.
[ > n_i:=transform[tallyinto](X,limits):
n_i := [8.5 .. 8.75, Weight( 8.75 .. 9.00, 3 ), Weight( 9.00 .. 9.25, 6 ), Weight( 9.25 .. 9.50, 2 ),
   Weight( 9.50 .. 9.75, 6), Weight( 9.75 .. 10.00, 7), 10.00 .. 10.25, 10.25 .. 10.50]
```

Calculate the observed *relative* frequency in each cell. Do this by dividing the observed f
requency by N using the transform[scaleweight[ ]] ( ) command.

```
> f_hat:=transform[scaleweight[1/N]](n_i);
```

$$f\_hat := \left[\text{Weight}\left(8.5 .. 8.75, \frac{1}{27}\right), \text{Weight}\left(8.75 .. 9.00, \frac{1}{9}\right), \text{Weight}\left(9.00 .. 9.25, \frac{2}{9}\right), \right.$$

$$\text{Weight}\left(9.25 .. 9.50, \frac{2}{27}\right), \text{Weight}\left(9.50 .. 9.75, \frac{2}{9}\right), \text{Weight}\left(9.75 .. 10.00, \frac{7}{27}\right),$$

$$\left. \text{Weight}\left(10.00 .. 10.25, \frac{1}{27}\right), \text{Weight}\left(10.25 .. 10.50, \frac{1}{27}\right)\right]$$

In one graph plot the theoretical frequency and the observed frequency histogram.

```
> plot(f(x),x=Xmin..(Xmin+M*deltax),color=black,thickness=2):
  histogram(f_hat, color=gray):
  plots[display](%,%%,labels=[`ln(Q)`,`frequency`],
    labeldirections=[HORIZONTAL,VERTICAL],
    labelfont=[TIMES,ROMAN,16]);
```

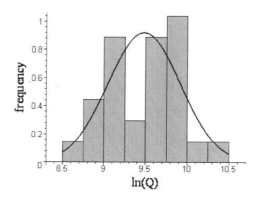

The log-transformed data appears to be more symmetric about its mean, but the graph itself does not seem to suggest a good fit. Again, these observations are subjective.

The Empirical Cumulative Distribution Function

We have seen that a frequency histogram may be built from observed data. Similarly, a cumulative empirical distribution function may be built if desired. It is a staircase function with jump discontinuities at the values of the random variable where frequency data is observed. In other words, the value of the empirical distribution function at $X=x$ is equal to the number of data points that are less than or equal to x divided by the total number of data points, N. If we construct an empirical cumulative distribution function from an observed frequency histogram with M class intervals of length ΔX, each with n_i data points, then

$$F_j = \frac{\sum_{i=1}^{j} n_i}{N}, \quad j=1, 2, ..., M \tag{7.2}$$

where

F_j=ordinate of empirical cumulative distribution function at the j-th
     class interval

Clearly, at j=1 the numerator in equation (7.2) is minimum, or zero, and F_j
is minimum or zero. At j=M the numerator in equation (7.2) is equal to
N and F_j=1. Thus, the empirical cumulative distribution function satisfies
the rules of the expected, or theoretical, cumulative distribution function
of random variables described in Chapter 3 (see properties 1 through 5,
below equation (3.3)). When the class interval length is changed, the
shape of the empirical cumulative distribution function does not change
as drastically as that of the frequency histogram. The most important
change is the specific location of the jump discontinuity. Yet, a
comparison between the empirical cumulative distribution function with
that of a candidate distribution does not offer more insight than a
comparison between corresponding frequency histograms.

There is a modification of the cumulative distribution that leads to a
practical method of evaluating a candidate distribution's worthiness as a
model. This method is based on calculating the values of F_j based on
ranked-ordered data and plotting them on probability paper. Observing
the form of equation (7.2), the value of the empirical cumulative
distribution at j is based on the summation of values up to that cell, Σn_i.
This creates a monotonically increasing staircase function as j increases.
This suggests that Σn_i may be related, by analogy, to the relative
position of a data point with respect to the rest of the data ranked in
increasing order of magnitude as X increases. Thus, if we ranked the
observed data in decreasing order of magnitude and assign a rank m=1 to
the highest in magnitude data, and m=N to the lowest in magnitude, then
$P(X>x_j) \sim m/N$. That way, when m=1 (the highest value of X) the
probability that the value is exceeded is minimum, and when m=N (the
lowest value of X) the probability that the value is exceeded is maximum
and equal to 1. If we take into consideration the fact that the sample is
always incomplete, we can create empirical formulae that allow the
estimation of probabilities. Thus, if we order the data in decreasing order
of magnitude while the value of the rank, m, increases, then

$$F_j = P(X \leq x_j) = 1 - \frac{m}{N+1}, \quad P(X > x_j) = \frac{m}{N+1} \qquad (7.3)$$

where

F_j=empirical cumulative distribution position of the j-th value of X

As the sample size, N, grows, equations (7.3) become more accurate. The so-called *return period*, *T*, which is a concept very common in uncertainty studies of extreme occurrences of events, is defined as the inverse of the probability of exceeding: $T=1/P(X>x_j)=(N+1)/m$. The return period is the *average* time interval between successive occurrences of events of a given probability. Equations (7.3), and many others developed for similar purposes, permit the calculation of the empirical cumulative distribution function from the ordered data itself, rather than from a frequency histogram with a given class interval width. The function thus obtained may be plotted as a staircase function and compared to the corresponding theoretical cumulative distribution function of a candidate distribution.

Alternatively, one can plot the former on a modified version of the theoretical curve, called *probability paper*. Probability paper of a theoretical distribution results after a modification of the probability scale of its cumulative distribution function. The modification is such that the theoretical cumulative distribution function would plot as a straight line, rather than as the normal elongated "S" curve. For instance, probability paper for the Standard Normal distribution may be constructed by adjusting the probability scale in such a way that the cumulative distribution function plots as a straight line. This paper, called *Normal probability paper* or simply *Normal paper*, can be graphically constructed (see for example, Haan, 1977), purchased at an art supply store, or downloaded from the Internet. Unfortunately, as of this writing, most spread sheet and computer programs do not offer probability paper as an option to be selected when plotting data.

Exponential probability paper is produced by a similar procedure. Not all probability distributions may be arranged in this manner to produce probability paper. Normal, Lognormal, Gumbel, Weibull, and Exponential random variables are examples of some probability distributions for which probability paper may be produced. Therefore this graphical procedure is limited in scope in that sample data may be tested for only a few probability distributions. In addition, Except for Normal probability paper, the commercial availability of probability papers is limited. In spite of these difficulties, and the limitations inherent to the subjective graphical fitting of curves, the procedure is so simple in its application that it is worth the effort for preliminary studies in engineering. With probability paper of a distribution available, one can plot the empirical cumulative distribution function of a sample data. If the cloud of points can be reasonably fitted to a straight line, there is reason to believe that the underlying distribution of the paper may be a good model for the data. As with any visual graphical procedure, this observation is subjective: How disperse should the cloud be before we reject the above hypothesis? How many points should be allowed to be far apart from the fitted line before we reject the candidate's distribution? How far is too far? Professional experience is paramount at this time. The recommended procedure is to plot the data in various probability papers in order to reject the distributions that offer a poor fit and select the best fit for further study of its worthiness.

The procedure to test a distribution's worthiness as a possible candidate to model experimental data of a random variable X is as follows:

1. Rank the data in decreasing order of magnitude.

2. Assign a rank, m, to each data point, with $m=1$ to the maximum value x_j, $m=2$ to the second greater, etc.

3. Calculate the empirical cumulative distribution function: apply equation (7.3) to each data point.

4. Obtain probability paper of a selected distribution.

5. Plot the empirical cumulative distribution on the selected probability paper.

6. If a straight line may be fitted through the points, the underlying theoretical distribution may be set aside for further consideration as a model for the data.

Example 7.4
In Example 7.1, (1) use Normal probability paper to determine if the data follows a Lognormal probability distribution. (2) Assuming the Lognormal distribution is a good model, use the fitted line as a design criterion and estimate the flow with a return period of 50 years.

Solution
(1) Following steps 1 through 6 above, the third column in Table 7.1 contains the rearranged flow values in decreasing order of magnitude. The fourth column has the natural logarithm of the rearranged flows. In the fifth column a rank, m , has been assigned to each flow. The sixth column has the return period, T, associated with each flow. Alternatively, one can use the cumulative distribution value according to equation (7.3). To test the suitability of the Lognormal distribution using Normal

probability paper, one must plot $Ln(Q_j)$ versus T in Normal probability paper (see Figure 7.3). If Lognormal paper is available, then one must plot Q_j versus T. Note that Figure 7.3 shows the cumulative distribution function rotated $90°$ counter clockwise. Finally, a straight line is fitted through the points. There is reason to believe that the Lognormal distribution might be a good model.

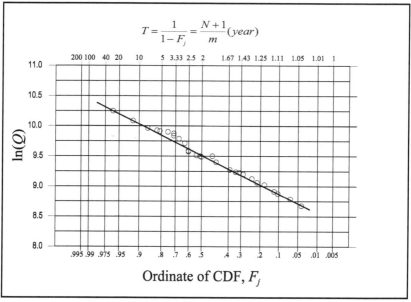

Figure 7.3: Probability Paper for Example 7.4

(2) Entering Figure 7.3 with $T=50 year$, $\ln(Q)=10.31$, and $Q=30,000\,m^3/s$. Note that extrapolating the line may lead to large errors as small errors in logarithmic readings may translate into errors in flow rates of the order of thousands of m^3/s.

An obvious improvement of the graphical procedure consists in designing a computer program that alters the scale of a given cumulative distribution function, plots the data on the resulting probability "paper," and fits a linear regression line through the points. The statistics of the line may be used to objectively assess the quality of the fit to the underlying density function.

Example 7.5: Computer Program to Plot Data on Normal Paper
Historical annual water yield, Q ($m^3\times10^6$), for a watershed is recorded in the following vector: $Q:=[61.7, 44.4, 52.4, 35.6, 39.2, 49.8,$

24.2, 45.6, 40.0, 49.7, 65.3, 44.0, 51.6, 56.5, 44.6, 71.4, 51.8, 43.8, 45.5, 58.2]. Write a Maple program that plots the data on Normal probability paper and fits a linear regression line. State wether or not the data follows a Normal density function.

Solution

```
Initialize; call statistics module; enter data
> restart: with(stats[statevalf]):
   Q:=[61.7, 44.4, 52.4, 35.6, 39.2, 49.8, 24.2, 45.6, 40.0,
        49.7, 65.3, 44.0, 51.6, 56.5, 44.6, 71.4, 51.8, 43.8,
        45.5, 58.2]:
   N:=describe[count](Q):            #Number of data points
   Qorden:=sort(Q):                  #sort series (ascending order)
   F:=[seq(j*100./(N+1),j=1..N)]:    #values of CDF
Create tickmarks of cumulative distribution function 0.1% to 99.0%
> invF:=icdf[normald]:
   YTicks1:=[seq(evalf(invF(i/100))=convert(i,string),
            i=[.1,.5,1.0,5.0,10.0,50.0,90.0,95.0,99.0])]:
   YTicks2:=[seq(evalf(invF(i/100))="",
            i=[.2,.3,.4,2,3,4,20,30,40,60,70,80,96,97,98])]:
   YTicks:=[op(YTicks1),op(YTicks2)]:
Transform the probability scale such that it plots as a straight line
> Fy:=map(invF,F/100.):
Organize data into coordinates [x.y]
> Data:=zip((x,y)->[x,y],Qorden,Fy)[]:
Fit a linear regression line to the data
> FIT:=unapply(rhs(fit[leastsquare[[x,y]]]([Qorden,Fy])),x):
   mu:=solve(FIT(x)=0);              #calculate the mean
   sigma:=solve(FIT(x)=1)-mu;        #standard deviation
   rho:=describe[linearcorrelation](Qorden,Fy);
```

$$\mu := 48.76500000$$
$$\sigma := 12.28761545$$
$$\rho := 0.9789693945$$

```
Plot data and fitted regression on Normal paper
> plots[display](
        plot(FIT,30..70,color=black,thickness=2),
        PLOT(POINTS(Data),AXESTICKS(DEFAULT,YTicks),AXESSTYLE(BOX)),
        symbol=circle,labels=[`Q (m^3x10^6)`,`F (%)`],
        labeldirections=[HORIZONTAL,VERTICAL],
        labelfont=[TIMES,ROMAN,16]  );
```

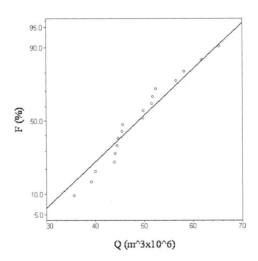

The program creates tick marks of the Normal cumulative distribution function adjusted to plot as a straight line, fits a linear regression line, and plots it along with the data. Note the sample correlation coefficient is about 0.979, indicating a good fit. Since the data appears to follow a straight line, there is reason to believe that the Normal probability distribution is a good model to describe the random variable annual yield volume, *Q*. Further tests, such as the *Chi-Squared goodness-of-fit test* should be performed on the data to evaluate the assumption of normality.

Example 7.6: Program to Test Data for the Lognormal Distribution
Modify the program on Example 7.5 to test for lognormality.

Solution
Minor changes to the program are needed. Essentially, the flow series is transformed into a logarithmic one, and tick marks for the horizontal, log-transformed, scale are created.

```
[ Initialize; call statistics module; enter data
[ > restart: with(stats[statevalf]):
    Q1:=[61.7, 44.4, 52.4, 35.6, 39.2, 49.8, 24.2, 45.6, 40.0,
         49.7, 65.3, 44.0, 51.6, 56.5, 44.6, 71.4, 51.8, 43.8,
         45.5, 58.2]:
    Q:=map(log,Q1):
    N:=describe[count](Q):     #Number of data points
    MinQ:=min(seq(Q[i],i=1..N)):
    MaxQ:=max(seq(Q[i],i=1..N)):
    Qorden:=sort(Q):           #sort series in ascending order
    F:=[seq(j*100./(N+1),j=1..N)]:  #values of CDF
[ Create tickmarks of cumulative distribution from 0.1% to 99.0%
[ > invF:=icdf[normald]:
    XTicks:=[seq(evalf(log(i))=convert(i,string),
                 i=[10,20,30,40,50,60,70,80,90,100])]:
    YTicks1:=[seq(evalf(invF(i/100))=convert(i,string),
                  i=[.1,.5,1.0,5.0,10.0,50.0,90.0,95.0,99.0])]:
    YTicks2:=[seq(evalf(invF(i/100))="",
                  i=[.2,.3,.4,2,3,4,20,30,40,60,70,80,96,97,98])]:
    YTicks:=[op(YTicks1),op(YTicks2)]:
[ Transform the cumulative probability such that it plots as a straight line
[ > Fy:=map(invF,F/100.):
[ Organize data into coordinates [x,y]
[ > Data:=zip((x,y)->[x,y],Qorden,Fy)[]:
[ Fit a linear regression line to the data
[ > FIT:=unapply(rhs(fit[leastsquare[[x,y]]]([Qorden,Fy])),x):
    solve(FIT(x)=0): mu:=exp(%):
    solve(FIT(x)=1): sigma:=exp(%)-mu;
    rho=describe[linearcorrelation](Qorden,Fy):
                        μ := 47.56087531
                        σ := 15.26092130
                        ρ = 0.9542395470
[ Plot data and fitted regression on Normal paper
[ > plots[display](
        plot(FIT,MinQ..MaxQ,color=black,thickness=2),
        PLOT(POINTS(Data),AXESTICKS(XTicks,YTicks),AXESSTYLE(BOX)),
        symbol=circle,labels=[`Q (m^3x10^6)`,`F (%)`],
        labeldirections=[HORIZONTAL,VERTICAL],
        labelfont=[TIMES,ROMAN,16]    );
```

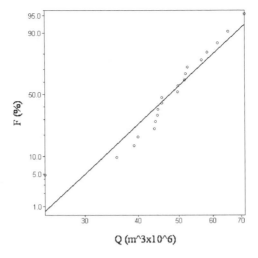

Q (m^3x10^6)

The data also appears to follow a Lognormal distribution, but not as well as a Normal distribution. Notice the value of the correlation coefficient has decreased. Again, further statistical tests would be required to evaluate which distribution would be more suitable. Other probability distributions lend themselves to be plotted on probability paper, such as the Exponential and the Weibull (see Example 9.7, Chapter 9).

7.2 STATISTICAL TESTS FOR GOODNESS OF FIT

In section 7.1, we learned to compare frequency histograms and empirical cumulative distribution functions with the corresponding expected frequency histograms of theoretical probability distributions. This was done with the purpose of studying the possibility of using the theoretical curve as a possible general model for the data. We concluded that while the visual comparison of a sample versus a theoretical curve is useful for a preliminary selection of candidate distributions for further study, such comparison is subjective as a sole estimation means by itself. The question is: how good is the fit between the sample and the theoretical curves? To answer this question objectively, we need an index or a numeral measure that assesses how close the fit is. In addition, it would be desirable that an assessment method took in consideration the fact that we are comparing a limited sample, with its own measurement uncertainty with a curve derived from a population. One of these methods is the *Chi-Squared goodness-of-fit test*.

The Chi-Squared Goodness-of-Fit Test
This test answers the question: does a sample follow a Normal (or

Binomial, Lognormal, Poisson, or any other candidate) distribution? The method requires the calculation of a test statistic derived   from a comparison of the observed frequencies with respect to the expected theoretical frequencies of the candidate's distribution. As with other statistical inference methods studied in previous chapters, the test statistic approximately follows a *Chi Squared probability density function*. The hypothesis that a set of data follows a selected theoretical curve is accepted if the statistic falls within a critical region delimited by abscissa values of the Chi-Squared distribution. The procedure involves the preparation of the data in cells and the calculation of their frequency as if we were going to plot a frequency histogram (see section 7.1). Here is a summary of the procedure:

1. Divide the sample data of size N into M cells and record the observed *absolute frequency* of each cell i, n_i. Each cell has a width, ΔX_i, which does not need to be equal to the rest. If there are cells with $n_i < 4$, increase their width and adjust their frequencies

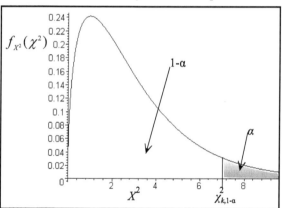

Figure 7.4: The Chi Squared Density Function
Showing the Critical Region Delimited by $\chi^2_{k,1-\alpha}$

2. Select a candidate probability density function to test as a possible model to describe the data. Calculate the *expected absolute frequency* for each cell, $e_i = f_i \times N$, where f_i is the *expected relative frequency* given by equation (7.1). If there are cells with $e_i < 4$, change their width ΔX_i and adjust their frequencies.

3. Calculate the test statistic

$$R = \sum_{i=1}^{M} \frac{(n_i - e_i)^2}{e_i}, \quad M \geq 5 \qquad (7.4)$$

R approximately follows a Chi Squared distribution with k degrees of freedom and α level of significance, $\chi^2_{k,1-\alpha}$, where $k=M-p-1$, with p the number of parameters of the theoretical distribution.

4. State the null hypothesis H_0: the sample comes from a population with the assumed probability density function; and the alternative hypothesis H_1: the sample does not come from this population.

5. If $R<\chi^2_{k,1-\alpha}$, accept H_0 (see Figure 7.4). If $\chi^2_{k,1-\alpha}<R$, reject H_0 and implicitly accept H_1. On the other hand, if $R<\chi^2_{k,\alpha}$ state that there is reason to believe the data was rigged since the fit is too good.

Example 7.7: Tension Strength of Concrete Cylinders

100 concrete cylinders were tested to failure to tension. The stress at failure, X (*psi*), along the axis of the cylinders is recorded in Table 7.4. Does the population of X follow a Normal Probability density function?

Table 7.4: Sample Tension Stress at Failure for Concrete Cylinders (*psi*)

320	380	340	410	380	340	360	350	320	370
350	340	350	360	370	350	380	370	300	420
370	390	390	440	330	390	330	360	400	370
320	350	360	340	340	350	350	390	380	340
400	360	350	390	400	350	360	340	370	420
420	400	350	370	330	320	390	380	400	370
390	330	360	380	350	330	360	300	360	360
360	390	350	370	370	350	390	370	370	340
370	400	360	350	380	380	360	340	330	370
340	360	390	400	370	410	360	400	340	360

Solution

$N=100$, the Normal density function has $p=2$ parameters, μ_X and σ_X^2. The calculated sample mean is $\bar{X}=364.7psi$, and the sample standard deviation is $S_X=26.7psi$. Following steps 1 through 5 above, we organize the data in Table 7.5 with $\Delta X=10psi$. We did not complete the last column since there are several cells with expected absolute frequency $e_i<4$. Hence, we combine into one cells 1, 2, and 3, and cells 12, 13, 14, and 15, respectively (see Table 7.6). Since the candidate's distribution is Normal, the expected absolute frequency was calculated as $e_i=f_X(X_i)\Delta X_i \times N$, where $f_X(X_i)$ is the Normal density function evaluated at the cell mid point, X_i (see Equation (7.1)), and the cell width ΔX_i is now variable. We now state the hypotheses. H_0: data comes from a Normal population, $X \sim N(\mu_X, \sigma_X)$. H_1: data does not come from a Normal Population. $k=M-p-1=10-2-1=7$. We arbitrarily choose a level of significance $\alpha=0.05$, since it has not been given. From Table A.3, Appendix A, $\chi^2_{7,0.95}=14.0671$, and $\chi^2_{7,0.05}=2.1673$. Since $R<\chi^2_{7,0.95}$ we accept H_0. Since $\chi^2_{7,0.05}<R$ we do not suspect the data of being rigged.

Table 7.5: Frequency Data Preparation for Example 7.7

Cell, i	Cell Limits (psi)	Cell Mid Point X_i(psi)	Observed Absolute Frequency, n_i	Expected Absolute Frequency, e_i	$\dfrac{(n_i-e_i)^2}{e_i}$
1	295 - 304	300	2	0.793	
2	305 - 314	310	0	1.832	
3	315 - 324	320	4	3.679	
4	325 - 334	330	6	6.421	
5	335 - 344	340	11	9.740	
6	345 - 354	350	14	12.840	
7	355 - 364	360	16	14.712	
8	365 - 374	370	15	14.650	
9	375 - 384	380	8	12.679	
10	385 - 394	390	10	9.537	
11	395 - 404	400	8	6.235	
12	405 - 414	410	2	3.543	
13	415 - 424	420	3	1.750	
14	425 - 434	430	0	0.751	
15	435 - 444	440	1	0.280	
$M=15$			$\Sigma n_i=N=100$		

Table 7.6: Data Preparation for Example 7.7 with Variable Cell Width

Cell, i	Cell Limits (psi)	Cell Mid Point X_i(psi)	Observed Absolute Frequency, n_i	Expected Absolute Frequency, e_i	$\dfrac{(n_i-e_i)^2}{e_i}$
1	295 - 324	310	6	5.497	0.046
2	325 - 334	330	6	6.421	0.028
3	335 - 344	340	11	9.740	0.163
4	345 - 354	350	14	12.840	0.105
5	355 - 364	360	16	14.712	0.113
6	365 - 374	370	15	14.650	0.008
7	375 - 384	380	8	12.679	1.727
8	385 - 394	390	10	9.537	0.023
9	395 - 404	400	8	6.235	0.500
10	405 - 444	425	6	4.666	0.381
$M=10$			$\Sigma n_i=N=100$		$R=\Sigma 3.094$

Example 7.8

In Example 7.7 write a computer program to plot in one graph the observed and the expected relative frequency histograms of cylinder strength.

Solution

```
⌐ Input data, failure stress, X(psi) and basic statistics.
⌐ > with(stats): with(stats[statplots]):
    X:=[320,380,340,410,380,340,360,350,320,370,350,
        340,350,360,370,350,380,370,300,420,370,390,
        390,440,330,390,330,360,400,370,320,350,360,
        340,340,350,350,390,380,340,400,360,350,390,
        400,350,360,340,370,420,420,400,350,370,330,
        320,390,380,400,370,390,330,360,380,350,330,
        360,300,360,360,360,390,350,370,370,350,390,
        370,370,340,370,400,360,350,380,380,360,340,
        330,370,340,360,390,400,370,410,360,400,340,
        360]:
    N:=describe[count](X):
    p:=2: deltax:=10:
    X_bar:=describe[mean](X):
    S_X:=describe[standarddeviation[1]](X):
⌐ Identify the minimum/maximum observed values with the min( ) and max( ) commands.
⌐ > Xmin:=min( seq(X[i],i=1..N) )-deltax/2:
    Xmax:=max( seq(X[i],i=1..N) )+deltax/2:
⌐ Define number of cells and cell limits.
  M:=15:
  limits:=[ seq( Xmin+i*deltax..Xmin+(i+1)*deltax, i=0..(M-1) ) ]:
⌐ Define observed absolute frequency.
⌐ > n_i:= transform[tallyinto](X,limits): transform[statsort](%):
⌐ Define the observed relative frequency.
⌐ > f_hat:=transform[scaleweight[1/N]](n_i):
⌐ In one graph plot the observed relative frequency histogram and the theoretical Normal density.
⌐ > histogram(f_hat,color=gray):
    plot(statevalf[pdf,normald[X_bar,S_X]],Xmin..Xmax,
        color=black,thickness=2,
        labelfont=[TIMES,ROMAN,16]):
    plots[display](%,%%,labels=[`X(psi)`,`fi`]);
```

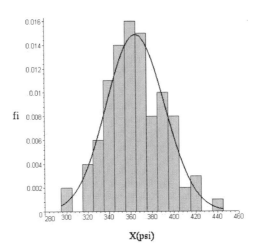

This program plots the observed relative frequency histogram with constant width cells. Remember that Maple re-scales the histograms (i.e., automatically applies equation (7.1)) to make them comparable to the expected frequency.

Example 7.9: Programming of a Goodness of Fit Test
Write a program to solve Example 7.7.

Solution

```
[ Input data, failure stress, X(psi), calculate basic statistics.
[ > restart: with(stats): with(stats[statplots]):
    X:=[320,380,340,410,380,340,360,350,320,370,350,
        340,350,360,370,350,380,370,300,420,370,390,
        390,440,330,390,330,360,400,370,320,350,360,
        340,340,350,350,390,380,340,400,360,350,390,
        400,350,360,340,370,420,420,400,350,370,330,
        320,390,380,400,370,390,330,360,380,350,330,
        360,300,360,360,360,390,350,370,370,350,390,
        370,370,340,370,400,360,350,380,380,360,340,
        330,370,340,360,390,400,370,410,360,400,340,360]:
    N:=describe[count](X): p:=2: alpha:=0.05:
    X_bar:=describe[mean](X):
    S_X:=describe[standarddeviation[1]](X):
    describe[range](X);
```
$$300..440$$
```
[ Define number of cells and cell limits. This time we enter them manually.
[ limits:=[295,325,335,345,355,365,375,385,395,405,445]:
[ M:=describe[count](limits)-1:
[ Define the interval, or ranges of values, in each cell.
[ > ranges:=[ seq( limits[i]..limits[i+1], i=1..M ) ]:
[ Calculate the value of each cell width, deltaxl.
[ > deltax:=[ seq( (limits[i+1]-limits[i]), i=1..M) ]:
[ Calculate the value of X at each cell's mid point.
[ > X_i:=[ seq( limits[i]+deltax[i]/2, i=1..M) ]:
[ Define observed absolute frequency.
[ > Oi:=transform[tallyinto](X,ranges):
    n_i:= transform[frequency](Oi):
[ Calculate the absolute expected relative frequency.
[ > e_i:=[ seq( statevalf[pdf,normald[X_bar,S_X]](X_i[i])
                            *deltax[i]*N, i=1..M ) ]:
[ Calculate the R statistic.
[ > R := sum( (n_i[i]-e_i[i])^2/e_i[i], i=1..M);
```
$$R := 3.008522434$$
```
[ Ho: Data comes from a Normal population.
[ H1: Data does not come form a Normal population.
[ Find the Chi Squared values.
[ > k:=M-p-1:
    chi2_up:=statevalf[icdf,chisquare[k]](1-alpha);
    chi2_lo:=statevalf[icdf,chisquare[k]](alpha);
```
$$chi2\_up := 14.06714045$$
$$chi2\_lo := 2.167349909$$
```
[ Make a decision.
[ > if (R<chi2_up) then `Accept Ho`
    elif (R>chi2_up) then `Accept H1`
    elif (R<chi2lo) then `We suspect data is rigged` end if;
```
$$Accept Ho$$

Most commands in this program do not show the partial output. However, it is a good idea to retype the commands with a semicolon at the end and execute them one by one. That way, one can see the partial output and correct potential errors. Notice the cell limits are entered manually. This provides a good means to observe variations in the test statistic which is not possible with a completely automatic program, especially since the test is sensitive to the choice of class intervals.

Example 7.10: A Self-Contained Goodness-of-Fit Program

Modify the program in Example 7.9 to automatically adjust cells, such that the observed absolute frequency is greater than 4 values.

Solution

The first block of commands in the following program is the only part where the user enters data or defines parameters. The rest of the program automatically consolidates the cells having less than 4 data points, and calculates the width, limits, ranges, and abscissa value at mid point for each cell. Then the Chi-Squared test is done as in Example 7.9.

```
[ Enter data, failure stress X(psi), for Chi Squafred test for Normality.
[ > restart: with(stats): with(stats[statplots]):
  X:=[320,380,340,410,380,340,360,350,320,370,350,
      340,350,360,370,350,380,370,300,420.,370,390,
      390,440,330,390,330,360,400,370,320,350,360,
      340,340,350,350,390,380,340,400,360,350,390,
      400,350,360,340,370,420.,420.,400,350,370,330,
      320,390,380,400,370,390,330,360,380,350,330,
      360,300,360,360,360,390,350,370,370,350,390,
      370,370,340,370,400,360,350,380,380,360,340,
      330,370,340,360,390,400,370,410,360,400,340,360]:
  N:=describe[count](X): p:=2: alpha:=0.05: M:=15:
  X_bar:=describe[mean](X):
  S_X:=describe[standarddeviation[1]](X):
  r:=describe[range](X):
  dx:=ceil((op(2,r)-op(1,r))/M):
[ This block defines limits, width, and ranges of values for each cell.
[ Xmin:=op(1,r)-dx/2:
  Xmax:=op(2,r)+dx/2:
  limits[1]:=Xmin:
  limits[2]:=limits[1]+dx:
  for i from 3 to M do
  limits[i]:=limits[i-1]+dx:
  end do:
  limits[M+1]:=Xmax:
  seq(limits[i],i=1..M+1):
  deltax:=[ seq( (limits[i+1]-limits[i]), i=1..M) ]:
  ranges:=[ seq( limits[i]..limits[i+1], i=1..M ) ]:
[ Define intial observed absolute frequency.
[ > Oi:=transform[tallyinto](X,ranges):
  n_i:= transform[frequency](Oi):
```

The following block consolidates cells with less than 4 values, and redefines width, limits, ranges, and observed frequency for each cell. Ignore the error message.

```
> j:=1: tot:=0: m:=0:
  for k from 1 to M-1 do
  for i from j while tot<4 do
  tot:=tot+n_i[i]:
  end do:
  m:=m+1:
  newdeltax[k]:=sum(deltax[l],l=j..i-1);
  j:=i:
  tot:=0:
  end do:
  if(n_i[M-1]<4) then newdeltax[m]:=newdeltax[m]+deltax[M-1]:
  end if:
  M:=m:
  deltax:=[seq(newdeltax[p],p=1..M)]:
  limits[1]:=Xmin:
  for i from 2 to M do
  limits[i]:=limits[i-1]+deltax[i-1];
  end do:
  limits[M+1]:=Xmax:
  ranges:=[seq(limits[i]..limits[i+1],i=1..M )]:
  Oi:=transform[tallyinto](X,ranges):
  n_i:= transform[frequency](Oi):
Error, invalid subscript selector
```

Calculate the value of X at each cell's mid point.

```
> X_i:=[seq( limits[i]+deltax[i]/2, i=1..M)]:
```

Calculate the absolute expected relative frequency.

```
> e_i:=[ seq( statevalf[pdf,normald[X_bar,S_X]](X_i[i])
                          *deltax[i]*N, i=1..M-1 ) ]:
```

Calculate the R statistic.

```
> R := sum( (n_i[l]-e_i[l])^2/e_i[l], l=1..M-1);
```

$$R := 2.686863039$$

Ho: Data comes from a Normal population.
H1: Data does not come form a Normal population.
Find the Chi Squared values.

```
> k:=M-p-1:
  chi2_up:=statevalf[icdf,chisquare[k]](1-alpha);
  chi2_lo:=statevalf[icdf,chisquare[k]](alpha);
```

$$chi2\_up := 14.06714045$$
$$chi2\_lo := 2.167349909$$

Make a decision.

```
> if (R<chi2_up) then `Accept Ho`
  elif (R>chi2_up) then `Accept H1`
  elif(R<chi2lo) then `We suspect data is rigged` end if;
```

$$Accept\ Ho$$

Example 7.11: Assessing Weigh Scale Effectiveness in a Highway

A portion of the I-75 highway has a weigh scale in operation. Experience shows that after being open for about 15 *min*, the truckers have warned each other and are taken alternate routes. Since the advent of a more powerful C.B. radio, it is suspected that the 15-*min* interval is no longer valid. An engineer was hired to assess this premise. She designed an experiment consisting of observing 70 time intervals (each 15min long) when the station was open and recorded the number of truck overloads (Table 7.7). The observed absolute frequency, n_i, is simply the observed number of 15-min periods with k overloads. Determine whether the data is random or affected by another deterministic cause, such as truckers avoiding the check point. Make a recommendation.

Table 7.7: Truck Overload frequency in 70 15-min Periods for Example 7.11

Cell, i	Number of Truck Overloads in an Interval, k_i	Observed Absolute Frequency, n_i
1	0	20
2	1	36
3	2	8
4	3	4
5	4	1
6	5	0
7	6	1
8	7	0

Solution

Here we have an application of probabilistic modeling and statistical data analysis in decision management. This is a case of modeling arrivals. In Chapter 3 we introduced the Poisson random variable as a practical distribution to model arrivals. The underlying assumption here is that if the data is random (i.e., follows a Poisson distribution), then the overloads are governed by the probabilistic laws of arrivals and the 15-min intervals are adequate. With this premise, let us perform a Chi-Squared goodness-of-fit test of the data to a Poisson distribution. Note this is a discrete random variable and the class "intervals" have no width. To satisfy the minimum cell-frequency requirements, we combine the last 5 cells as shown in Table 7.8.

Table 7.8: Data Preparation for Example 7.11

Cell, i	Number of Overloads, k_i	Observed Absolute Frequency, n_i	Expected Absolute Frequency, e_i	$(n_i-e_i)^2/e_i$
1	0	20	24.32	0.77
2	1	36	25.71	4.12
3	2	8	13.59	2.30
4	3, 4, 5, 6, 7	6	6.39	0.02
M=4		$N=\Sigma n_i$=70		$R=\Sigma$7.21

Assume α=0.05. The Poisson distribution has p=1 parameter, ω. The number of intervals is N=70 15-min periods. The total number of overloads is $N_o = \sum_{7} k_i \times n_i = 74$ trucks. The overload arrival rate in the Poisson distribution is $\omega = N_o / N = 74/70 = 1.057$ trucks/period (see equation (3.22)). The expected absolute frequency is calculated form the Poisson distribution (see equation (7.1)):

$$e_i = P(k \text{ overloads in } t=1 \text{ period}) \times N = \frac{\omega^k}{k!} e^{-\omega} \times N$$

Note that for cell $i=4$, we added the expected absolute frequencies of $k=3$, 4, 5, 6, and 7 overloads (i.e., $6.39 = (P(3)+P(4)+P(5)+P(6)+P(7)) \times 70$. We now state the hypotheses. H_0: data comes from a Poisson population; overloaded truck arrivals are random. H_1: data does not come from a Poisson Population; truckers are being warned. $k=M-p-1=4-1-1=2$. From Table A.3, Appendix A, $\chi^2_{2,0.95} = 5.9915$. Since $\chi^2_{2,0.95} < R$ we accept H_1. The engineer should recommend that the 15-*min* interval be shortened to ensure random arrivals. Note this conclusion apparently contradicts a visual comparison of observed versus expected absolute frequency histograms (Figure 7.5). Yet, the numerical calculations are more accurate. There are two problems that might affect the above conclusion. One is that the engineer must ensure the experiment was performed correctly according to the rules of probabilistic experimentation. In other words the chosen 70 periods of observation must be identical in nature. She should question whether weekend observations have the same features as weekday observations, etc. The second problem relates to the fundamental assumption; it is possible that the data does not fit a Poisson distribution and yet it is random. In other words, it is possible that the overload arrivals fit another probability distribution. In the absence of more data, the previous conclusion seems reasonable.

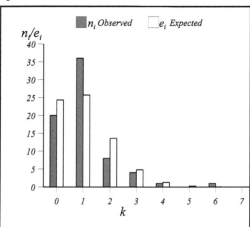

Figure 7.5: Histogram of Example 7.11

The previous examples indicate that the Chi-Squared goodness-of-fit test of data to probability distributions is sensitive to the selection of the class interval width. The selection of a class interval with cells having little or no observed or expected frequency produces large values in the R statistic, and thus a tendency to reject the null hypothesis, even though the data may actually follow the candidate's distribution. It may take some trial and error to select the best class interval width. The test is also insensitive to the tails of the distribution. In other words, extremely high, or extremely low, values of the random variable may follow a distribution different from the rest of the data. In many engineering fields it is a well-known fact that extreme values follow a distribution completely different from that of the regular values. For instance, in hydrologic engineering it is known that while the mean annual streamflow values may follow a Normal or Gamma distribution, the maximum annual (i.e., flood) and the minimum (i.e., drought) values follow a Gumbel, Lognormal, or special distributions such as the Log-Pearson family of curves. Depending on the problem in question, care should be taken to assess if extreme values participate with the same regular series.

In this chapter we have seen that the problem of fitting distributions to data is not as direct and objective as one might wish. Having performed a correct experiment, the modeler needs to assume possible candidate distributions to test. The comparison of a data histogram with that of a candidate distribution, after a correct selection of the class interval width, may help to narrow the selection and eliminate obviously wrong candidates. Plotting the data on probability paper may further refine the pool of possibilities. Lastly, performing a goodness-of-fit test of the finalists distributions will point the modeler toward a correct choice. It is possible that more than one distribution will fit the data, but there is usually a best choice. There are other goodness-of-fit tests the engineer may attempt, besides the Chi Square test. The *Kolmogorov-Smirnov test*, the *W test*, and the *E test* are but a few examples of accurate procedures developed to test for specific distributions. The reader interested in deepening in this subject should consult the appropriate literature (i.e., Devore, 1995; Han and Shapiro, 1975).

PROBLEMS

7.1 In Example 7.1, we observe that a symmetric (i.e., zero skewness) distribution, such as the Normal distribution, may not be a good model. It is suspected that a Lognormal distribution may work. Assess this possibility by constructing a histogram for the logarithm of the flows and write your conclusions.

7.2 Table 7.9 contains frequency data on the life (*years*), L (i.e., time before failure), of the plastic water pump of 120 washing machines. The company is seeking a model to define the appropriate length of the warranty and suspects that a Weibull distribution with $m=1.5$, and $c=1.5$ might be a good one. Compare the observed frequency histogram with respect to the theoretical expected one for a class interval $\Delta L=0.25 years$. State your conclusion.

Table 7.9: Frequency Data for Problem 7.2

Cell, i	L Cell Range	Center Value L_i	Observed Absolute Frequency, n_i
1	0.125 - 0.374	0.25	16
2	0.375 - 0.624	0.50	26
3	0.625 - 0.874	0.75	18
4	0.875 - 1.124	1.00	11
5	1.125 - 1.374	1.25	13
6	1.375 - 1.624	1.50	6
7	1.625 - 1.874	1.75	5
8	1.875 - 2.124	2.00	5
9	2.125 - 2.374	2.25	4
10	2.375 - 2.624	2.50	4
11	2.625 - 2.874	2.75	2
12	2.875 - 3.124	3.00	4
13	3.125 - 3.374	3.25	2
14	3.375 - 3.624	3.50	2
15	3.625 - 3.874	3.75	2

7.3 Modify the program in Example 7.3 to produce a histogram with a different class interval, $\Delta \ln(Q)$, and thus different number of cells, M. Make sure all data are used in the plot. Try for instance $\Delta L=2 years$. Do you think a better fit is obtained?

7.4 The regulations of the board of health in a state specify that the fluoride concentration in drinking water must not exceed 1.5*ppm*. The following daily measurements were taken early in the morning at random in the water distribution system (source: Mendenhall and Ott, 1976): 0.75, 0.94, 0.88, 0.72, 0.81, 0.86, 0.89, 0.78, 0.92, 0.85, 0.84, 0.84, 0.77, 1.05, 0.97, 0.85, 0.83, 0.76, 0.94, 0.93, 0.97, 0.89, 0.82, 0.83, 0.79. Using probability paper, investigate the Normal distribution as a possible model. State your conclusion.

7.5 Redo Example 7.7 *by hand* (i.e., do not use a computer program), this time with a class interval width of 20*psi* in the internal cells.

7.6 Modify the program in Example 7.9 to produce a histogram with a class interval of 5*psi*, constant. State your conclusions on the goodness of fit.

7.7 Modify the program in Example 7.9 to perform the Chi-Squared goodness-of-fit test of Example 7.7 when the internal class interval width is 20*psi*.

7.8 Modify the program in Example 7.10 to change the internal class interval width to 6*psi*, instead of 10*psi*. Do you reach a different conclusion?

Table 7.10: Frequency Data for Problem 7.9

Cell, i	Cell Limits	Cell Mid Point Y_i	Observed Absolute Frequency, n_i
1	1.275 - 1.324	1.300	1
2	1.325 - 1.374	1.350	5
3	1.375 - 1.424	1.400	6
4	1.425 - 1.474	1.450	13
5	1.475 - 1.524	1.500	8
6	1.525 - 1.574	1.550	17
7	1.575 - 1.624	1.600	14
8	1.625 - 1.674	1.650	7
9	1.675 - 1.724	1.700	1
10	1.725 - 1.774	1.750	3

7.9 Table 7.10 has sample data of a random variable, Y, already organized. Test at the 5% level of significance that this data comes from a Normal population. Hint: for the calculation of the mean and the standard deviation use the moments' formulae in Chapter 3.

7.10 Solve Problem 7.9 by modifying the program in Example 7.9 to perform a Chi-Squared goodness-of-fit test of data already organized in class intervals.

Table 7.11: Frequency Data for Problem 7.11

Cell, i	Cell Limits	Observed Absolute Frequency, n_i
1	41.50 - 41.99	4
2	42.00 - 42.49	11
3	42.50 - 42.99	15
4	43.00 - 43.49	14
5	43.50 - 43.99	23
6	44.00 - 44.49	20
7	44.50 - 44.99	12
8	45.00 - 45.49	2

7.11 Table 7.11 shows the results of an experiment on tensile strength of steel sheeting, $T(kg/mm^2)$. Test for normality at the 5% level of significance.

"The Universal Intelligence endowed us with the ability to run repetitive statistical tests on our relationships and life experiences. In this context, the physical world is the means of evolution. As the ancient alchemists taught, your body is your experimental crucible. Your alchemical laboratory is the external world. Your words and deeds in relation to others are the alchemical dissolution and distillation processes. Your emotions are the alchemical Mercury, that dissolves and extracts the essence of our experiences and improves our personality. Your thoughts and reasoning, which improve with experience, make the alchemical Sulphur. As this process continues, the body, or the alchemical Salt, will decay and eventually die. This marks the transmutation from a raw and heavy (lead) personality to an enlightened one (gold), eventually. The body is thus the vehicle to this evolution."

Sergio E. Serrano ("The Three Spirits. Applications of Huna to Health, Prosperity, and Personal Growth" SpiralPress, Ambler, Pennsylvania, 2007)

8 REGRESSION ANALYSIS

In this chapter we explore the fundamentals of linear regression models and some nonlinear cases reducible to linear regression. These are a wide class of problems in everyday engineering practice related to the prediction of the value of one variable based on that of another. For instance, we may be interested in predicting tomorrow's air temperature, Y, based on today's air pressure, X. In order to do it, we must establish a statistical relationship between the two variables. This relationship is accomplished after an experiment whereby the N measurements of the two variables are simultaneously recorded under identical conditions. From the N samples of X and Y, (X_i, Y_i), $i=1, 2, ..., N$ we attempt to fit the simplest possible mathematical relationship that may be used in future interpolation of Y values, given current measurements of X.

A word of caution concerning the conclusions and applications of regression models is in order. The fact that we derive a good regression line between the two variables does not necessarily mean that there is a causal relationship between them. For example, if we find a good relationship between energy consumption at a factory and net company's profit, it would be a misuse of statistics to predict that spending a great deal of energy would translate into higher profits. There is at least one intermediate step of using the energy for products that are actually sold that connects the expenditure of energy to net profits. The statistics literature is filled with these type of erroneous conclusions, many of them to manipulate public opinion for commercial or political purposes.

8.1 STATISTICAL MEASURES BETWEEN TWO RANDOM VARIABLES

The Sample Covariance and Sample Correlation Coefficient

In Chapter 5 we introduced the concept of covariance between two random variables, C_{XY}. Equation (5.18) characterized the *population* covariance as a measure of the linear relationship between two random variables X and Y. If a limited sample of size N is available from the two variables, the *sample covariance*, $\hat{C}_{XY}$ (the "hat" denotes "sample") is given as

$$c_{XY} = \hat{C}_{XY} = \frac{1}{N-1} \sum_{i=1}^{N} \left(X_i - \bar{X} \right)\left(Y_i - \bar{Y} \right) \tag{8.1}$$

where

c_{XY}=sample covariance between X and Y (lowercase c)
$\bar{X}$=the sample mean of X (see equation (6.1))
$\bar{Y}$=the sample mean of Y

The *population* correlation coefficient, ρ_{XY}, was also introduced in Chapter 5, equation (5.20), as a normalized version of the covariance between two random variables. Similarly, the *sample correlation coefficient* is given as

$$r_{XY}=\hat{\rho}_{XY}=\frac{c_{XY}}{S_X S_Y} \tag{8.2}$$

where

r_{XY}=sample correlation coefficient between X and Y
S_X=sample standard deviation of X (see below equation (6.2))
S_Y= sample standard deviation of Y

The sample correlation coefficient has the properties described below equation (5.20). These sample statistics are fundamental measures of the linear relationship between trials of two random variables. As other sample statistics, the sample covariance and the sample correlation coefficient depend on the sample size and are themselves random variables. As the sample size increases, the sample statistics approach in magnitude the population parameters (i.e., the population covariance, C_{XY}, and population correlation coefficient, ρ_{XY}). The population parameters are deterministic constants.

8.2 THE LEAST-SQUARES STRAIGHT LINE

Fitting a Straight Line through the Scatter Diagram

Referring again to a sample of two random variables X and Y, a *scatter diagram* is a two-dimensional graph of the sample points with coordinates (X_i, Y_i). It usually indicates the shape of the curve relating X to Y. If a straight line may be attempted through the cloud of points, how do we know where to draw the line to best predict Y given X? Thus, we need to establish a measure of the fit of the line to a given set of data. One way to approach the problem is to minimize the sum of the squared errors, d_i, as shown in Figure 8.1. The i-th data point has a deviation, d_i, equal to the difference between its vertical coordinate, Y_i, and the corresponding vertical coordinate of the best fitted line, $\hat{Y}_i$ (with the "hat" denoting "estimate" of Y_i). Thus, the best line has the equation $\hat{Y}=\hat{a}X+\hat{b}$, where the slope and the intercept, $\hat{a}$ and $\hat{b}$, respectively, are to be found by minimizing $\varepsilon = \sum_{i=1}^{N} d_i^2$.

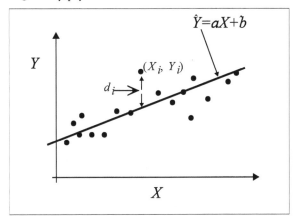

Figure 8.1: A Scatter Diagram with a Fitted Line

Expanding,

$$\varepsilon = \sum_{i=1}^{N} \left(Y_i - \hat{Y}_i \right)^2 = \sum_{i=1}^{N} \left[Y_i - \left(\hat{a} X_i + \hat{b} \right) \right]^2 \qquad (8.3)$$

To obtain a minimum, we differentiate with respect to $\hat{a}$ and $\hat{b}$, respectively, and equate to zero:

$$\frac{\partial \varepsilon}{\partial \hat{a}} = 2 \sum_{i=1}^{N} \left[Y_i - (\hat{a} X_i + \hat{b}) \right] X_i = 0$$

$$\frac{\partial \varepsilon}{\partial \hat{b}} = 2 \sum_{i=1}^{N} \left[Y_i - (\hat{a} X_i + \hat{b}) \right] = 0 \qquad (8.4)$$

Solving simultaneously equations (8.4),

$$\hat{a} = \frac{N \sum_{i=1}^{N} X_i Y_i - \sum_{i=1}^{N} X_i \sum_{i=1}^{N} Y_i}{N \sum_{i=1}^{N} X_i^2 - \left(\sum_{i=1}^{N} X_i \right)^2}$$

$$\hat{b} = \frac{\sum_{i=1}^{N} X_i^2 \sum_{i=1}^{N} Y_i - \sum_{i=1}^{N} X_i \sum_{i=1}^{N} X_i Y_i}{N \sum_{i=1}^{N} X_i^2 - \left(\sum_{i=1}^{N} X_i \right)^2} \qquad (8.5)$$

From equations (6.2) and (8.1), it can be shown that equations (8.5) may be written as

$$\hat{a}=\frac{c_{XY}}{S_X^2}, \qquad \hat{b}=\bar{Y}-\hat{a}\bar{X} \tag{8.6}$$

Similarly, the best line $\hat{Y}=\hat{a}X+\hat{b}$ may be written as

$$\frac{\hat{Y}-\bar{Y}}{S_Y}=r_{XY}\frac{X-\bar{X}}{S_X} \tag{8.7}$$

which provides a simple formula for the regression line in terms of familiar sample statistics.

Table 8.1: Results from Experiment in Example 8.1

Reading Errors in 30 *min*	Hours on the Job
26	1.0
28	1.5
32	2.5
28	2.0
30	3.5
40	4.5
36	4.0
36	5.5
40	6.0
33	4.5
35	3.5
37	7.0
39	8.0
25	1.0
27	1.5
38	5.5

Example 8.1: Quality Control of Microchip Manufacturing

Workers at a microchip production line inspect each unit individually for defects through a microscope. The quality control engineer suspects that the number of reading errors (i.e., number of defective chips being accepted) increases throughout the day and is probably due to visual fatigue as measured by the number of hours a worker is on the job. Table 8.1 shows the results of an experiment measuring the number of reading errors in several 30-*min* periods, and the number of hours the workers in the shift had been on the job at the start of the time period. (1) Fit a linear regression model, plot the best line with the scatter diagram, and state your conclusions. (2) Predict the number of errors in 30 *min* at five hours of work in a shift.

Solution

(1) Let X be the number of hours workers have been on the job at the beginning of a 30-*min* period and Y the number of reading errors in the same lapse. The following Maple worksheet illustrates the application.

```
[ (1) Initialize and enter data.
[ > restart: with(stats): with(stats[statplots]):
    X:=[1.0,1.5,2.5,2.0,3.5,4.5,4.0,5.5,6.0,4.5,3.5,
        7.0,8.0,1.0,1.5,5.5]:
    Y:=[26,28,32,28,30,40,36,36,40,33,35,37,39,25,27,38]:
[ Calculate sample statistics, and line slope and intercept.
[ The subcommand [1] means "sample" statistic is calculated.
[ > N:=describe[count](X):
    X_bar:=describe[mean](X):
    Y_bar:=describe[mean](Y):
    S_X:=describe[standarddeviation[1]](X):
    S_Y:=describe[standarddeviation[1]](Y):
    c_XY:=describe[covariance[1]](X,Y):
    r_XY:=describe[linearcorrelation](X,Y);
    a:=c_XY/S_X^2; b:=Y_bar-a*X_bar;       #equation (8.6)
    y_fit:=x->a*x+b;    #best fit line defined as a function
```
$$r\_XY := 0.8864763953$$
$$a := 1.988699011$$
$$b := 25.48093818$$
```
[ For comparison, use the Maple regression procedure to find the slope and the intecept.
[ Note the resulting slope and intercept estimates are slightly different, because the fit command
    calculates "population" statistics.
[ > fit[leastsquare[[x,y]]]([X,Y]);
```
$$y = 24.97133407 + 2.121278942\,x$$
```
[ Re-define the above line as a functional for later plotting.
[ > y1:=unapply(rhs(%),x);
```
$$y1 := x \rightarrow 24.97133407 + 2.121278942\,x$$
```
[ Plot the data and the two regression lines derived in one graph.
[ > g1:=scatterplot(X,Y,color=black,symbol=box):
    g2:=plot(y_fit(x),x=0..8,color=black,thickness=2):
    g3:=plot(y1(x),x=0..8,color=black,linestyle=3,thickness=2):
    plots[display](g1,g2,g3,labels=[` X `,` Y `],
        labelfont=[TIMES,ROMAN,16]);
```

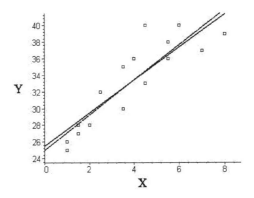

```
⌐ (2) Predict with both regression lines.
| > y_fit(5); y1(5);
|
|                                  35.42443324
|                                  35.57772878
⌐
```

There is a difference between the line derived from equation (8.6) and that produced by Maple. The former uses the sample standard deviation in the calculations (i.e., N-1 degrees of freedom), whereas Maple assumes population values in its procedure, which is not the case in this example. As the sample size increases, the two lines become indistinguishable. The sample correlation coefficient is about 0.886. That is about 88.6% of the value of Y is explained by that of X. Therefore there is reason to believe that as the shift working time increases, the errors in reading increase linearly.

Nonlinear Curves Reducible to Straight Lines

Sometimes the scatter diagram of data suggests a nonlinear curve rather than a straight line. In this situation there are several types of curves which can be reduced to straight lines by a simple linear transformation. We will study three types in particular.

Exponential Curve: $y = ba^x$

Taking logarithms on both sides of the equation,

$$\log(y) = \log(b) + \log(a^x) = (\log(a))x + \log(b)$$
$$\Rightarrow Y = AX + B, \quad Y = \log(y), \quad A = \log(a), \quad B = \log(b)$$

$$(8.8)$$

One-Term Polynomials: $y = ax^n$

Taking logarithms,

$$\log(y) = \log(a) + n\log(x)$$
$$\Rightarrow Y = AX + B, \quad Y = \log(y), \quad X = \log(x), \quad A = n, \quad B = \log(a)$$

$$(8.9)$$

Hyperbola: $y = 1/(ax + b)$

This could be written as

$$\frac{1}{y} = ax + b \quad \Rightarrow Y = ax + b, \quad Y = \frac{1}{y}$$

$$(8.10)$$

Example 8.2

A prototype fruit press operates with air pressure. Table 8.2 shows the result of an experiment that recorded the pressure, P (dyn/cm^2), and the corresponding volume, V (cm^3). From thermodynamics, the engineer expects to obtain a relationship of the form $PV^\beta = c$, where β and c are constants. (1) Fit an appropriate regression model. (2) Plot the scatter diagram of the data along with the fitted model. (3) Find the optimal values of the constants and predict the value of the pressure when the volume is $1000\,cm^3$.

Table 8.2: Results from the Experiment in Example 8.2

Volume, V (cm^3)	Air Pressure, P (dyn/cm^2)
890.0	42.2
1013.0	34.1
1186.6	25.9
1453.8	19.6
1943.9	13.2
2360.2	7.0
1702.1	17.2
1245.3	23.8
2210.4	12.0
851.2	37.8
2093.4	11.2
1100.7	31.0

Solution

The theoretical (i.e., ideal) relationship $PV^\beta = c$ suggests a one-term polynomial model. Taking logarithms on both sides of this equation,

$$\log(P) = \log(c) - \beta\log(V)$$

$$\Rightarrow Y = AX + B, \quad Y = \log(P), \quad X = \log(V), \quad A = -\beta, \quad B = \log(c)$$

Thus, by taking logarithms to the V and P data, we may fit a linear model from which the parameters of the nonlinear equation may be found. The following Maple worksheet illustrates the calculations. Notice the correlation coefficient of the log-transformed data suggests an excellent relationship. However, a sample correlation coefficient of a log-transformed (linear) data does not necessarily mean a good fit for the original (nonlinear) data. Logarithmic transformations of high magnitude values are compressed into low ranges and thus small errors in the transformed domain may translate into large ones in the original quantity.

(1) Initialize program and enter data.

```
> restart: with(stats): with(stats[statplots]):
  V:=[890.0,1013.0,1186.6,1453.8,1943.9,2360.2,
      1702.1,1245.3,2210.4,851.2,2093.4,1100.7]:
  P:=[42.2,34.1,25.9,19.6,13.2,7.0,17.2,23.8,12.0,
      37.8,11.2,31.0]:
  N:=describe[count](V):
  min(seq(V[i],i=1..N)): Vmin:=floor(%):
  max(seq(V[i],i=1..N)): Vmax:=ceil(%):
```

Transform data and fit a line to linear model.

```
> N:=describe[count](V):
  Y:=[ seq(log[10](P[i]),i=1..N) ]:
  X:=[ seq(log[10](V[i]),i=1..N) ]:
  X_bar:=describe[mean](X):
  Y_bar:=describe[mean](Y):
  S_X:=describe[standarddeviation[1]](X):
  S_Y:=describe[standarddeviation[1]](Y):
  c_XY:=describe[covariance](X,Y):
  r_XY:=describe[linearcorrelation](X,Y);
  a:=c_XY/S_X^2: b:=Y_bar-a*X_bar:
```

$$r\_XY := -0.9782256756$$

Define nonlinear model.

```
> alpha:=-a; c:=10^b; p:=v->c*v^(-alpha):
```

$$\alpha := 1.381734452$$

$$c := 454578.7279$$

(2)

```
> g1:=scatterplot(V,P,color=black,symbol=box):
  g2:=plot( p(v),v=Vmin..Vmax,color=black,thickness=2):
  plots[display](g1,g2,tickmarks=[4,default],
      labels=["V (cm^3)","P (dyn/cm^2)"],
      labeldirections=[HORIZONTAL,VERTICAL] );
```

(3)

```
> p(1000);
```

$$32.53911621$$

8.3 CONFIDENCE INTERVALS OF THE REGRESSION MODEL

The regression line obtained with a sample of size N yields an *estimate* of the line, $\hat{Y}=\hat{a}X+\hat{b}$. Clearly $\hat{Y}$, $\hat{a}$, and $\hat{b}$ depend on the particular *sample* and its size. Thus, if we take k sets of samples, each one would yield a different line $\hat{Y}_i=\hat{a}_iX+\hat{b}_i$, $i=1$, 2, ..., k, and each would be an *estimate* of the true population least-squares line. The particular values of $\hat{a}$ and $\hat{b}$ obtained with any sample set are *point estimators* of the true *population parameters* $\hat{a}$ and $\hat{b}$, respectively. Actually, they are *unbiased* estimators, because $E\{\hat{a}\}=a$ and $E\{\hat{b}\}=b$. In addition, for any given $X=x$, the estimate $\hat{Y}$ for each of the k sample sets will differ from the true population value $Y=aX+b$. Therefore, while Y, a, and b are theoretical population constants, the corresponding estimates $\hat{Y}$, $\hat{a}$, and $\hat{b}$ are random variables with their own probability density functions.

Often the modeler only has one sample from which the above estimates are calculated. The question is: how can one evaluate the quality of the model? To answer this question, we can use confidence intervals or hypothesis testing procedures, such as those discussed in Chapter 6. In this chapter we will mainly focus on the practical calculation of the confidence intervals of the slope, the intercept, and the predicted values of the regression model.

The classical theory on the confidence intervals of the slope, the intercept, and the prediction values of linear regression models rest on the fundamental assumption that the random variable X, follows a Normal distribution, that is, $X\sim N(\mu_X, \sigma_X)$. For the general case of other distributions, the following concepts are only approximations.

Case 1: Confidence Interval of the Slope a, σ_X^2 is not Known, $N<60$

We are seeking the $100(1-\alpha)\%$ confidence interval (α is the confidence level) of the population slope, a, of a linear regression model. This case applies to the situation when the variance of X is not known (i.e., the usual scenario), and the number of data points in the sample, N, is less than 60. The test statistic is given by

$$R=\frac{(\hat{a}-a)\sqrt{\frac{(N-1)(N-2)}{N}}\,S_X}{S_{XY}}, \qquad S_{XY}=\sqrt{\frac{1}{N}\sum_{i=1}^{N}\left(Y_i-\hat{Y}_i\right)^2} \qquad (8.11)$$

where the statistic S_{XY} is based on the summation of the square of the differences between the measured Y_i and the corresponding predicted $\hat{Y}_i$. R follows a Student's t density function with $N-2$ degrees of freedom.

Following a procedure similar to that in section 6.3, Chapter 6, the confidence limits are given by

$$L = \hat{a} - t_{N-2,1-\alpha/2} \sqrt{\frac{N}{(N-1)(N-2)}} \frac{S_{XY}}{S_X}$$

$$U = \hat{a} + t_{N-2,1-\alpha/2} \sqrt{\frac{N}{(N-1)(N-2)}} \frac{S_{XY}}{S_X}$$

(8.12)

where L and U are the lower and upper confidence limits, respectively; and $t_{N-2,1-\alpha/2}$ is the abscissa of the Student's t distribution, with N-2 degrees of freedom, corresponding to a cumulative area of $1-\alpha/2$ (see Table A.2, Appendix A).

Case 2: Confidence Interval of the Slope a, σ_X^2 is not Known, $60 \le N$

We are seeking the $100(1-\alpha)\%$ confidence interval of the population slope, a, of a linear regression model. This case applies to the situation when the variance of X is not known and the number of data points in the sample, N, is greater or equal to 60. The test statistic, R, is given by equation (8.11). It now follows a Standard Normal density function (i.e., $R \sim N(0, 1)$). The confidence limits are given by

$$L = \hat{a} - Z_{1-\alpha/2} \sqrt{\frac{N}{(N-1)(N-2)}} \frac{S_{XY}}{S_X}$$

$$U = \hat{a} + Z_{1-\alpha/2} \sqrt{\frac{N}{(N-1)(N-2)}} \frac{S_{XY}}{S_X}$$

(8.13)

where $Z_{1-\alpha/2}$ is the abscissa of the Standard Normal distribution corresponding to a cumulative area of $1-\alpha/2$ (see Table A.1, Appendix A). Note that equation (8.13) is the same as equation (8.12) except that $t_{N-2,1-\alpha/2}$ is replaced by $Z_{1-\alpha/2}$.

Case 3: Confidence Interval of Predicted Y, σ_X^2 is not Known, $N < 60$

We are seeking the $100(1-\alpha)\%$ confidence interval of the population predicted value Y given $X=x$ of a linear regression model. In this case the variance of X is not known and the number of data points in the sample, N, is less than 60. The test statistic is given by

$$R = \frac{(\hat{Y}-Y)\sqrt{N-2}}{S_{XY}\sqrt{N+1+\left(\dfrac{N(x-\bar{X})^2}{(N-1)S_X^2}\right)}}$$

(8.14)

where $\hat{Y}$ and Y are evaluated at $X=x$. R follows a Student's t density function with $N-2$ degrees of freedom. The confidence limits are given by

$$L=\hat{Y}-t_{N-2,1-\alpha/2}\frac{S_{YX}}{\sqrt{N-2}}\sqrt{N+1+\left(\frac{N(x-\bar{X})^2}{(N-1)S_X^2}\right)}$$

$$U=\hat{Y}+t_{N-2,1-\alpha/2}\frac{S_{YX}}{\sqrt{N-2}}\sqrt{N+1+\left(\frac{N(x-\bar{X})^2}{(N-1)S_X^2}\right)}$$

(8.15)

Case 4: Confidence Interval of Predicted Y, σ_X^2 is not Known, $60 \leq N$

We are seeking the $100(1-\alpha)\%$ confidence interval of the population predicted value Y given $X=x$ of a linear regression model. In this case the variance of X is not known and the number of data points in the sample, N, is greater or equal to 60. The test statistic is given by equation (8.14). It now follows a Standard Normal density function (i.e., $R\sim N(0,\ 1)$). The confidence limits are given by

$$L=\hat{Y}-Z_{1-\alpha/2}\frac{S_{YX}}{\sqrt{N-2}}\sqrt{N+1+\left(\frac{N(x-\bar{X})^2}{(N-1)S_X^2}\right)}$$

$$U=\hat{Y}+Z_{1-\alpha/2}\frac{S_{YX}}{\sqrt{N-2}}\sqrt{N+1+\left(\frac{N(x-\bar{X})^2}{(N-1)S_X^2}\right)}$$

(8.16)

where $Z_{1-\alpha/2}$ is the abscissa of the Standard Normal distribution corresponding to a cumulative area of $1-\alpha/2$. Note that equation (8.16) is the same as equation (8.15) except that $t_{N-2,1-\alpha/2}$ is replaced by $Z_{1-\alpha/2}$.

Case 5: Confidence Interval of Intercept b, σ_X^2 is not Known, $N<60$

We are seeking the $100(1-\alpha)\%$ confidence interval of the population intercept of the regression line, b, which reduces to that of Case 3 the confidence interval of the predicted value Y given $X=0$ of a linear regression model. In this case the variance of X is not known and the number of data points in the sample, N, is less than 60. The test statistic and the confidence limits are given by equations (8.14) and (8.15), respectively, except that $\hat{Y}$ and Y are evaluated at $X=0$.

Case 6: Confidence Interval of Intercept b, σ_X^2 is not Known, $60 \leq N$

We are seeking the $100(1-\alpha)\%$ confidence interval of the population intercept of the regression line, b, which reduces to that of Case 4 the confidence interval of the predicted value Y given $X=0$ of a linear regression model. In this case the variance of X is not known and the number of data points in the sample, N, is greater or equal to 60. The test statistic and the confidence limits are given by equations (8.14) and (8.16), respectively, except that $\hat{Y}$ and Y are evaluated at $X=0$.

Example 8.3
In Example 8.1, find the 95% confidence interval for the slope, a.

Solution
Required is the confidence interval for the slope a. $\alpha=0.05$, σ_X^2 is unknown, and $N<60$. Use Case 1. The following is a modification of the Maple worksheet of Example 8.1. It uses equations (8.11) and (8.12).

```
[ Initialize, enter data, and calculate sample statistics.
 > restart: with(stats): with(stats[statplots]):
   alpha:=0.05:
   X:=[1.0,1.5,2.5,2.0,3.5,4.5,4.0,5.5,6.0,4.5,3.5,
       7.0,8.0,1.0,1.5,5.5]:
   Y:=[26,28,32,28,30,40,36,36,40,33,35,37,39,25,27,38]:
   N:=describe[count](X):
   X_bar:=describe[mean](X):
   Y_bar:=describe[mean](Y):
   S_X:=describe[standarddeviation[1]](X):
   S_Y:=describe[standarddeviation[1]](Y):
   c_XY:=describe[covariance](X,Y):
   r_XY:=describe[linearcorrelation](X,Y):
   a:=c_XY/S_X^2: b:=Y_bar-a*X_bar:   #equation (8.6)
   y_fit:=x->a*x+b:  #best fit line defined as a function
   S_XY:=sqrt( sum(
        (Y[i]-y_fit(X[i]))^2,i=1..N )/N ): #equation (8.11)
   T:=statevalf[icdf,studentst[N-2]](1-alpha/2):
[ Calculate confidence limits using equation (8.12).
 > a-T*sqrt(N/((N-1)*(N-2)))*S_XY/S_X: L:=evalf(%);
   a+T*sqrt(N/((N-1)*(N-2)))*S_XY/S_X: U:=evalf(%);
```

$L := 1.349391382$

$U := 2.628006640$

We are 95% confident that the slope a is between 1.349 and 2.628.

Example 8.4
In Example 8.1 calculate and draw the "envelope" 90% confidence interval of the predicted workers reading errors. On the same graph plot the regression line.

Solution
Required is the confidence interval for the predicted reading errors, Y, for various values of the number of hours on the job, X. $\alpha=0.1$, σ_X^2 is

unknown, and *N*<60. Use Case 3. The following is a modification of the Maple worksheet of Example 8.1. It uses equation (8.15).

```
[ (1) Initialize, enter data, calculate sample statistics, line slope, and intercept..
[ > restart: with(stats): with(stats[statplots]):
    alpha:=0.1:
    X:=[1.0,1.5,2.5,2.0,3.5,4.5,4.0,5.5,6.0,4.5,3.5,
        7.0,8.0,1.0,1.5,5.5]:
    Y:=[26,28,32,28,30,40,36,36,40,33,35,37,39,25,27,38]:
    N:=describe[count](X):
    Xmin:=min(seq(X[i],i=1..N)):
    Xmax:=max(seq(X[i],i=1..N)):
    X_bar:=describe[mean](X):
    Y_bar:=describe[mean](Y):
    S_X:=describe[standarddeviation[1]](X):
    S_Y:=describe[standarddeviation[1]](Y):
    c_XY:=describe[covariance](X,Y):
    r_XY:=describe[linearcorrelation](X,Y):
    a:=c_XY/S_X^2: b:=Y_bar-a*X_bar:   #equation (8.6)
    Y_hat:=x->a*x+b:   #best fit line defined as a function
    S_XY:=sqrt( sum(
        (Y[i]-Y_hat(X[i]))^2,i=1..N )/N ): #equation (8.11)
    T:=statevalf[icdf,studentst[N-2]](1-alpha/2):
[ Define the upper and lower confidence limits (equation (8.15)) as functions of x.
[ > L:=x->Y_hat(x)-T*S_XY/sqrt(N-2)
            *sqrt(N+1+(N*(x-X_bar)^2)/((N-1)*S_X^2)):
    U:=x->Y_hat(x)+T*S_XY/sqrt(N-2)
            *sqrt(N+1+(N*(x-X_bar)^2)/((N-1)*S_X^2)):
[ Plot the data, the regression line and the confidence lines in one graph.
[ > g1:=scatterplot(X,Y,color=black,symbol=box):
    g2:=plot(Y_hat(x),x=Xmin..Xmax,color=black,
        thickness=2):
    g3:=plot(L(x),x=Xmin..Xmax,color=black,
        linestyle=3,thickness=2):
    g4:=plot(U(x),x=Xmin..Xmax,color=black,
        linestyle=3,thickness=2):
    plots[display](g1,g2,g3,g4,labels=[` X `,` Y `]);
```

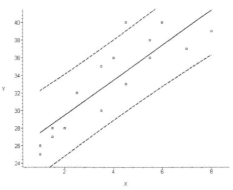

Example 8.4 illustrates several important concepts regarding the quality of our model. In this case, the 90% upper and lower confidence lines occupy a wide band around the fitted regression line because the correlation coefficient is low and *N* is small. This suggests a poor model.

As the correlation coefficient approaches 1.0 and the sample size increases, the confidence band becomes narrower, thus indicating a better model. The narrower the confidence band, the more confident we are that the fitted regression line is closer to the true population one. Also, as the confidence level α increases, we are less confident (i.e., $100(1-\alpha)\%$) of our results but the confidence band becomes narrower. Another feature, not too apparent in Example 8.4, is that the envelope confidence limits are not straight lines, but concave curves around the regression line. Thus, the confidence line is narrowest near the point with coordinates $(\bar{X}, \bar{Y})$, (i.e., the scatter diagram center of mass). As we move away from this point, the confidence band becomes wider, thus indicating a less reliable regression line. This supports the concept that extrapolating fitted lines beyond the region of data points is risky and inappropriate.

Example 8.5

In Example 8.1, find the 95% confidence interval for the intercept b.

Solution

Required is the confidence interval of the population intercept of the regression line, b, $\alpha=0.05$, σ_X^2 is not known, and $N<60$. Use Case 5, which reduces to Case 3 with equation (8.15) evaluated at $X=0$. Hence, we modify the Maple worksheet of Example 8.4.

```
[ Initialize, enter data, calculate sample statistics, line slope, and intercept..
 > restart: with(stats): alpha:=0.05:
   X:=[1.0,1.5,2.5,2.0,3.5,4.5,4.0,5.5,6.0,4.5,3.5,
       7.0,8.0,1.0,1.5,5.5]:
   Y:=[26,28,32,28,30,40,36,36,40,33,35,37,39,25,27,38]:
   N:=describe[count](X):
   X_bar:=describe[mean](X):
   Y_bar:=describe[mean](Y):
   S_X:=describe[standarddeviation[1]](X):
   S_Y:=describe[standarddeviation[1]](Y):
   c_XY:=describe[covariance](X,Y):
   r_XY:=describe[linearcorrelation](X,Y):
   a:=c_XY/S_X^2: b:=Y_bar-a*X_bar:   #equation (8.6)
   Y_hat:=x->a*x+b:   #best fit line defined as a function
   S_XY:=sqrt( sum(
             (Y[i]-Y_hat(X[i]))^2,i=1..N )/N ): #equation (8.11)
   T:=statevalf[icdf,studentst[N-2]](1-alpha/2):\
[ Define the upper and lower confidence limits (equation (8.15) evaluated at x=0).
 > Y_hat(0)-T*S_XY/sqrt(N-2)*sqrt(N+1+(N*X_bar^2)
                       /((N-1)*S_X^2)):   L:=evalf(%);
   Y_hat(0)+T*S_XY/sqrt(N-2)*sqrt(N+1+(N*X_bar^2)
                       /((N-1)*S_X^2)):   U:=evalf(%);
                       L := 19.41382794
                       U := 31.54804842
```

We are 95% confident that the intercept of the regression line, b, is between 19.414 and 31.548. Again, the reasons for such a wide interval are low values of r and N. Also the intercept is at one extreme of the scatter diagram where the confidence band is widest.

In the case of nonlinear curves that reduce to linear ones by one of the transformations discussed before, we can only apply Cases 1 through 6 to the transformed, linearized, model. The confidence intervals thus found can then be expressed in terms of the original quantities via the inverse (e.g., logarithmic) transformation.

Example 8.6

In Example 8.2, (1) calculate and plot the 95% confidence band of the predicted pressure of the linearized model; and (2) plot the regression line and confidence band of the nonlinear model.

Solution

(1) Required is the confidence interval for the predicted log-pressure, Y, for various values of log-volume, X. $\alpha=0.05$, σ_X^2 is unknown, and $N<60$. Use Case 3. The following is a modification of the Maple worksheet of Example 8.2.

```
[ Initialize, enter data, and fit a line to a linear model.
[ > restart: with(stats): with(stats[statplots]):
   alpha:=0.05:
   V:=[890.0,1013.0,1186.6,1453.8,1943.9,2360.2,1702.1,
       1245.3,2210.4,851.2,2093.4,1100.7]:
   P:=[42.2,34.1,25.9,19.6,13.2,7.0,17.2,23.8,12.0,
       37.8,11.2,31.0]:
   N:=describe[count](V):
   min( seq(V[i],i=1..N) ): Vmin:=floor(%):
   max( seq(V[i],i=1..N) ): Vmax:=floor(%):
   Y:=[ seq(log[10](P[i]),i=1..N) ]:
   X:=[ seq(log[10](V[i]),i=1..N) ]:
   X_bar:=describe[mean](X):
   Y_bar:=describe[mean](Y):
   S_X:=describe[standarddeviation[1]](X):
   S_Y:=describe[standarddeviation[1]](Y):
   c_XY:=describe[covariance](X,Y):
   r_XY:=describe[linearcorrelation](X,Y):
   a:=c_XY/S_X^2: b:=Y_bar-a*X_bar:
   Y_hat:=x->a*x+b:
   S_XY:= sqrt( sum( (Y[i]-Y_hat(X[i]))^2,i=1..N )/N ):
   T:=statevalf[icdf,studentst[N-2]](1-alpha/2):
   L:=x->Y_hat(x)-T*S_XY/sqrt(N-2)
              *sqrt(N+1+(N*(x-X_bar)^2)/((N-1)*S_X^2)):
   U:=x->Y_hat(x)+T*S_XY/sqrt(N-2)
              *sqrt(N+1+(N*(x-X_bar)^2)/((N-1)*S_X^2)):
[ Plot linear regression and confidence band.
[ > g1:=scatterplot(X,Y,color=black,symbol=box):
   g2:=plot(Y_hat(x),x=log[10](Vmin)..log[10](Vmax),
       color=black,thickness=2):
   g3:=plot(L(x),x=log[10](Vmin)..log[10](Vmax),color=black,
       linestyle=3,thickness=2):
   g4:=plot(U(x),x=log[10](Vmin)..log[10](Vmax),color=black,
       linestyle=3,thickness=2):
   plots[display](g1,g2,g3,g4,labels=[` X `,` Y `]);
```

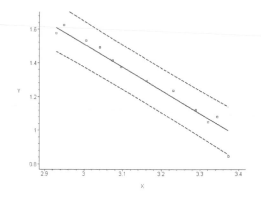

[(2) Define nonlinear model of regression line and confidence lines.
[> beta:=-a: c:=10^b: p:=v->c*v^(-beta) :
  Lo:=v->10^L(log[10](v)) :
  Up:=v->10^U(log[10](v)) :
[Plot nonlinear data, regression, and confidence band.
[> G1:=scatterplot(V,P,color=black,symbol=box) :
  G2:=plot(p(v),v=Vmin..Vmax,color=black,thickness=2) :
  G3:=plot(Lo(v),v=Vmin..Vmax,color=black,linestyle=3,
      thickness=2) :
  G4:=plot(Up(v),v=Vmin..Vmax,color=black,linestyle=3,
      thickness=2) :
  plots[display](G1,G2,G3,G4,tickmarks=[4,default],
      labels=["V (cm^3)","P (dyn/cm^2)"],
      labeldirections=[HORIZONTAL,VERTICAL]) ;

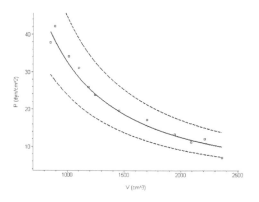

We have seen that a single graph combining the scatter diagram of the data, the fitted regression line, and the confidence band provides a simple, clear, means to assess and report an engineering regression model. In this chapter we have emphasized the use of confidence intervals since they constitute the most popular way to present regression results in engineering. However, we may also design hypothesis tests concerning each of the regression parameters. As explained in Chapter 6, the procedure reduces to the calculation of a test statistic, which follows a specific probability density function, the formulation of a null and an alternative hypothesis concerning the value or the range of values of the regression parameter, and the acceptance or rejection of the null hypothesis depending on whether or not the calculated statistic falls in the critical region delimited by the statistic density function.

The subject of correlation and regression constitutes a statistical field on its own. We have barely touched upon some of its practical applications. In the regression line $\hat{Y}=\hat{a}X+\hat{b}$ we said that $\hat{X}$ and $\hat{Y}$ are single (scalar) random variables. In the general case, they may be column vectors of several random variables. In this case the "slope" $\hat{a}$ is a square matrix, and the "intercept" $\hat{b}$ is a column matrix. Thus, the regression equation is a matrix equation governed by the rules of linear algebra. This is the field of multiple linear regression that permits the consideration of several input and output variables. For an introduction to these concepts the reader is referred to Devore (1995) and Rosenkrantz (1997).

PROBLEMS

8.1 Table 8.3 shows the values of a pressure valve width opening and the corresponding tail-water level. (1) Fit a linear regression model between the two variables (i.e., find the slope and the intercept). (2) Calculate the correlation coefficient and interpret it. (3) Plot the regression line along with the scatter diagram of the data. (4) Predict the level when the width is 3.50cm.

Table 8.3: Results of the Experiment in Problem 8.1

Valve Width, W (cm)	Water Level, L (m)
0.68	12.45
0.72	9.93
1.27	6.64
2.01	10.14
2.63	8.93
3.06	13.34
3.15	11.56
4.00	16.72
4.03	19.62
4.50	15.03

8.2 Propane consumption, G ($m^3 \times 10^6$), in an area over the last 25 years is recorded in Table 8.4. Planning engineers believe the increase over time is exponential, that is $G = ba^t$, where t is time in *years* after given origin, and a and b are constants. (1) Find the value of the constants of an appropriate regression model. (2) Plot the scatter diagram of the data with the fitted nonlinear curve. (3) Predict the consumption of propane at $t=30$ years.

Table 8.4: Results of the Experiment in Problem 8.2

Time, t (*years*)	Propane Consumption, G ($m^3 \times 10^6$)
1	0.25
2	0.35
4	0.40
6	0.50
8	0.85
10	1.25
12	1.65
14	2.50
16	3.75
18	5.30
20	7.60
22	11.10
25	19.10

8.3 In Problem 8.1, calculate 95% confidence interval of the slope of the regression line.

8.4 In Problem 8.1, calculate and draw the "envelope" 75% confidence interval of the predicted water level. On the same graph plot the regression line.

8.5 In Problem 8.1, find the 90% confidence interval for the intercept b.

8.6 In Problem 8.2, (1) calculate and plot the 95% confidence band of the predicted gas consumption of the linearized model; and (2) plot the regression line and confidence band of the nonlinear model.

9 RELIABILITY OF ENGINEERING SYSTEMS

9.1 THE CONCEPT OF RELIABILITY

How long will a machine component last? Is an operation safe? Will a system or a project perform according to its design specifications? These are typical questions engineers in project management and design fields must answer. In Chapter 1 we discussed the philosophical implications of prediction. Naturally, one cannot predict the future performance of a system or a system component in an objective deterministic sense.

One approach to the problem is the extrapolation of historical conditions. In other words, one might expect that a system that has performed satisfactorily in the past may continue to do so in the future. There are several problems with this approach. In many engineering fields, the collection of historical data is expensive, difficult, and subject to errors. Examples abound in the environmental fields. For example, time series systems that require long-term historical records in order to predict future events face the problem of the total absence of past records. Even if adequate historical records are available, as in the case of a system component that has performed admirably in past projects, what guarantees that it will continue to do so? How do we know that historical conditions, those that are reflected in past records, will prevail in the future?

These are difficult questions to answer. The best-quality component eventually and inexplicably fails, sometimes without warning, sometimes after a symptomatic malfunction that gives us time for repair or replacement. "Failure" may be defined as either poor performance at a specified level, as well as complete breakdown. Once a system or a system component has failed, we can disassemble it, investigate its process, and find a deterministic reason for its failure. For example, we may determine the cause of fractured iron or concrete components in a structural element as excessive shear stress. We can take an amplifier apart and find that a condenser is burned. We can deduce that the failure of an airport traffic control system is due to overload. However, we can conclude about the possible deterministic cause of failure only *after* the failure event occurred. This inductive type of research is very useful in the planning of strategies to improve and modify existing designs, and in the identification of individuals or organizations to hold responsible for a failure. For instance, the investigation of a chemical container spill in an aquifer may produce improved spill prevention container designs and specifications for the aquifer restoration project. In addition, this post failure investigation may help identify the individuals or the company responsible at the time of the event. This will facilitate the implementation of fines and penalties for any infringed environmental laws or property damage. The post failure investigation also constitutes the basis for the statistical data necessary for design improvements. For example, if a large number of tail-wing pins of a particular airplane model fail, the company may use this as an indication that the component needs to be improved. In a sense, the performance of mass-produced industry products constitutes a statistical "experiment" from which failure probabilities may be estimated. If a product line is manufactured by the same system, and produced for a specific consumer under more or less similar conditions, then we are reasonably satisfying the requirements of statistical experimentation of "repetitive experimentation under identical conditions."

This leads us to the problem of predicting failure. Recognizing once again that we cannot predict failure in a deterministic sense, we are reassured that we indeed may predict the future in a broad statistical sense. Based on a conscientious statistical experimentation on the frequency of failure of a system or system component, we may estimate the statistical properties and even the probability density function of failure if sufficient information is available. For example, from historical and laboratory tests a manufacturer may assure us that the *mean* life time of a machine is X number of months, with a *standard deviation* of σ_X months. While not being able to predict the precise date when the machine will fail, this is valuable information for the engineer in charge of planning the number of replacement machines in stock and the timing of the purchase. For the manufacturer himself, the failure statistics are extremely valuable, not only for the future machine design improvements, but also for the implementation of support and warranty systems that will assure customer satisfaction and ultimately the maximization of profits. Thus, the concept of reliability is nothing else but an application of the laws of probabilistic events studied in previous chapters. Reliability theory constitutes a field of study by itself. In the following sections we offer the fundamentals only. To probe further, we suggest to consult Leemis (2009), Singpurwalla (2006), O'Connor (2002), Barlow (1998), and Barlow and Proschan (1987, 1975).

9.2 TIME RELIABILITY

With the concepts of the last section, let us define T a random variable representing the time to failure of a system component after being placed in service, and $R(t)$ the *reliability* of the system component or the probability that the system component operates t units of time after being placed in service. From equation (3.28), and noting that time is a positive quantity, if $f_T(t)$ is the probability density function of the time to failure, T, then the probability that the system fails before t is

$$P(T \le t) = F_T(t) = \int_0^t f_T(\tau) d\tau \qquad (9.1)$$

where
   T=time to failure
   $f_T(t)$=probability density function of the time to failure
   $F_T(t)$=cumulative distributions function of T

From equation (2.18), the probability that system operates after t is

$$P(T > t) = R(t) = 1 - F_T(t) \qquad (9.2)$$

where
   $R(t)$=system reliability

From equations (3.30), (9.1), and (9.2), the mean time or *expected time to failure* is

$$E\{T\} = \int_0^\infty t f_T(\tau) d\tau = \int_0^\infty t \cdot \frac{dF_T(t)}{dt} dt = \int_0^\infty t \frac{d}{dt}(1 - R(t)) dt = -\int_0^\infty t \cdot \frac{dR(t)}{dt} dt \qquad (9.3)$$

Applying integration by parts with $u=t$ and $dv=(dR(t)/dt)dt$, and noting that $R(\infty)=0$, then equation (9.3) becomes

$$E\{T\} = \int_0^\infty R(t) dt \qquad (9.4)$$

Example 9.1: Reliability of Electrical Motors
   A company produces electrical motors for vacuum cleaners. After extensive measurements, quality control engineers found that the main core has a probability density function of time to failure in years given by

$$f_T(t) = \frac{a}{t+a}, \qquad 0 \le t \le 10, \ a = 0.2766$$

(1) Find the cumulative distribution of time to failure. (2) Derive the reliability function. (3) Calculate the probability that a unit (any unit) works after 6 years. (4) estimate the expected time to failure.

Solution

(1) From equation (9.1),

$$F_T(t) = \int_0^t f_T(\tau) d\tau = \int_0^t \frac{a d\tau}{\tau + a}$$

Let $z = \tau + a$, from which $dz = d\tau$. The limits of integration are: as $\tau \to 0$, $z \to a$; as $\tau \to t$, $z \to t + a$. Thus, the cumulative distribution function is given by

$$F_T(t) = \int_a^{t+a} \frac{a\,dz}{z} = a\big[\ln(z)\big]_a^{t+a} = a\big(\ln(t+a) - \ln(a)\big) = a\ln\left(\frac{t+a}{a}\right)$$

(2) From equation (9.2),

$$R(t) = 1 - F_T(t) = 1 - a\ln\left(\frac{t+a}{a}\right)$$

(3) From equation (9.2),

$$P(T>6) = R(6) = 1 - 0.2766 \times \ln\left(\frac{6 + 0.2766}{0.2766}\right) = 0.137$$

There is a 13.7% chance that a unit will last longer than 6 years. Remember that reliability is a measure of probability and therefore, as a check, $0 \le R(t) \le 1$.

(4) From equation (9.4),

$$E\{T\} = \int_0^{10} R(t) dt = \int_0^{10} \left[1 - a\ln\left(\frac{t+a}{a}\right)\right] dt = 10 - a\int_0^{10} \ln\left(\frac{t+a}{a}\right) dt$$

Substituting $z = (t+a)/a$ and solving,

$$E\{T\}=10-a^2[z\ln(z)-z]_1^{\frac{10+a}{a}}=10-a^2\left[\frac{10+a}{a}\right]\left[\ln\left(\frac{10+a}{a}\right)-1\right]-a^2=2.49 years$$

In many practical situations the question of reliability is formulated as in questions (3) and (4) in Example 9.1. In other words, we are interested in finding the probability that the typical system component under scrutiny will fail after the warranty period, the expected life of the unit, etc. These questions usually arise prior to the unit being placed in operation. Alternatively, if the unit has been operating properly for a period of time we may wonder how much longer it will work. Mathematically, this question can be posed by using the conditional probability concepts discussed in Chapter 2. For instance, suppose that the system is working properly at a time $t>0$. What is the probability that it will fail before a time τ, with $\tau>t$? If we were interested in finding the probability that the system fails at a time τ, without any condition, then we would simply use equation (9.1): $P(T\leq\tau)=F_T(\tau)$. However, there is an additional condition here: the system is operating at a time t. Therefore, the proper question should be the following: what is the probability of $T\leq\tau$ *given* that $T>t$? This is found from the conditional cumulative distribution function of the time to failure: $P(T\leq\tau|T>t)=F_T(\tau|T>t)$. From the formula for conditional probability, equation (2.19),

$$F_T(\tau|T>t)=\frac{P(T\leq\tau\cap T>t)}{P(T>t)}, \qquad \tau>t \qquad (9.5)$$

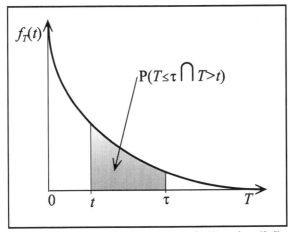

Figure 9.1: A Graph of Numerator in Equation (9.5)

Figure 9.5 shows a graph of the probability density function of the time to failure of a system component. The shaded area is equal to the numerator in equation (9.5). It can be calculated as the difference between the cumulative distribution function evaluated at τ and that at t, respectively. For the denominator of equation (9.5), we use equation (9.2):

$$F_T(\tau \mid T > t) = \frac{F_T(\tau) - F_T(t)}{1 - F_T(t)} = \frac{F_T(\tau) - F_T(t)}{R(t)}, \qquad \tau > t \qquad (9.6)$$

Differentiating with respect to τ,

$$\frac{dF_T(\tau \mid T > t)}{d\tau} = f_T(\tau \mid T > t) = \frac{f_T(\tau)}{R(t)}, \qquad \tau > t \qquad (9.7)$$

Thus, we found that the conditional probability density function of the time to failure is given as the ratio of the (marginal) failure density evaluated at τ over the reliability function evaluated at t.

Now, let us write $\tau = t + \Delta t$, where Δt is a small time interval. Equation (9.7) becomes

$$f_T(t + \Delta t \mid T > t)\Delta t = \frac{f_T(t + \Delta t)\Delta t}{R(t)} \approx \frac{f_i}{R(t)} \approx h(t) \qquad (9.8)$$

where
  f_i=expected relative frequency of $f_T(t)$ in the interval of size Δt
  $h(t)$=*failure rate* at time t

In the numerator of equation (9.8) we have used equation (7.1), Chapter 7. Recall that the expected relative frequency in an i-th interval of size Δt, f_i, is approximately equal to the area under the probability density function, $f_T(t + \Delta t)\Delta t$. We used this concept to plot frequency histograms of empirical distribution functions. The expected relative frequency f_i in this case is equal to the number of system components failing during an interval Δt divided by the total number of system components (see Chapter 7). As $\Delta t \to 0$, in the limit, equation (9.8) becomes

$$h(t) = \frac{f_T(t)}{R(t)} \qquad (9.9)$$

The *failure rate* is then defined as the failure density function divided by the reliability function.

Let us now differentiate equation (9.2) with respect to t.

$$\frac{dR(t)}{dt} = \frac{d}{dt}\left(1 - F_T(t)\right) = -f_T(t) \qquad (9.10)$$

In words, the time derivative of the reliability function equals the negative of the failure density function. Substituting equation (9.10) into equation (9.9),

$$h(t) = -\frac{\dfrac{dR(t)}{dt}}{R(t)} = -\frac{d}{dt}\left[\ln(R(t))\right] \qquad (9.11)$$

Integrating equation (9.11) and noting that $R(0)=1$ and $h(0)=0$,

$$-\int_0^t h(\tau)d\tau = \ln\left[R(t)\right] \qquad (9.12)$$

or

$$R(t) = e^{-\int_0^t h(\tau)d\tau} \qquad (9.13)$$

Equation (9.13) yields a general expression for the reliability function when the system failure rate is variable with time. This is the common scenario, which requires the system monitoring over time. In the special case when the failure rate $h(t)=k$, where k is a constant, then equation (9.13) reduces to

$$R(t) = e^{-kt} \qquad (9.14)$$

We have arrived at a result commonly used in system failure analysis: the reliability of a system component decreases exponentially with time.

Example 9.2
    In Example 9.1 derive the reliability and the failure rate. On a single graph plot these functions along with the failure density and the failure cumulative distribution functions.

Solution

The following Maple worksheet illustrates the procedure. We have lumped most commands without showing the output. However, it is strongly advised that the reader executes and displays each command. That way, if an error is present, or if Maple is unable to solve an integral, one may find its precise location.

```
[ Derive reliability functions.
[ > f:=t->a/(t+a) : int(f(tau),tau=0..t) : F:=unapply(%,t):
    R:=t->1-F(t) :         #equation (9.2)
    h:=t->f(t)/R(t) :      #equation (9.9) :
[ Plot functions in a single graph.
[ > a:=0.2766:
    G1:=plot(f(t) ,t=0..10,color=black,thickness=2,
        legend="f(t)") :
    G2:=plot(F(t) ,t=0..10,color=black,linestyle=2,thickness=2,
        legend="F(t)") :
    G3:=plot(R(t) ,t=0..10,color=black,linestyle=3,thickness=2,
        legend="R(t)") :
    G4:=plot(h(t) ,t=0..9,color=black,linestyle=4,thickness=2,
        legend="h(t)") :
    plots[display] (G1,G2,G3,G4,labels=[`t (years)`,``],
        labelfont=[TIMES,ROMAN,16]) ;
```

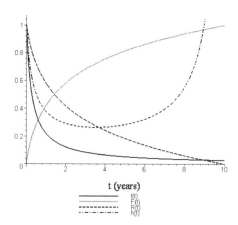

t (years)

The probability density function of the time to failure, $f_T(t)$, is a decreasing function of time indicating that there are fewer chances of having a unit with high time to failure, T. The cumulative distribution function, $F_T(t)$, has a value of zero at $T=0$ and gradually increases to 1 at $T=10 years$. This means that there is a zero probability that the unit will fail prior to $T=0$, and the chances that the unit will fail prior to a given time $T=t$ increase as t increases.

The failure rate, $h(t)$, is very high at the beginning of the life time of a unit. This corroborates the common belief that undetected defective devices that fail usually do so during the early periods. As time increases, the failure rate becomes steady. As time approaches 10*years*, the failure rate becomes infinity. This is due to equation (9.11) where the reliability function approaches zero at this limiting time. The engineering interpretation is that most units are expected to fail as time approaches this limit, due to wear out.

Finally, the reliability function is maximum at $T=0$, and gradually decreases with time until it becomes zero at the end of the service life of the unit. This simply means that as time increases, the chance that a unit will last longer than such time decreases. In other words, as time increases, we are less and less confident that the unit will continue to work. The shape of these functions is typical of systems that decay or wear with time.

Example 9.3

A device has a probability density function of the time to failure T given by $f_T(t)=te^{-t}$, $t \geq 0$ in *months*. Derive the reliability and the failure rate. In a single graph plot these functions along with the failure density and the failure cumulative distribution function.

Solution

Define density function; integrate and factorize exp(-t) to obtain the cumulative; derive the reliability and the failure rate functions; simplify.

```
> restart: f:=t->t*exp(-t);
  int(f(tau),tau=0..t); collect(%,exp(-t)): F:=unapply(%,t);
  simplify(1-F(t)): R:=unapply(%,t);
  f(t)/R(t): h:=unapply(%,t);
  G1:=plot(f(t),t=0..10,color=black,thickness=2,
    legend="f(t)"):
  G2:=plot(F(t),t=0..10,color=black,linestyle=3,thickness=2,
    legend="F(t)"):
  G3:=plot(R(t),t=0..10,color=black,linestyle=3,thickness=2,
    legend="R(t)"):
  G4:=plot(h(t),t=0..10,color=black,linestyle=4,thickness=2,
    legend="h(t)"):
  plots[display](G1,G2,G3,G4,labels=[`T (months)`,``],
    labelfont=[TIMES,ROMAN,16]);
```

$$f := t \to t\,e^{(-t)}$$

$$-e^{(-t)} - e^{(-t)}\,t + 1$$

$$F := t \to 1 + (-1-t)\,e^{(-t)}$$

$$R := t \to (1+t)\,e^{(-t)}$$

$$h := t \to \frac{t}{1+t}$$

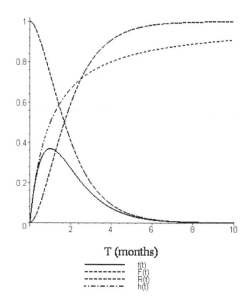

T (months)

——————   f(t)
- - - - - - - -   F(t)
- - - - - - - -   R(t)
· — · — · — ·   h(t)

Here we have a situation of a positively-skewed time to failure density function. The reliability function decreases with time, but the rate of decrease is small at the beginning. In contrast with Example 9.2, the failure rate always increases with time, reaching a value of 1 at infinity.

9.3 RELIABILITY OF SYSTEMS

Systems in general do not have one, but many components. When studying the reliability of a whole system, one must study the reliability of the system components as well as the type of system in question. How the system components are arranged to produce an output is important in determining the effect of possible failure of one or more components on that of the entire structure. According to the way their components are arranged systems may be classified as systems in series, systems in parallel, and hybrid systems.

Systems in Series
A system in series is one in which its components are arranged in a sequential manner. As shown in Figure 9.2, the output from component i becomes the input to component $i+1$. The output from component $i+1$ becomes the input to component $i+2$, etc. Because of this sequential arrangement of components, there exists a serial performance dependency. For component i to be successful in executing its mission, component $i-1$ must also be successful as well. Similarly, component i

must be successful if component $i+2$ is to be successful. As a generalization, the overall success of a system depends on the success of all of the N system components. Therefore, the failure of only one of the components implies the failure of the entire system. Often this is the case along the *critical path* of PERT operational models. The critical path is constituted by the most important tasks in a project. The failure or the delay in one of its components will determine a failure or a delay in the entire line. For the calculation of system reliability we apply the rules of the probability of several events, taking into account the dependency or independency of the individual events, as outlined in Chapter 2.

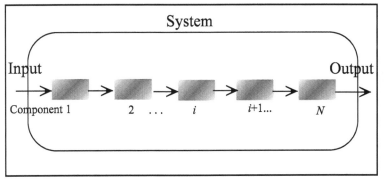

Figure 9.2: A Representation of a System in Series

Example 9.4: Fundamental Processes in a Water Treatment Plant

An oversimplification of the fundamental processes of a water treatment plant conceives the treatment process as composed of three main phases arranged in series: T_1 =time to failure of sedimentation, T_2 = time to failure of filtration, and T_3 = time to failure of disinfection. From historical records of system performance, it is known that $R_1(t)=P(T_1>t)=0.95$; $R_2(t)=P(T_2>t)=0.89$; and $R_3(t)=P(T_3>t)=0.91$. It has also been observed that the processes are independent (i.e., a malfunction of a structure in sedimentation does not affect the malfunction in filtration or disinfection). Calculate the overall reliability of the plant to deliver clean water.

Solution

We know that the system components are statistically independent, which implies that the probability of sedimentation *and* filtration working properly is equal to the product of the individual probabilities (see equation (2.24)). We also know that system components are arranged in series, that is water passes through the sedimentation process, then it is filtrated, and then it is disinfected. This implies that for the entire plant to fulfill its objective (i.e., for the water output to be properly treated), the three fundamental processes must be successfully executed (i.e., water must be sedimented *and* filtered *and* disinfected). Since these events are independent, the reliability of the plant, $R_s(t)$, is given as

$$R_s(t) = P(T_p > t) = P\big((T_1 > t) \cap (T_2 > t) \cap (T_3 > t)\big) = 0.95 \times 0.89 \times 0.91 = 0.77$$

As a generalization, if the components of a system with time to failure T_i, $i = 1, 2, ..., N$ are arranged in series, and if its components are independent of each another, then the overall system reliability if given by

$$R_s = P(T_s > t) = \prod_{i=1}^{N} P(T_i > t) = \prod_{i=1}^{N} R_i(t) \tag{9.15}$$

where
  $\prod$ = the product operator
  R_s = reliability of a system with components arranged in series
  T_s = time to failure of entire system
  T_i = time to failure of an individual component i
  R_i = reliability of component i
  N = number of components

In words, the reliability of a system with independent components arranged in series is equal to the product of the reliability of the individual components. It is interesting to note that for a system in series,

$$R_s(t) \leq R_i(t), \qquad i = 1, 2, ..., N \tag{9.16}$$

That is, the reliability of a system in series is less than or equal to that of any component. If some of the components in the system are not independent, then equation (9.15) must be modified according to the principles of conditional probabilities studied in section 2.5, Chapter 2. The evaluation of the probability of failure of one component given that another fails requires the simultaneous historical recording of failures of the components in question.

To calculate the overall system failure rate, take logarithms on both sides of equation (9.15):

$$\ln(R_s(t)) = \sum_{i=1}^{N} \ln(R_i(t)) \tag{9.17}$$

Differentiating with respect to t and using equation (9.11),

$$h_s = \sum_{i=1}^{N} h_i(t) \tag{9.18}$$

That is, the system failure rate is equal to the summation of failure rates of each component.

Systems in Parallel

A system in parallel is one in which its components operate simultaneously. In other words, component or task i operates at the same time as component $i+1$, $i+2$, (see Figure 9.3) etc.

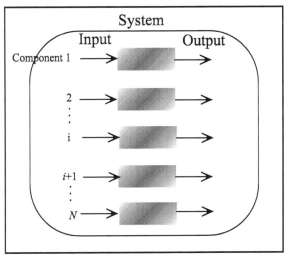

Figure 9.3: A Representation of a Parallel System

In a parallel system, the failure of component, i, does not affect that of another, $i+1$, or that of any other, which continues to operate. Thus, all components must fail for the whole system to fail. For this reason parallel systems are expected to be more reliable, but more expensive, than serial systems.

Example 9.5

An urban county is supplied with electricity by two power plants, 1 and 2, each having a capacity to supply the entire area. If the reliability of plant 1 is $R_1 = 0.87$, and that of plant 2 is $R_2 = 0.78$, estimate the reliability of power supply for the county. Assume independence between 1 and 2.

Solution

Let us define the probability of failure of plant 1 as $F_1 = P(T_1 \le t)$, with T_1 the time to failure of plant 1, and $F_2 = P(T_2 \le t)$, with T_2 the time to failure of plant 2. From equations (2.23) and (9.2), if each plant is independent of the other, the cumulative distribution of the failure of the system in parallel $F_p(t)$, is

$$F_p(t) = F_1 \times F_2 = \left(1 - R_1(t)\right) \times \left(1 - R_2(t)\right)$$

Thus, the reliability of the entire county, $R_p(t)$, is

$$R_p(t) = 1 - F_p(t) = 1 - \left((1 - 0.87) \times (1 - 0.78)\right) = 0.97$$

As a generalization, the overall reliability of a system with independent components arranged in parallel is given by

$$R_p = 1 - \prod_{i=1}^{N} \left(1 - R_i(t)\right) \tag{9.19}$$

where
   R_p = reliability of a system with components arranged in parallel
   R_i = reliability of component i
   N = number of components

It is interesting to note that for a system in parallel,

$$R_p(t) \ge R_i(t), \qquad i = 1, 2, \dots, N \tag{9.20}$$

That is, the reliability of a system in parallel is greater than or equal to that of any component.

Hybrid Systems

A system with some of its components arranged in series and some in parallel is called a hybrid system. The derivation of the reliability functions for such a system can be accomplished by the application of the principles in the previous two categories. The system is resolved into serial subsystems or parallel subsystems. For each subsystem the reliability functions can be derived according to its type. The resulting subsystems are either in series or in parallel with one another, and the process continues. This approach does not work in feedback systems or in hybrid systems possessing bridges amongst some of its components. Even in systems that can be easily decomposed into equivalent subsystems in series and in parallel, the underlying assumption suggests that the principle of superposition of the various components is valid. The principle of superposition states that a system's output is equivalent to the addition of the isolated effect of each component on the system response. In other words, it is assumed that the system is linear. Many natural and engineering systems are nonlinear in its response to combined components, and the above approach constitutes an approximation.

9.4 ENGINEERING MODELS OF FAILURE

Until now we have seen the applications of failure models to the calculation of reliability functions once the probability density function of the time to failure T, $f_T(t)$, is known. A fundamental requisite for this analysis is the availability of such density. As we stated in section 1, this requires the collection of historical data about the time to failure of the system or system component performing under identical conditions. This data must be fitted to a theoretical probability distribution by using the methods described in Chapter 7. Indeed many functions have been found to fit experimental failure data. In this section we briefly mention the Weibull distribution which has been found to adequately describe the several phases of failure frequency.

The Weibull Failure Model

In Chapter 3 we introduced the Weibull random variable and suggested its use in extreme events' analysis and in reliability studies. Applied to the modeling of time to failure, its probability density function is given by equation (3.52) as

$$f_T(t) = \frac{g}{c}\left(\frac{T}{c}\right)^{g-1} e^{-(T/c)^g}, \qquad T,\ g,\ c > 0 \qquad (9.21)$$

where
  g=a shape parameter
  c=a parameter called "the characteristic time"

As seen in Figure 9.4, by altering the values of the parameters g and c, a density function with a different shape, scale, and skewness results. This makes the Weibull distribution a versatile density to model a wide range of random phenomena. In the case of failure models, by adjusting the value of g a wide range of failure models may be fitted. For instance, when g=1, the Weibull distribution produces an exponential probability density of time to failure, $f_T(t)$, which results in a constant failure rate,

$h(t)$. Recall that a constant failure rate is typical of the middle portion in the life of a system component, after the "break-in" period (i.e., when $g<1$), which is marked by a decreasing failure rate with time. When $g>1$, the failure rate is an increasing function of time. This is representative of the "wear-out" period at the end of the life of a component. Thus, g is referred to as a *shape parameter*. The parameter c is called the *characteristic time*.

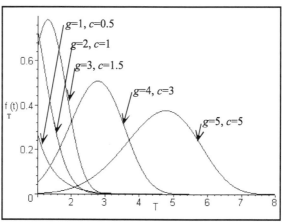

Figure 9.4: The Weibull Density for Various Values of g and c.

From equations (3.30), (331), and (9.21) the expected time to failure and its variance of a Weibull model are given by

$$E\{T\}=c\Gamma\left(1+\frac{1}{g}\right), \qquad Var(T)=c^2\left[\Gamma\left(1+\frac{2}{g}\right)-\Gamma^2\left(1+\frac{1}{g}\right)\right] \qquad (9.22)$$

where $\Gamma(\ )$ is the gamma function defined by equation (3.47). From equations (9.1) and (9.21) the cumulative distribution function of the Weibull model is given by

$$F_T(t)=P(T\le t)=1-e^{-(T/c)^g}, \qquad T\ge0 \qquad (9.23)$$

From equations (9.23) and (9.2) the reliability function is given by

$$R(t)=e^{-(t/c)^g}, \qquad T\ge0 \qquad (9.24)$$

Finally, from equations (9.9), (9.21), and (9.24), the failure rate of a Weibull model is given by

$$h(t)=\frac{g}{c}\left(\frac{t}{c}\right)^{g-1}, \qquad t>0 \qquad (9.25)$$

Figure 9.5 shows the failure rate with respect to time for $c=5$ and various values of g according to equation (9.25). As stated earlier, for values of $g<1$ the failure rate decreases with time, which illustrates the typical "break-in" period of new products. When $g=1$, the failure rate is constant, which illustrates the typical failure of devices within their normal life. For values of $g>1$, we have linear and nonlinear increase in the failure rate with respect to time, which illustrates the final "wear-out" period at the end of service of a device.

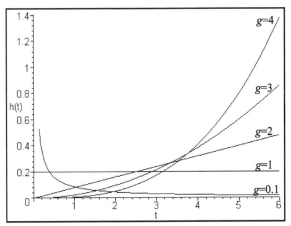

Figure 9.5: Failure Rate of a Weibull Model for $c=5$
and Various values of g.

Fitting Data to a Weibull Failure Model

The procedure to fit data to probability distributions was outlined in Chapter 7. Among the methods discussed was the plotting of the empirical cumulative distribution of experimental data on probability paper. The Weibull density function lends itself to an easy transformation of its cumulative distribution that produces a straight line when plotted. Taking logarithms on both sides of equation (9.23),

$$-\ln\left(1-F_T(t)\right)=\left(\frac{t}{c}\right)^g \tag{9.26}$$

Upon taking logarithms again,

$$\ln\left[-\ln\left(1-F_T(t)\right)\right]=g\ln(t)-g\ln(c) \tag{9.27}$$

Equation (9.27) is of the form $y=ax+b$ with $y=\ln\left[-\ln\left(1-F_T(t)\right)\right]$, $a=g$, $x=\ln(t)$, and $b=-\ln(c)$. That is, the theoretical Weibull cumulative distribution function plots as a straight line after the above transformation.

We can use this feature to fit a Weibull model to the empirical cumulative distribution function collected from data. This is done by first setting $F_j = F_T(t)$, where F_j is the ordinate of the empirical cumulative distribution function at the j-th failure time corresponding to a sample failure time t_j. In other words, $F_j = P(T \le t_j)$ is calculated from either equation (7.2) or (7.3). Next, we calculate the j-*th* ordinate, $y_j \approx \ln(-\ln(1 - F_j))$, and plot it with respect to its abscissa $x_j \approx \ln(t_j)$, for all sample points j. If a straight line may be reasonably fitted, then there is reason to believe that the Weibull density function is a good model for the failure data. If this is the case, an estimate of g, $\hat{g}$, is obtained from the slope of the fitted line, and an estimate of c, $\hat{c}$, is obtained from the ordinate intercept of the fitted line, $\hat{b}$, as $\hat{c} = e^{\hat{b}/\hat{g}}$.

If we use a rank-ordered data as a basis for the evaluation of the empirical cumulative distribution function (see equation (7.3)), then equation (9.27) applied to the j-th plotting position becomes

$$y_j = \hat{a}x_j + \hat{b}, \qquad y_j = \ln\left[-\ln\left(\frac{m}{N+1}\right)\right] \tag{9.28}$$

$$\hat{a} = \hat{g}, \qquad x_j = \ln(t_j) \qquad \hat{b} = -\hat{g}\ln(\hat{c})$$

where

y_j = j-th plotting ordinate
x_j = j-th plotting abscissa
$\hat{a}$ = estimate of the line slope
$\hat{b}$ = estimate of the line ordinate intercept
m = ranks of data points in failure series arranged in decreasing order
N = number of data points
t_j = failure time of j-th data point
$\hat{g}$ = estimate of g
$\hat{c}$ = estimate of c

Let us summarize the procedure to fit a Weibull model to a series of failure data:

1. Collect N sample values of failure times, t_j, $j = 1, 2,..., N$ of the system under consideration.

2. Rank failure times in decreasing order of magnitude, assigning a rank $m = 1$ to the longest failure time, $m = 2$ to the second longest, etc.

3. For each point j, calculate its abscissa, x_j, and its ordinate, y_j, from equation (9.28).

4. Plot x_j versus y_j for all j on linear paper and fit a straight line $y=\hat{a}x+\hat{b}$, where $\hat{a}$ is the best-line slope and $\hat{b}$ is the best-line ordinate intercept. Alternatively, fit a linear regression through the scatter diagram of points.

5. Estimate the Weibull parameters from equation (9.28): $\hat{g}=\hat{a}$, and $\hat{c}=e^{-\hat{b}/\hat{g}}$.

Table 9.1: Data and Calculations for Example 9.6

Sample, j	Failure Age, t_j $(km\times10^3)$	Rearranged Age, t_j $(km\times10^3)$	Rank, m	x_j	y_j
1	6.107	8.558	1	2.147	1.234
2	8.144	8.531	2	2.144	1.008
3	7.808	8.378	3	2.126	0.848
4	7.806	8.320	4	2.119	0.717
5	5.859	8.267	5	2.112	0.601
6	7.631	8.144	6	2.097	0.496
7	7.122	8.026	7	2.083	0.398
8	8.558	8.012	8	2.081	0.304
9	7.738	7.911	9	2.068	0.213
10	8.531	7.808	10	2.055	0.124
11	8.378	7.806	11	2.055	0.036
12	6.038	7.738	12	2.046	-0.052
13	7.656	7.704	13	2.042	-0.140
14	7.269	7.656	14	2.036	-0.230
15	6.645	7.631	15	2.032	-0.320
16	6.992	7.553	16	2.022	-0.413
17	7.342	7.368	17	1.997	-0.510
18	8.026	7.342	18	1.994	-0.610
19	8.320	7.299	19	1.988	-0.714
20	7.299	7.269	20	1.984	-0.825
21	8.267	7.122	21	1.963	-0.943
22	7.368	7.097	22	1.960	-1.070
23	4.848	7.041	23	1.952	-1.209
24	7.041	6.992	24	1.945	-1.363
25	7.553	6.645	25	1.894	-1.537
26	5.530	6.107	26	1.809	-1.738
27	7.704	6.038	27	1.798	-1.979
28	8.012	5.859	28	1.768	-2.285
29	7.097	5.530	29	1.710	-2.708
30	7.911	4.848	30	1.579	-3.418

Example 9.6: A Model of Tire Failure

A company has tested a new prototype of steel-belted tire for thread separation failure. All the samples were taken on the same type of tire equipped on the same type of vehicle running at 120 $km/hour$. Table 9.1 shows the "age" of the tire in km covered at the time of failure. Fit a Weibull failure model to the data.

Solution

N=30. We follow steps 1 through 5 above, summarizing the calculations in Table 9.1. x_j is calculated from equation (9.28) as $x_j = \ln(t_j)$. y_j is calculated from equation (9.28) as $y_j = \ln(-\ln(m/(N+1)))$, where m is the rank. The coordinates, (x_j, y_j), of the arranged cumulative distribution function are plotted in Figure 9.6 and a straight line $y = \hat{a}x + \hat{b}y$ is fitted, where $\hat{a} = 8.179$ is the slope, and $\hat{b} = -16.787$ is the intercept (not shown). The straight line indicates that a Weibull distribution might be a good model for the probability density function of tire failure. Now from equation (9.28), the parameters of the Weibull model are $\hat{g} = \hat{a} = 8.179$, and $\hat{c} = e^{-\hat{b}/\hat{g}} = e^{16.787/8.179} = 7.787$.

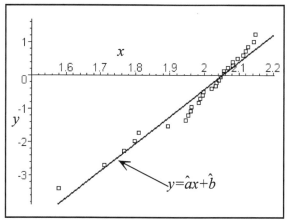

Figure 9.6: Tire Failure Cumulative Distribution
Function for Example 9.6.

Example 9.7: A Program for Weibull Probability Paper

Write a Maple program that solves Example 9.6. For that purpose, modify the program in Example 7.5 (Chapter 7) to implement equation (9.28), plot the data on Weibull paper, and estimate the parameters g and c.

Solution

```
[ Initialize; call statistics module; enter data.
[ > restart: with(stats[statevalf]):
    T:=[6.107,8.144,7.808,7.806,5.859,7.631,7.122,8.558,
         7.738,8.531,8.378,6.038,7.656,7.269,6.645,6.992,
         7.342,8.026,8.320,7.299,8.267,7.368,4.848,7.041,
         7.553,5.530,7.704,8.012,7.097,7.911]:
    N:=describe[count](T):              #number of points
    X:=map(log,T):                      #equation (9.28)
    r:=describe[range](X):              #range of values
    MinX:=op(1,r):                      #minimum Abscissa
    MaxX:=op(2,r):                      #maximum abscissa
    Xorden:=sort(X,`>`):                #sort series
    F:=[seq(j*100./(N+1),j=1..N)]:      #values of CDF
[ Create tickmarks of CDF 0.1% to 99.0%
[ > XTicks:=[seq(evalf(log(i))=convert(i,string),
    i=[.1,.2,.3,.4,.5,.6,.7,.8,.9,1,2,3,4,5,6,7,8,9,10,
       20,30,40,50,60,70,80,90,100,200,300,400,500,600,700,
       800,900,1000,15000,20000,25000,3000,4000,5000,6000,
       7000,8000,9000,10000,20000,30000,40000,50000,60000,
       70000,80000,90000,100000] ) ]:
    YTicks1:=[seq(evalf(log(log(100/(100-i))))=convert(i,string),
    i=[.1,.2,.3,.4,.5,1,2,3,4,5,10,20,30,40,50,60,70,80,90,95,96,
       97,98,99] ) ]:
    YTicks:=[op(YTicks1)]:
[ Transform the F such that it plots as a straight line.
[ > Fy:=[seq(log(-log(F[i]/100.)),i=1..N)]: #Equation (9.28)
[ Organize data into coordinates [x,y].
[ > Data:=zip((x,y)->[x,y],Xorden,Fy)[]:
[ Fit a linear regression line to the data.
[ > FIT:=unapply(rhs(fit[leastsquare[[x,y]]]([Xorden,Fy])),x):
    g:=op(2,FIT(x))/x;                  #Weibull parameters
    c:=exp(-op(1,FIT(x))/g);            #Equation (9.28)
```

$$g := 8.179887709$$

$$c := 7.786063346$$

```
[ Plot data and fitted regression on Webull paper.
[ > plots[display](
    plot(FIT,MinX..MaxX,color=black,thickness=2),
    PLOT(POINTS(Data),AXESTICKS(XTicks,YTicks),AXESSTYLE(BOX)),
       symbol=box,labels=[`Age (km*10^3)`, `F (%)`],
       labelfont=[TIMES,ROMAN,16] );
```

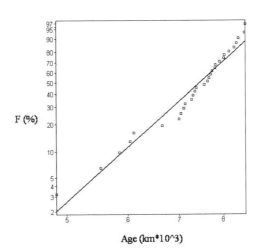

Age (km*10^3)

PROBLEMS

9.1 Assume that the failure probability density function for a device is given by $f_T(t) = \omega e^{-\omega t}$, with t in hours. (1) Find the cumulative distribution of time to failure. (2) Derive the reliability function. (3) Calculate the probability that the device works after 2000 hours if $\omega = 0.001$ $hour^{-1}$. (4) estimate the expected time to failure.

9.2 In Problem 9.1, find the conditional probability density function for a time $\tau > t$.

9.3 In Problem 9.1, derive the reliability and the failure rate functions. Plot the reliability function.

9.4 Consider a system with a time to failure given by a Uniform density function between 0 and $10 days$. Derive and plot the cumulative distribution, the reliability and the failure rate functions.

9.5 The launch of a satellite requires the following sequence of operations with the associated probabilities of success: 1. Ignition of booster rocket, $R_1 = 0.99$. 2. Bolts holding satellite to main rocket blow, $R_2 = 0.98$. 3. Gas jets spin satellite for stability, $R_3 = 0.965$. 4. Booster rocket shutdown at desired final velocity, $R_4 = 0.97$. These tasks occur in series. Assuming independence among them, estimate the probability of mission success, or reliability.

9.6 In Problem 9.5, suppose that the system component reliability is exponential, $R_i = e^{-\omega_i t}$, instead of constant, with $\omega_1 = \omega_2 = 0.01$, $\omega_3 = \omega_4 = 0.02 hour^{-1}$. Calculate (1) the reliability of the system; and (2) the mean time to failure of the entire system.

9.7 A system has two components, 1 and 2, operating in parallel. If the reliability of each component is exponential with parameters ω_1 and ω_2, respectively, calculate (1) the reliability of the system; and (2) the mean time to failure of the entire system.

9.8 Table 9.2 lists the annual maximum flows of a river, Q. Assuming that extremely high flow rate in a river (i.e., floods) may be treated as "failure" data, modify the program in Example 9.7 to investigate the Webull distribution as a possible model. Specifically: (1) have the program read the data, calculate the transformed coordinates of the Weibull cumulative distribution function, fit a straight line using a least-squares regression model, plot the scatter diagram of the data along with the fitted line, and estimate the parameters of the Weibull distribution. (2) Assuming the

Weibull distribution is a good model, estimate the flow with a return period of 50 years.

Table 9.2: Flood Frequency Analysis of Problem 9.7

Year	Q (m^3/s)	Year	Q (m^3/s)
1984	16200	1997	17000
1985	6500	1998	10300
1986	20300	1999	19900
1987	14700	2000	13600
1988	20500	2001	5700
1989	10100	2002	8800
1990	7200	2003	9400
1991	28300	2004	19200
1992	18300	2005	6800
1993	24300	2006	14700
1994	17900	2007	8400
1995	12100	2008	20100
1996	10900	2009	10400

9.9 Modify the program in Example 9.7 to test the sensitivity of the Weibull parameter estimation method to the sample size, N. First generate $N=50$ Weibull random numbers with $g=10$ and $c=8$. Assume this is your time to failure data, t_j, and estimate the Weibull parameters $\hat{g}$ and $\hat{c}$. How close are they to the true values? Repeat for $N=100, 500,$ and 1000, respectively. State your conclusions.

"We go through life employing a psychic stochastic filter that allows us to use the memory of past experiences and the feedback from present circumstances to correct our mistakes, make small adjustments to our decisions, change the course of future actions, and modify personal goals."

Langid

10 DESIGN OF ENGINEERING EXPERIMENTS

10.1 THE CONCEPT OF STATISTICAL EXPERIMENT DESIGN

In this chapter we cover the practical issues of experimental design in engineering. Throughout this book we have referred to the "experiment" as a means to acquire data for the purpose of estimating the possible value of a population parameter, which is subsequently used to study the relationship with other variables or to fit probabilistic models for simulation or engineering design. The engineering experiment is the foundation of statistical and probabilistic analysis.

The experiment consists in the repetitive measurement of a variable. We have emphasized that the fundamental rule for the statistical experiment requires that the measurements be performed under identical conditions. This implies a systematic repetition of the same measurement procedure by following the same experimental protocol, utilizing the same (class of) measurement device manufactured by the same company, and repeating the observation under the same spatial and temporal environmental conditions. By repeating the experiment under identical conditions, we are assuring the data analyst or modeler that the samples are independent of each another, which is an important statistical assumption. In addition, experimentation under identical conditions facilitates the elimination of the effect of other intervening variables that are not components of the uncertainty being quantified. If an experiment is performed with different protocols, the sample values reflect not only the effect of random variability, but also that introduced by different instrumentation or procedure. This is why it is difficult to compare data bases obtained by different groups in different historical events.

We are not concerned here with the observational procedure associated with a particular engineering operation. This is the experimental protocol dictated by the relevant engineering testing organization. For example, the detailed procedure for measuring the module of elasticity of iron bars, X, is outlined in the corresponding ASTM testing norm. The norm describes the length of the bar sample, the type of tensile machine, the collocation of a deformation meter, etc. This is what the engineer learned early in his/her career as the "experiment" itself. The *statistical experiment* is the sequential repetition of the same test on several bar samples. By now the reader has learned that performing the test only once does not yield the "true" deterministic value, but rather a sample of what truly is a random process. To quantify this uncertainty, the engineer designs a statistical experiment consisting in repeating the same test on N bar samples.

A fundamental question in experimental design is: how large should N be? We have seen in previous chapters that the sample statistics are random variables that depend on the sample size (see Chapter 6). As the sample size increases, the calculated statistics approach the true population parameter values. In other words, as N increases $\bar{X} \to \mu_X$, $S_X \to \sigma_x$, $Var(\bar{X}) \to 0$, and $Var(S_X) = 0$. A conservative approach to the problem is to take a very large set of observations. This might be appropriate in the quality control of an industrial operation that massively produces an item. The routine selection of a large number of samples is done easily and inexpensively. However, in many engineering projects the execution of a test requires large expenditures involving equipment, personnel, and time. For example, the measurement of the transmissivity value of an aquifer is done via a pumping test in the field. This requires the construction of a deep well followed by the pumping operation itself. The results must be analyzed with the aid of a mathematical model that simulates groundwater flow. The experiment and its subsequent analysis are expensive.

At this point the question on the value of N emerges as an important engineering decision with economical restrictions. Let us rephrase the question: how large should N be in order to *accurately* evaluate a parameter with the *least investment* of resources? With the foregoing thoughts, we define the design of a statistical experiment as the determination of the minimum sample size of a random variable that will yield a specified maximum error in the estimation of a population parameter. Thus, the design of an experiment is essentially a plan for the purchase of a specified quantity of information at the lowest cost possible. In the following sections we offer a basic introduction to the problem of experimental design. First we study the estimation of the population mean from limited sampling and then the problem of determination of the sample size itself. Experimental design is a vast subject in statistics with a level well beyond the scope of the present book. The fundamental concepts are indeed simple and should be mastered by all engineering professionals. For a more in depth treatment consult Antony (2003), Barrentine (1999), Hicks (1999), and Mendenhall (1968). For a very readable treatment about experimental design, sampling planning, and monitoring with emphasis in environmental problems see Gilbert (1987). Principles of sampling with applications are covered in Cochran (1977).

10.2 ESTIMATING THE POPULATION MEAN FROM LIMITED SAMPLING

In this section we explore the effect of the sample size on the accuracy of the sample mean of a random variable. Let us suppose that the size of the population of a random variable X is N_p. In many circumstances N_p may be essentially infinite (e.g., the number of 10-g soil samples under a foundation field). In certain cases, however, it is possible to have a finite number N_p (e.g., the number of hours in a day to test for air quality). In order to select a sample size $N<N_p$, one may use *random sampling*. If N_p is finite and refers to an ordinal or temporal set, then the procedure consists in selecting N random numbers from 0 to N using a random number generator or a table of random numbers. The generated numbers refer to the location in the set (e.g., the days when the sample will be taken).

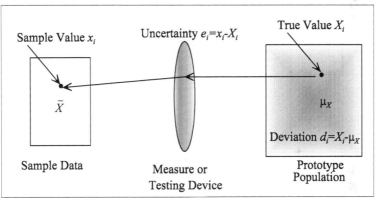

Figure 10.1: A Representation of the Measurement of a Sample

With N samples of X, x_i, $i=1, 2,..., N$, collected with identical protocol, one can calculate the *sample* mean, $\bar{X}$, from equation (6.1) and the sample variance, S_X^2, from equation (6.2). Each sample value x_i is a measurement of a true population value X_i (see Figure 10.1). The difference $e_i=x_i-X_i$ may be conceived as measurement uncertainty. In addition, each true value X_i may be expressed as the difference between the population mean, μ_X, and a deviation, d_i. Hence, a simple model for the measurement of x_i is

$$x_i=\mu_X+d_i+e_i=X_i+e_i \qquad (10.1)$$

where

x_i=sample measurement of the i-th population unit of X

d_i=amount the i-th true value differs from the population mean (i.e., $d_i=X_i-\mu_X$)

μ_X=true population mean of X calculated over N_p units
X_i=true population value of the i-th population unit
e_i=x_i-X_i=amount by which the i-th measurement of X, x_i, differs from the true X_i

If there are no systematic (i.e., deterministic) measurement, sample collection, or handling biases, then the mean uncertainty calculated over the entire population of N_p units is zero. Thus, the *population* mean of X, μ_X, is given by equation (3.6), and the *population* variance, σ_X^2, is given by equation (3.10). If we take only a small sample $N<N_p$, the population parameters are usually unknown. Unbiased sample estimates of these parameters, $\overline{X}$ and S_X^2, may be computed from equations (6.1) and (6.2), respectively. Since $\overline{X}$ depends on the sample size, N, it is a random variable in itself (see Chapter 6). Therefore, the variance of $\overline{X}$, $Var(\overline{X})$, is a measure of the random sampling error:

$$Var(\overline{X}) = \frac{1}{N}(1-f)\sigma_X^2 \qquad (10.2)$$

where
$f=N/N_p$=the sampling fraction
N_p=Number of population units

The *sampling fraction* is simply the portion of the N_p population actually measured. The quantity $(1-f)$ in equation (10.2) is called the *population correction factor*. Equation (10.2) is valid when sampling without replacement from a finite population of N_p units. If sampling with replacement is used, $f=0$. We also note that if N_p is very large, f is also zero.

An unbiased estimator of $Var(\overline{X})$ computed from the N data is

$$S_{\overline{X}}^2 = \frac{1}{N}(1-f)S_X^2 \qquad (10.3)$$

where
$S_{\overline{X}}^2$=estimate of $Var(\overline{X})$, S_X^2=sample variance of X

The *standard error* of $\overline{X}$ is the square root of equation (10.3):

$$S_{\overline{X}} = \left(\sqrt{\frac{1}{N}(1-f)}\right)S_X \qquad (10.4)$$

where
$\quad S_{\bar{x}}$=standard error of the sample mean
$\quad S_{X}$=sample standard deviation of X

Example 10.1: Effect of Sample Size on the Estimation Error

A contractor was hired to estimate the cyanide concentration, C ($\mu g/L$), at a location in a stream in March 2010. He took 10 samples c_i, i=1, 2 ..., 10, each representative of a 24-*hour* duration during the month of March. The N=10 sampling days were selected at random from the N_p = 31 potential days by choosing 10 two-digit numbers between 1 and 31 from a random number generator. (1) If the concentration measurements were 1450, 1510, 1480, 1680, 2130, 2110, 1010, 1290, 1150, and 1300 $\mu g/L$, respectively, calculate the standard error. (2) If the 10 measurements are considered representative of March of all years, estimate the standard error.

Solution
(1) From equation (3.6) the population mean would be given by

$$\mu_C = \frac{1}{31}\sum_{i=1}^{31} C_i$$

were C_i is the true concentration on day i. This true mean is estimated from equation (6.1) as

$$\bar{C} = \hat{\mu}_C = \frac{1}{10}\sum_{i=1}^{10} c_i = 1511 \mu g/L$$

From equation (3.10) the population variance is

$$\sigma_C^2 = \frac{1}{31}\sum_{i=1}^{31}(C_i - \mu_C)^2$$

which is estimated with the sample variance, equation (6.2), as

$$S_C^2 = \hat{\sigma}_C^2 = \frac{1}{10-1}\sum_{i=1}^{10}(c_i - \bar{C})^2 = 138876.966(\mu g/L)^2$$

From equation (10.2), the variance of the sample mean is

$$Var(\bar{C}) = \frac{1}{10}\left(1 - \frac{10}{31}\right)\sigma_C^2$$

which is estimated from equation (10.3) as

$$S_{\bar{C}}^2 = \frac{1}{10}\left(1-\frac{10}{31}\right)S_C^2 = 9407.795(\mu g/L)^2$$

From equation (10.4) the standard error is $S_{\bar{C}} = 96.994 \mu g/L$. Since the target population was defined to consist of only 31 unknown daily concentrations for a month, then the factor $(1-f)$ is needed when computing the standard error in equation (10.4).

(2) If the 10 measurements are considered representative of concentrations that would arise during March of all years, then N_p may be so large that f is essentially zero. Setting $f=0$ in equation (10.4) gives $S_{\bar{C}} = 117.846 \mu g/L$, which represents an increase of 21.5% with respect to the standard error calculated when $f=10/31$. Clearly, the conceptualization and the resulting size of the target population affect f and the value of the standard error.

10.3 ESTIMATING THE REQUIRED NUMBER OF MEASUREMENTS

In this section we study two practical methods of determining the number of samples to take, N. Each approach is concerned with choosing N such that the estimated mean achieves some *pre specified accuracy*.

Pre-Specified Variance

Suppose that the true (i.e., population) mean of a random variable X, μ_X, is estimated by the sample mean, $\bar{X}$, and it is decided that the variance of the sample mean, $Var(\bar{X})$, must be less than a specified value V. Setting $V=Var(\bar{X})$ in equation (10.2) and solving for N,

$$N = \frac{\sigma_X^2}{V+\dfrac{\sigma_X^2}{N_p}} \tag{10.5}$$

If N_p is very large then equation (10.5) becomes

$$N = \frac{\sigma_X^2}{V} \tag{10.6}$$

Thus, if the population variance is known, it is easy to calculate N for any selected V. As the selected V increases (i.e., the required accuracy decreases), the required number of measurements N decreases. In practice the value of V is selected according to budgetary and practical restrictions.

If the population variance, σ_X^2, is not known, it may be estimated by taking an initial set of measurements N_1 and computing the sample variance, S_X^2, from equation (6.2). This initial number of measurements may be compiled from previous studies. It is easy to show (Gilbert, 1987) that additional samples should be taken such that the final number of measurements is given by

$$N = \frac{S_X^2}{V}\left(1 + \frac{2}{N_1}\right) \tag{10.7}$$

where
N_1 = initial (tentative) number of measurements

The above estimate assumes that the sample mean, $\overline{X}$, is normally distributed (Cochran, 1977), and therefore it is an approximate result in most cases.

Example 10.2: Samples Required to Verify a Contractor's Data
Lead contamination of soils near an industrial plant is under investigation. An environmental contractor has sampled the area extensively and reported that the variance of the soil-lead concentration is about $100\,(mg/kg)^2$. The government wants to take a reduced set of independent samples to verify the reported mean concentration. Based on project objectives and available budget, the government decided to take measurements such that the variance of the sample mean is not greater than $4(mg/kg)^2$. (1) Calculate the number of samples to take. (2) If the government does not trust the previous reports and collects 15 preliminary samples that give a sample variance of $95(mg/kg)^2$, how many additional samples should be taken?

Solution
(1) Let X be the soil-lead concentration in mg/kg; $\sigma_X^2 = 100(mg/kg)^2$; $V = 4(mg/kg)^2$; and the number of soil possible samples in the plot N_p is very large. Thus, from equation (10.6),

$$N=\frac{\sigma_X^2}{V}=\frac{100}{4} = 25$$

(2) If the previous data is discarded, then σ_X^2 is unknown and must be estimated from the $N_1=15$ preliminary samples as $S_X^2=95(mg/kg)^2$. From equation (10.7) the total number of samples to take is

$$N=\frac{S_X^2}{V}\left(1+\frac{2}{N_1}\right)=\frac{95}{4}\left(1+\frac{2}{15}\right)=27$$

Thus, 27-15=12 additional samples must be taken.

Pre-Specified Margin of Error

Instead of specifying V, the variance of $\overline{X}$, one may specify the absolute margin of error, d, to be tolerated along with an acceptable small probability, α, of exceeding that error. In other words, one can choose N such that

$$P(|\overline{X}-\mu_X|\geq d)\leq\alpha \tag{10.8}$$

Inequality (10.8) reminds us of the definition statements of confidence intervals. For example, if $d=10$ and $\alpha=0.05$, the number of samples to take, N, must be such that there is a $100\alpha=5\%$ chance that the absolute difference, positive or negative, between the sample mean, $\overline{X}$, calculated from N collected samples and the population mean μ_X is greater than or equal to 10. When this conception is used, V in equations (10.5) and (10.6) is replaced by $(d/Z_{1-\alpha/2})^2$, where $Z_{1-\alpha/2}$ is the abscissa value under the Standard Normal distribution corresponding to a cumulative area of $1-\alpha/2$. Thus, equations (10.5), and (10.6) become,

$$N=\frac{(Z_{1-\alpha/2}\sigma_X/d)^2}{1+\frac{(Z_{1-\alpha/2}\sigma_X/d)^2}{N_p}} \tag{10.9}$$

and

$$N=\left(\frac{Z_{1-\alpha/2}\sigma_X}{d}\right)^2 \tag{10.10}$$

for small and large N_p, respectively. The values of $Z_{1-\alpha/2}$ may be obtained from Table A.1 (Appendix A) or a computer program. Table

10.1 lists values of $Z_{1-\alpha/2}$ interpolated from Table A.1 for typical values of the upper tail α.

Table 10.1: Typical Standard Normal Abscissa Values

α	0.20	0.10	0.05	0.01
$Z_{1-\alpha/2}$	1.2816	1.6449	1.960	2.5758

We remark again that equations (10.9) and (10.10) assume that the sample mean, X, is normally distributed, which occurs when the random variable X is normally distributed, or when N is sufficiently large. If X is approximately Normal, but the population variance, σ_X^2, is not known, then the sample variance, S_X^2, is used instead of the population and the Student's t distribution is used instead of the Normal. That is, $Z_{1-\alpha/2}$ is replaced by $T_{N-1,1-\alpha/2}$, which is the abscissa value under the Student's t distribution corresponding to a cumulative area of $1-\alpha/2$, and N-1 degrees of freedom. Thus, equations (10.9) and (10.10) become

$$N = \frac{(T_{N-1,1-\alpha/2}S_X/d)^2}{1 + \dfrac{T_{N-1,1-\alpha/2}S_X/d)^2}{N_p}} \tag{10.11}$$

and

$$N = \left(\frac{T_{N-1,1-\alpha/2}S_X}{d}\right)^2 \tag{10.12}$$

for small and large N_p, respectively. The values of $T_{N-1,1-\alpha/2}$ may be obtained from Table A.2 (Appendix A) or a computer program. Since $T_{N-1,1-\alpha/2}$ depends on N, an iterative procedure must be used. First, we use equations (10.9) or (10.10), depending on the size of N_p, to determine a tentative value of N. We call this initial value N_1. Next find $T_{N_1-1,1-\alpha/2}$ in Table A.2 and use it to find a new value of N, called N_2, in equation (10.11) or (10.12). Next find $T_{N_2-1,1-\alpha/2}$ in Table A.2 and use it to find a new value of N, called N_3, in equation (10.11) or (10.12). This iteration is repeated several times until the value of N converges to a constant. Ideally this procedure should be implemented in a computer program.

Example 10.3

In Example 10.2, assume that the population variance is $100 \, (mg/kg)^2$ and that the government agency decides to accept a 10% chance that the difference between the sample mean and the population mean is greater than or equal to $2mg/kg$. Calculate the number of samples to take.

Solution

$\sigma_X^2 = 100 \ (mg/kg)^2$, $\alpha = 0.10$, $d = 2mg/kg$, $P(|\bar{X} - \mu_X| \geq 2) \leq 0.10$, and N_p is very large. We use equation (10.10) and Table 10.1:

$$N = \left(\frac{Z_{1-\alpha/2}\sigma_X}{d}\right)^2 = \left(\frac{1.6449 \times 10}{2}\right)^2 = 67.64 \approx 68$$

Note that from equation (10.10) the number of samples may be decreased if we are willing to accept a greater margin of error, d, or if we are willing to increase the chances that this margin, α, is exceeded.

Example 10.4

Repeat Example 10.3 if the government assumes the population variance is unknown and instead uses a sample variance of $95(mg/kg)^2$ calculated from limited preliminary sampling.

Solution

σ_X^2 is unknown, $S_X^2 = 95(mg/kg)^2$, $\alpha = 0.10$, $d = 2mgh/kg$, $P(|\bar{X} - \mu_X| \geq 2) \leq 0.10$, and N_p is very large. We first use equation (10.10) and Table 10.1 to calculate the initial sample size:

$$N_1 = \left(\frac{Z_{1-\alpha/2}S_X}{d}\right)^2 = \left(\frac{1.6449 \times \sqrt{95}}{2}\right)^2 = 64.26 \approx 65$$

From equation (10.12) and Table A.2, an improved estimate of the sample size is

$$N_2 = \left(\frac{T_{N_1-1,1-\alpha/2}S_X}{d}\right)^2 = \left(\frac{T_{64,0.95}\sqrt{95}}{2}\right)^2 = \left(\frac{1.6690 \times 9.7468}{2}\right)^2 = 66.16 \approx 67$$

Again from equation (10.12), Table A.2, and N_2,

$$N_3 = \left(\frac{T_{66,0.95}\sqrt{95}}{2}\right)^2 = 66.10 \approx 67$$

Similarly,

$$N_4 = \left(\frac{T_{66,0.95}\sqrt{95}}{2}\right)^2 = 66.10 \approx 67$$

and $N \rightarrow 67$ samples.

PROBLEMS

10.1 Write a Maple program that solves Example 10.1.

10.2 A concrete plant guarantees that the standard deviation of concrete of a specified strength is 250*psi*. If a construction engineer wants to estimate the mean strength to compression of a delivered batch such that the variance of the sample mean is not greater than $2500 \, (psi)^2$, how many samples should he take from the batch?

10.3 In Problem 10.2, the engineer does not trust the manufacturer recommendation and decides to take a few samples from the batch and test for failure strength with the following results in *psi*: 3205, 3129, 2847, 3350, 2955, 2990, 2833, and 2878. How many more samples should he take?

10.4 In Problem 10.2, assume that the population standard deviation of concrete strength is indeed 250*psi*. If the engineer is willing to accept a 5% chance that the margin of error in the mean strength is greater than or equal to 100 *psi*, estimate the number of samples he needs to collect.

10.5 Write a Maple program that solves Example 10.4.

10.6 In Problem 10.2, suppose that the population variance is not known and that the sample variance is calculated from the following preliminary samples taken by the engineer: 3205, 3129, 2847, 3350, 2955, 2990, 2833, and 2878 *psi*. If the engineer is willing to accept a 5% chance that the margin of error in the mean strength is greater than or equal to 100*psi*, estimate the number of samples he needs to collect.

"Are love and hate mutually exclusive binomial random processes? Can we love and simultaneously hate someone? Humans seem to be able to experience a myriad of concurrent emotions when dealing with others. Our middle self, or conscious self, may rationalize a relationship as productive and positive, while at the same time our lower self or unconscious may experience feelings of distrust and repulsion towards the same subject. Most of our complexes, fixations, and unhappiness derive from these unresolved conflicts between different parts of our psyche. The mathematics of human emotions has not yet been developed."

Sergio E. Serrano ("The Three Spirits. Applications of Huna to Health, Prosperity, and Personal Growth." SpiralPress, Ambler, PA, 2007.

11 EXPERIMENTS AND TESTS FOR TWO OR MORE POPULATIONS

11.1 COMPARISON OF PARAMETERS OF TWO POPULATIONS

Many engineering studies include, as part of their analyses, the investigation of possible differences in the means of two populations. As an example, we might be interested in finding whether or not the mean sales of a product at a particular region are different from that at another. Alternatively, our project may require the determination of a possible difference in energy consumption at 8:00 A.M. and that at 8:00 P.M.. To study the possible differences between the two means, we take a number of samples from the two populations X_1 and X_2, N_1 and N_2 respectively, and perform a statistical test. The procedure is an extension of the tests about the mean and the variance performed on a single population in section 6.4, Chapter 6.

Case 1: Test for the Comparison of the Means of Two Populations, the Variances $\sigma_{X_1}^2$ and $\sigma_{X_2}^2$ are Known

The objective of this test is to assess whether or not the means of two Normal random variables X_1 and X_2, μ_{X_1} and μ_{X_2} respectively, are equal when the population variances, $\sigma_{X_1}^2$ and $\sigma_{X_2}^2$, are known. The test statistic is

$$R = \frac{(\bar{X}_1 - \bar{X}_2)}{\sqrt{\dfrac{\sigma_{X_1}^2}{N_1} + \dfrac{\sigma_{X_2}^2}{N_2}}} \tag{11.1}$$

where
$\bar{X}_1$ =the sample mean of population 1 based on a sample size N_1
$\bar{X}_2$ = the sample mean of population 2 based on a sample size N_2
$\sigma_{X_1}^2$ =the population variance of a Normal random variable X_1
$\sigma_{X_2}^2$ =the population variance of a Normal random variable X_2

R follows a Standard Normal random variable. We hypothesize μ_{X_1} to be greater, equal, or less than μ_{X_2}. Given a small level of significance, α, this test answers the following 3 questions about μ_{X_1} and μ_{X_2}:

Question 1: Is $\mu_{X_1} = \mu_{X_2}$ or is $\mu_{X_1} < \mu_{X_2}$?
The format of this question calls for a left-tail test with an area of α.

We would reject the hypothesis that $\mu_{X_1}=\mu_{X_2}$, and thus accept the alternative hypothesis that $\mu_{X_1}<\mu_{X_2}$, if the calculated R statistic is less than $-Z_\alpha$, which is equal to $Z_{1-\alpha}$ because of symmetry (see Figure 6.9).

The procedure is as follows:

(1) Calculate R from equation (11.1).

(2) State the null hypothesis H_0: $\mu_{X_1}=\mu_{X_2}$, and the alternative hypothesis H_1: $\mu_{X_1}<\mu_{X_2}$.

(3) If $-Z_\alpha<R$ accept H_0. If $R<-Z_\alpha$ accept H_1.

Question 2: Is $\mu_{X_1}=\mu_{X_2}$ or is $\mu_{X_1}\neq\mu_{X_2}$?
Here we have a two-tail test, each tail with an area of magnitude $\alpha/2$. The region of acceptance is bounded by the abscissas $-Z_{1-\alpha/2}$ and $Z_{1-\alpha/2}$. We would reject the hypothesis that $\mu_{X_1}=\mu_{X_2}$ if the statistic R falls in the shaded area of Figure 6.9.

The procedure is as follows:

(1) Calculate R from equation (11.1).

(2) State the null hypothesis H_0: $\mu_{X_1}=\mu_{X_2}$, and the alternative hypothesis H_1: $\mu_{X_1}\neq\mu_{X_2}$.

(3) If $-Z_{1-\alpha/2}<R<Z_{1-\alpha/2}$ accept H_0. If $R<-Z_{1-\alpha/2}$ or $Z_{1-\alpha/2}<R$ accept H_1.

Question 3: Is $\mu_{X_1}=\mu_{X_2}$ or is $\mu_{X_2}<\mu_{X_1}$?
We have a right-tail test with an area value of α. The region of acceptance is bounded by the abscissa $Z_{1-\alpha}$. We would reject the hypothesis that $\mu_{X_1}=\mu_{X_2}$ if the statistic R falls in the shaded area of Figure 6.10.

The procedure is as follows:

(1) Calculate R from equation (11.1).

(2) State the null hypothesis H_0: $\mu_{X_1}=\mu_{X_2}$, and the alternative hypothesis H_1: $\mu_{X_2}<\mu_{X_1}$.

(3) If $R<Z_{1-\alpha}$ accept H_0. If $Z_{1-\alpha}<R$ accept H_1.

Case 2: Test for the Comparison of the Means of Two Populations, $\sigma_{X_1}^2$ and $\sigma_{X_2}^2$ are not Known, $\sigma_{X_1}^2 = \sigma_{X_2}^2$, $60 \le N_1$, and $60 \le N_2$

This test is similar to that in Case 1 in that it assesses whether or not the means of two Normal populations are equal, except that the population's variances are unknown but equal to one another, and the sample sizes are greater than 60 data values. The test statistic is

$$R = \frac{(\overline{X}_1 - \overline{X}_2)}{S_w \sqrt{\dfrac{1}{N_1} + \dfrac{1}{N_2}}} \tag{11.2}$$

The statistic S_w^2 is the pooled estimator of the common population variance $\sigma^2 = \sigma_{X_1}^2 = \sigma_{X_2}^2$ and it is given by

$$S_w^2 = \frac{(N_1 - 1)S_{X_1}^2 + (N_2 - 1)S_{X_2}^2}{N_1 + N_2 - 2} \tag{11.3}$$

where
   $S_{X_1}^2$ = the sample variance based on a sample of size N_1
   $S_{X_2}^2$ = the sample variance based on a sample of size N_2

This test answers the same questions as those of Case 1 and the procedure is the same, except that now the statistic is calculated with equations (11.2) and (11.3).

Case 3: Test for the Comparison of the Means of Two Populations, $\sigma_{X_1}^2$ and $\sigma_{X_2}^2$ are not Known, $\sigma_{X_1}^2 \ne \sigma_{X_2}^2$, $60 \le N_1$ and $60 \le N_2$

This test assesses whether or not the means of two Normal populations are equal, the population's variances are unknown and different from each other, and the sample sizes are greater than, or equal to, 60 data values. The test statistic is

$$R = \frac{(\overline{X}_1 - \overline{X}_2)}{\sqrt{\dfrac{S_{X_1}^2}{N_1} + \dfrac{S_{X_2}^2}{N_2}}} \tag{11.4}$$

This test answers the same questions as those of Case 1 and the procedure is identical, except that now the statistic is calculated with equation (11.4).

Case 4: Test for the Comparison of the Means of Two Populations, $\sigma^2_{X_1}$ and $\sigma^2_{X_2}$ are not Known, $\sigma^2_{X_1} = \sigma^2_{X_2}$, and or $N_2 < 60$

This test assesses whether or not the means of two Normal populations are equal, the population's variances are unknown but equal to one another, and the sample size of X_1, or the sample size of X_2, is less than 60 data values. The test statistic is given by equations (11.2) and (11.3). This test answers the same questions as those of Case 1 and the procedure is the same, except that now the statistic is calculated with equation (11.2), and instead of the Standard Normal variate Z we use $t_{N_1 + N_2 - 2}$, that is the Student's t density function with $N_1 + N_2 - 2$ degrees of freedom.

Case 5: Test for the Comparison of the Means of Two Populations, $\sigma^2_{X_1}$ and $\sigma^2_{X_2}$ are not Known, $\sigma^2_{X_1} \neq \sigma^2_{X_2}$, and $N_1 < 60$ or $N_2 < 60$

This test assesses whether or not the means of two Normal populations are equal, the population's variances are unknown and different from one another, and the sample size of X_1, or that of X_2, is less than 60 data values. The test statistic is given by equation (11.4). The test answers the same questions as those of Case 1 and the procedure is the same, except that now the statistic is calculated with equation (11.4), and instead of the Standard Normal variate Z we use t_M, that is the Student's t density function with M degrees of freedom given by

$$M = \frac{\left(\dfrac{S^2_{X_1}}{N_1} + \dfrac{S^2_{X_2}}{N_2} \right)^2}{\dfrac{\left(\dfrac{S^2_{X_1}}{N_1} \right)^2}{(N_1 - 1)} + \dfrac{\left(\dfrac{S^2_{X_2}}{N_2} \right)^2}{(N_2 - 1)}} \qquad (11.5)$$

Example 11.1: Comparison of Two Population Means

A machine is used to package 4-*kg* boxes of a certain chemical powder. A modification is suggested to increase the speed of operation, but there is concern that the modified settings will cause the machine to fill the boxes with less mass of chemical than before. A test is conducted to assess this possibility. Fifty boxes are filled before and after modification with the results of $X_1 = 4.091kg$, and $X_2 = 4.075kg$, where X_1 and X_2 are the filled masses of chemical before and after machine modification, respectively. Assuming that chemical mass is Normally distributed with a standard deviation of 0.05*kg*, test at the 5% level of significance the hypothesis $\mu_{X_1} = \mu_{X_2}$ versus $\mu_{X_2} < \mu_{X_1}$.

Solution

We compare means, $N_1 = N_2 = 50$, $\bar{X}_1 = 4.091 kg$, $\bar{X}_2 = 4.075 kg$, $\sigma_{X_1} = \sigma_{X_2} = 0.05 kg$, and $\alpha = 0.05$. Use Case 1, Question 3. From equation (11.1).

$$R = \frac{(\bar{X}_1 - \bar{X}_2)}{\sqrt{\dfrac{\sigma_{X_1}^2}{N_1} + \dfrac{\sigma_{X_2}^2}{N_2}}} = \frac{(4.091 - 4.075)}{\sqrt{\dfrac{0.05^2}{50} + \dfrac{0.05^2}{50}}} = 1.60$$

H_0: $\mu_{X_1} = \mu_{X_2}$, H_1: $\mu_{X_2} < \mu_{X_1}$, and from Table A.1, $Z_{1-\alpha} = Z_{0.95} = 1.645$. Since $R < 1.645$, accept H_0. At the 5% level of significance, there is evidence to believe that the mean masses prior to and after modification are equal.

Example 11.2: Comparison of Two Population Means

An experiment to determine the viscosities, V_A and V_B, of two fluids Type A and Type B, respectively, yields the results of Table 11.1. Is there any significant difference in the mean viscosities of the two types at the 5% level of significance, assuming $V_A \sim N(\mu_A, \sigma_A)$ and $V_B \sim N(\mu_B, \sigma_B)$?

Table 11.1: Viscosity Measurements of Two Fluids ($N.s/m^2$)

Type A	10.28	10.27	10.30	10.32	10.27	10.27	10.28	10.29
Type B	10.31	10.31	10.26	10.30	10.27	10.31	10.29	10.26

Solution

A comparison of mean viscosities is required. $N_1 = N_2 = 8$, $\sigma_{X_1} \neq \sigma_{X_2}$ and unknown, and $\alpha = 0.05$. Use Case 5 Question 2. We use Maple:

```
> restart: with(stats): alpha:=0.05:
  A:=[10.28,10.27,10.30,10.32,10.27,10.27,10.28,10.29]:
  B:=[10.31,10.31,10.26,10.30,10.27,10.31,10.29,10.26]:
  NA:=describe[count](A):
  NB:=describe[count](B):
  Abar:=describe[mean](A):
  Bbar:=describe[mean](B):
  SA:=describe[standarddeviation[1]](A):
  SB:=describe[standarddeviation[1]](B):
  S1:=SA^2/NA:  S2:=SB^2/NB:
  R:=(Abar-Bbar)/sqrt(S1+S2);          #Equation (11.4)
  (S1+S2)^2/(S1^2/(NA-1)+S2^2/(NB-1)): #Equation (11.5)
  M:=round(%):
  t:=statevalf[icdf,studentst[M]](1-alpha/2);
                     R := -0.3721042036
                     t := 2.160368656
  Ho: mu_A=mu_B; H1: mu_a != mu_B
> if(R<t) and (R>-t) then `Accept Ho` else `Accept H1` end if;
                        Accept Ho
```

At the 5% level of significance, there is reason to believe that the means of the two viscosities are equal.

Case 6: Test for the Comparison of the Variances of Two Normal Populations

This test compares the difference of the variances, $\sigma_{X_1}^2$ and $\sigma_{X_2}^2$, of two Normal random variables, X_1 and X_2. The test statistic is given by

$$R = \frac{S_{X_1}^2}{S_{X_2}^2} \tag{11.6}$$

where

$S_{X_1}^2$ = the sample variance of X_1 based on a sample of size N_1

$S_{X_2}^2$ = the sample variance of X_2 based on a sample of size N_2

This test answers the following questions about $\sigma_{X_1}^2$ and $\sigma_{X_2}^2$.

Question 1: Is $\sigma_{X_1}^2 = \sigma_{X_2}^2$ or is $\sigma_{X_1}^2 < \sigma_{X_2}^2$?

This is a left-tail test. As before, we present a brief summary of the steps to follow.

(1) Calculate R from equation (11.6).

(2) State the null hypothesis, $H_0: \sigma_{X_1}^2 = \sigma_{X_2}^2$, and the alternative hypothesis, $H_1: \sigma_{X_1}^2 < \sigma_{X_2}^2$.

(3) If $F_{N_1-1,N_2-1,\alpha} < R$, accept H_0. If $R < F_{N_1-1,N_2-1,\alpha}$, accept H_1

where

$F_{N_1-1,N_2-1,\alpha}$ = abscissa value of the *F distribution* with degrees of freedom $N_1 - 1$ and $N_2 - 1$ corresponding to a cumulative area of α

The F probability distribution has two parameters (i.e., degrees of freedom) denoted in general by m_1 and m_2, respectively. The parameter m_1 is also called *the number of numerator degrees of freedom*, and m_2 is called *the number of denominator degrees of freedom*. These parameters are positive integers. The probability density function of the F random variable is given by

$$f_F(f) = \frac{\Gamma\left(\dfrac{m_1}{2}+\dfrac{m_2}{2}\right)\left(\dfrac{m_1}{m_2}\right)^{\frac{m_1}{2}} f^{\frac{m_1}{2}-1}}{\Gamma\left(\dfrac{m_1}{2}\right)\Gamma\left(\dfrac{m_2}{2}\right)\left[\left(\dfrac{m_1}{m_2}\right)f+1\right]^{\frac{m_1}{2}+\frac{m_2}{2}}}, \qquad f,\, m_1,\, m_2 > 0 \quad (11.7)$$

There is an important connection between the F and the Chi-Squared random variables. If X_1 and X_2 are two independent Chi-Squared random variables, with m_1 and m_2 degrees of freedom, respectively, then the ratio of the two random variables divided by their respective degrees of freedom,

$$F = \frac{X_1/m_1}{X_2/m_2} \qquad (11.8)$$

can be shown to follow an F distribution. Figure 11.1 shows a typical graph of the F density function. It is not symmetric and it would seem necessary to tabulate the upper and lower tail critical values. However, this is not necessary because of the following useful property:

$$F_{m_1,m_2,\alpha} = \frac{1}{F_{m_2,m_1,1-\alpha}} \qquad (11.9)$$

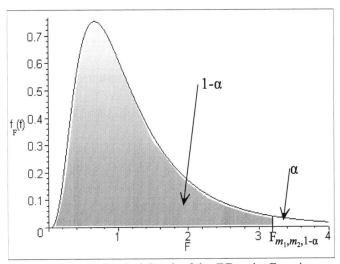

Figure 11.1: A Typical Graph of the F Density Function

Table A.4, Appendix A gives abscissa values of $F_{m_1,m_2,1-\alpha}$, for $\alpha=0.05$, various values of m_1 (in different columns of the table) and m_2 (different rows of the table). For example, $F_{6,10,0.95}=3.217$, and $F_{10,6,0.95}=4.060$. To obtain $F_{6,10,0.05}$, the abscissa value corresponding to a cumulative area of 0.05, with $m_1=6$ and $m_2=10$, we use equation (11.9): $F_{6,10,0.05}=1/F_{10,6,0.95}=1/4.060=0.246$. Tables A.5, A.6, and A.7 give abscissa values of $F_{m_1,m_2,1-\alpha}$, for $\alpha=0.025$, 0.01, and 0.005, respectively. Tables A.5 and A.7 are convenient when answering question 2, which requires tail areas of half of the level of significance. Abscissa values of F are also available in spreadsheet and computer algebra programs.

Question 2: Is $\sigma_{X_1}^2 = \sigma_{X_2}^2$ or is $\sigma_{X_1}^2 \neq \sigma_{X_2}^2$?

This is a two-tail test. Procedure:

(1) Calculate R from equation (11.6).

(2) State the null hypothesis, $H_0 : \sigma_{X_1}^2 = \sigma_{X_2}^2$, and the alternative hypothesis, $H_1 : \sigma_{X_1}^2 \neq \sigma_{X_2}^2$.

(3) $F_{N_1-1,N_2-1,\alpha/2} < R < F_{N_1-1,N_2-1,1-\alpha/2}$, accept H_0. If $R < F_{N_1-1,N_2-1,\alpha/2}$ or $F_{N_1-1,N_2-1,1-\alpha/2} < R$, accept H_1.

Question 3: Is $\sigma_{X_1}^2 = \sigma_{X_2}^2$ or is $\sigma_{X_2}^2 < \sigma_{X_1}^2$?

This is a right-tail test. Procedure:

1. Calculate R from equation (11.6).

2. State the null hypothesis, $H_0 : \sigma_{X_1}^2 = \sigma_{X_2}^2$, and the alternative hypothesis, $H_1 : \sigma_{X_2}^2 < \sigma_{X_1}^2$.

3. If $R < F_{N_1-1,N_2-1,1-\alpha}$, accept H_0. If $F_{N_1-1,N_2-1,1-\alpha} < R$, accept H_1.

Example 11.3: Comparison of Two Population Variances
In Example 11.1, the sample standard deviations of X_1 and X_2 are 0.12 and 0.08kg, respectively. The engineer suspects the variance has changed and may be different for the two groups. Assess this hypothesis at the 5% level of significance.

Solution

We are comparing variances. $N_1 = N_2 = 50$, $\bar{X}_1 = 4.091 kg$, $\bar{X}_2 = 4.075 kg$, $S_{X_1} = 0.12 kg$, $S_{X_2} = 0.08 kg$, and $\alpha = 0.05$. Use Case 6 Question 2. From equation (11.6),

$$R = \frac{S_{X_1}^2}{S_{X_2}^2} = \frac{0.12^2}{0.08^2} = 2.25$$

$H_0 : \sigma_{X_1}^2 = \sigma_{X_2}^2$; $H_1 : \sigma_{X_1}^2 \neq \sigma_{X_2}^2$. From Table A.5, interpolating, $F_{49,49,0.975} = 1.762$. From equation (11.9), $F_{49,49,0.025} = 1/F_{49,49,0.975} = 1/1.762 = 0.568$. Since $F_{49,49,0.975} < R$ we reject the null hypothesis and accept H_1. We say, at the 5% level of significance there is reason to believe that the variances of the two processes are different.

11.2 COMPARISON OF MEANS OF TWO OR MORE POPULATIONS: SINGLE FACTOR ANALYSIS OF VARIANCE (ANOVA)

Box Plots and the Logic Behind ANOVA

Methods for comparing the means of two populations are a simplification of real life situations when samples from several populations are investigated. For example, consider the experiment of investigating the effect of adding different amounts of sugar to a tank containing fermenting grain juice for the purpose of enhancing the production of ethanol for energy production. At a given date, the engineer collects several samples from a tank that received 1,000 kg of corn sugar, (called *treatment* 1), several samples from another tank that received 1,500 kg of sugar (called *treatment* 2), and several samples from a tank that received no sugar (called the *control group*). With an alcohol meter or a hydrometer the density of each sample is measured and the results are summarized in Table 11.2. The *response variable* in this experiment is the liquid density. The *explanatory variable* is the particular method (i.e., level of sugar added), which is an example of a nonnumerical variable called *a factor*. A particular value of the factor is called the *factor level* or *treatment*. For each treatment there are 8 samples. We say there are 8 *replicates* for each treatment. Since the treatments are distinguished from each other by the different levels of a single factor, the experiment is called a *single-factor experiment* or a *one-way* ANOVA (*analysis of variance*). In this chapter we introduce the basis of single-factor ANOVA. Design and analysis of multi factor experiments are covered in detail elsewhere (e.g., Rosenkrantz, 1997; Devore, 1995; Mendenhall, 1968;

Guenther, 1964; Kicks, 1964; Li, 1961). The goal of ANOVA is to study the mean of a response variable as a function of factor levels. In this sense, ANOVA is similar to regression analysis, except that the explanatory variable may be non-numeric.

On the bases of the sample means of the three treatments (see Table 11.2), one needs to decide whether the population means differ. Given the measurement errors involved in the experiment, one expects the sample means to differ. However, this does not necessarily mean that the population means are different. Even if the population mean densities were identical, the sample means most probably would differ. Hence, how does one decide whether the differences among the sample means are large enough to imply a difference among the corresponding population means? We shall answer this question with ANOVA.

Table 11.2: Comparison of Mean Density from Three Populations
(Small Amount of Within-Sample Variation)

Method Sample	1	2	3	4	5	6	7	8	Mean
Control Group, X_0	1011.5	1012.4	1010.3	1009.8	1008.8	1009.7	1010.4	1010.7	1010.5
Treatment 1, X_1	1040.2	1039.8	1024.7	1038.9	1040.5	1041.3	1042.1	1040.9	1038.6
Treatment 2, X_2	1061.0	1063.4	1058.8	1059.6	1060.7	1058.7	1059.9	1060.0	1060.3

Why the method is called analysis of variance can be seen easily by observing the data of Table 11.2. Do the data represent sufficient evidence to indicate a difference between population means? A visual analysis of the data suggests an affirmative answer. There is a clear differentiation among the sample mean densities: $X_0 = 1010.5$, $X_2 = 1038.6$, and $X_2 = 1060.3 gr/cm^3$, respectively for the Control group, Treatment 1, and Treatment 2. Additionally, there is very little variation within each population, that is the variation in the measurements within each sample is very small. In other words, the spread or variation among the sample means is so large *in comparison to within-sample variation* that we intuitively conclude that a real difference does exist among the population means.

Table 11.3: Comparison of Mean Density from Three Populations
(Large Amount of Within-Sample Variation)

Method Sample	1	2	3	4	5	6	7	8	Mean
Control Group, X_0	1040.5	980.5	1093.0	928.0	956.7	1064.3	985.8	1035.2	1010.5
Treatment 1, X_1	1013.9	1063.5	1070.2	1007.0	1019.7	1057.5	1071.9	1005.3	1038.6
Treatment 2, X_2	1085.0	1035.6	1028.7	1091.9	1079.2	1041.4	1027.0	1093.6	1060.3

How does our intuition work when a larger within-sample variation is observed? Table 11.3 shows data for the same experiment yielding identical sample means. However, in this case the variation among the sample means is not large relative to the variation within samples. Hence, it would be difficult to conclude that the samples were drawn from populations with different means.

The variations in the observations for the set of data, of Table 11.2 may be easily observed with a graph called a *box plot* (see Figure 11.2). Each box represents and summarizes the data from one population X_0, X_1, or X_2. The central line of the box indicates the *median* value of the sample. The left line indicates the *first quartile*. The right line of the box represents the *third quartile*. The two lines extending from the central box have a maximal length of 3/2 of the *interquartile* range but do not extend paste the *range* of the data. Points located outside the box represent *outliers* whose values extend beyond the extent of the previous elements.

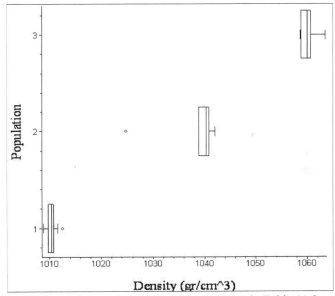

Figure 11.2: Side-by-Side Box Plot of Data in Table 11.2

Several new terms regarding variability of data need to be defined. The *median* of a data set X, also called the 50th *percentile*, is a number denoted by $\tilde{X}$ that divides the data set in half, so that at least half the data values are greater than or equal to $\tilde{X}$ and at least half the data values are less than or equal to $\tilde{X}$. From the cumulative distribution function of X, $F_X(x)$, $\tilde{X}$ may be found graphically or analytically as the number such that $F_X(\tilde{X})=0.5$. Clearly, for a symmetric probability density function,

$\tilde{X}=\overline{X}$.

The *first quartile* or the *lower quartile*, Q_1, is the number such that 25% of the data is less than or equal to it. In other words, $F_X(Q_1)=0.25$. The *third quartile* or the *upper quartile*, Q_3, is the number such that 75% of the data is less than or equal to it. In other words, $F_X(Q_3)=0.75$. The median, the first quantile, and the third quantile are the most common *quantiles* or *percentiles*. Other quantiles may be of interest (e.g., 10%, and 90% quantiles). They are defined similarly and can be determined from the empirical cumulative distribution function. The *interquartile range* is the interval of values in the data set included within the first and the third quartile. The *range* of a data set is the interval of values comprehended between the minimum value and the maximum value. Finally, an *outlier* is a data value with an unusually low or high magnitude (e.g., less than Q_1 or greater than Q_3).

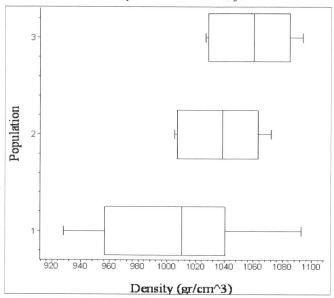

Figure 11.3: Side-by-Side Box lot of Data in Table 11.3

The above definitions characterize and summarize a data set, along with the moments and the empirical distribution function. They are conveniently summarized in a box plot, which can be produced by most computer programs at the call of an intrinsic function. Thus, Figure 11.2 represents the *side-by-side box plots* of the data in Table 11.2. It is easily seen that each of the data sets is clustered around its mean and that in all likelihood the three sets come from three different populations. In contrast, Figure 11.3 shows the side-by-side box plots of the data in Table

11.3. Although the samples' means are different, the large within-sample variation makes it difficult to conclude the sets come from different populations and may rather represent different (random) samples from a single population. Figures 11.2 and 11.3 indicate very clearly, then, what we mean by an analysis of variance. All differences in samples means are judged statistically significant or not significant by comparing them with a measure of the random variation (i.e., the variance) within the population data.

Example 11.4

Write the Maple commands needed to produce Figure 11.2.

Solution

```
> restart: with(stats): with(stats[statplots]):
  X0:=[1011.5,1012.4,1010.3,1009.8,1008.8,1009.7,1010.4,1010.7]:
  X1:=[1040.2,1039.8,1024.7,1038.9,1040.5,1041.3,1042.1,1040.9]:
  X2:=[1061.0,1063.4,1058.8,1059.6,1060.7,1058.7,1059.9,1060.0]:
  p:=boxplot(X0,X1,X2,shift=1,color=black,width=0.5,
     ytickmarks=3,labels=[`Density (gr/cm^3)`,`Population`],
     labelfont=[TIMES,ROMAN,16],
     labeldirections=[horizontal,vertical]):
  xyexchange(p);    #Show density in the abscissa
```

Case7: Test for the Comparison of the Means of More than Two Normal Populations

To explain the bases of ANOVA, we continue with the example of the previous section concerning the effect of sugar on liquid density. The results are reported in Table 11.2. We have $M=3$ factor levels or treatments, each containing $N=8$ samples. For each level, we calculate the sample mean and the sample variance as summarized in Table 11.4.

Table 11.4: Density Statistics from Three Treatments in Table 11.2

Statistic / Method	Sample Mean, $\bar{X}_i$ (gr/cm^3)	Sample Variance, S_i^2 $(gr/cm^3)^2$
Control Group, X_0	1010.5	1.2429
Treatment 1, X_1	1038.6	32.2457
Treatment 2, X_2	1060.3	2.2570

Intuitively, if we are interested in testing the equality of the population means, that is $\mu_{X_0} = \mu_{X_1} = \mu_{X_2}$ we might want to run all possible comparisons of population means. In other words, if we assume that the three distributions are approximately Normal, with a common variance

σ_X^2, we could run 3 versions of Case 4, section 11.1 of this chapter, each comparing two-population means. The 3 possible null hypotheses would be $\mu_{X_0} = \mu_{X_1}$, $\mu_{X_0} = \mu_{X_2}$, and $\mu_{X_1} = \mu_{X_2}$.

The first disadvantage of this approach is that it is tedious and time consuming. As the number of treatments increases, the number of combinations among them dramatically increases the required number of tests. However, the most important disadvantage of running multiple tests to compare means is that the probability of falsely rejecting at least one of the hypotheses increases as the number of tests increases. Although we may have the probability of Type I error fixed at $\alpha = 0.05$ (i.e., the level of significance) for each individual test, the probability of falsely rejecting at least one of these tests is greater than 0.05. In other words, the combined probability of Type I error for the set of 3 hypotheses would be greater than the 0.05 set for each individual test. Thus, we need a single test of the hypothesis "all three populations are equal," which is less tedious that the individual tests and which may be performed with the specified probability of Type I error.

Let us consider an extension of equation (11.3) applied to $M=3$ populations:

$$S_w^2 = \frac{(N_0-1)S_{X_0}^2 + (N_1-1)S_{X_1}^2 + (N_2-1)S_{X_2}^2}{N_0+N_1+N_2-3} \tag{11.10}$$

where
S_w^2=an estimator of the common variance of X_0, X_1, and X_2
N_0, N_1, N_2=sample sizes of treatments 0, 1, and 2 respectively
$S_{X_0}^2, S_{X_1}^2, S_{X_2}^2$=sample variances of X_0, X_1, and X_2 respectively

S_w^2 is used as an estimator of the common variance, σ_X^2, and measures the variability of the observations "within" the three populations. A generalization of equation (11.10) for several treatments is

$$S_w^2 = \frac{\displaystyle\sum_{i=0}^{M-1}(N_i-1)S_{X_i}^2}{\left(\displaystyle\sum_{i=0}^{M-1}N_i\right)-M}, \qquad M \geq 2 \tag{11.11}$$

where
S_w^2=an estimator of the common variance of X_i, $i=1, 2,..., M$
N_i=the sample size of treatment i
$S_{X_i}^2$=the sample variance of treatment i

M=number of treatments in the experiment

Now we consider a measure of the variability "between" or "among" the population means. If the null hypothesis $\mu_{X_0}=\mu_{X_1}=\mu_{X_2}$ is true, the three populations are identical with mean μ_X and variance σ_X^2. Drawing single samples from the three populations is equivalent to drawing three different samples from the same population. We need to decide the kind of variation to be expected for these sample means. If the variation is too great, the hypothesis that $\mu_{X_0}=\mu_{X_1}=\mu_{X_2}$ would be rejected. To assess the variation from sample mean to sample mean, we need to know the probability density function of the sample mean of 8 observations. From section 6.1, Chapter 6, recall that the density function of the sample mean, $\overline{X}$, is approximately Normal if the sample size is large with a mean equal to the same population mean $E\{\overline{X}\}=\mu_X$, and a variance equal to $Var(\overline{X})=\sigma_X^2/N_i=\sigma_X^2/8$. An estimate of the latter may be obtained by computing the sample variance of the sample means of the three treatments (see Rosenkrantz, 1997 for proof). In other words, Table 11.4 has M=3 sample means, $\overline{X}_0$, $\overline{X}_1$, and $\overline{X}_2$, one for each treatment, respectively. Using equation (6.2) we can use these three numbers to calculate a *sample variance of the sample means*. Thus, for M treatments in general,

$$\hat{\sigma}_{\overline{X}}^2=\frac{1}{M-1}\sum_{i=0}^{M-1}\left(\overline{X}_i-\hat{\mu}_X\right)^2 \tag{11.12}$$

where
$\hat{\sigma}_{\overline{X}}^2$=the sample variance of the sample means, that is an estimate of $Var(\overline{X})$, with the "hat" denoting "estimate"
$\overline{X}_i$=the sample mean of treatment i from equation (6.1) based on N_i observations
$\hat{\mu}_X$=sample mean of X from equation (6.1) based on all observations
M=number of treatments in the experiment

Since $\hat{\sigma}_{\overline{X}}^2$ is an estimate of $Var(\overline{X})=\sigma_X^2/N_i$, then we can define the quantity

$$S_B^2=N_i\hat{\sigma}_{\overline{X}}^2 \tag{11.13}$$

where
S_B^2=an estimate of the population variance, σ_X^2

The subscript B in equation (11.13) designates a measure of the variability among the sample means of the M populations. Equations

(11.12) and (11.13) may be generalized for the case of unequal sample sizes among the treatments:

$$S_B^2 = \frac{\sum_{i=0}^{M-1} N_i \bar{X}_i^2 - \frac{\left(\sum_{i=0}^{M-1} N_i \bar{X}_i\right)^2}{\sum_{i=0}^{M-1} N_i}}{M-1} \qquad (11.14)$$

We now have two estimates of the population variance σ_X^2, S_w^2 and S_B^2, respectively. We can use the ratio

$$R = \frac{S_B^2}{S_w^2} \qquad (11.15)$$

as a test statistic of the null hypothesis. I we repeat this experiment over and over again we would find that this ratio follows an F distribution with degrees of freedom $m_1 = M-1$ and $m_2 = \Sigma(N_i) - M$, respectively (see Guenther, 1964 for proof). If the null hypothesis is true, both S_w^2 and S_B^2 estimate σ_X^2 and R is expected to assume a value near $R=1$. When the null hypothesis is false, S_B^2 tends to be greater than S_w^2 due to the differences among the population means. Then we reject the null hypothesis in the upper tail of the distribution of $R = S_B^2 / S_w^2$. Given a significance level α equal to 0.01 or 0.05 (i.e., the probability of Type I error) we locate the rejection region for the right-tailed test using Tables A.4 through A.7, Appendix A.

In summary, the procedure for an ANOVA test about the equality of M populations means is as follows.

(1) Calculate the test statistic from equation (11.15), with S_w^2 from equation (11.11) and S_B^2 from equation (11.14).

(2) State the null hypothesis $H_0 : \mu_0 = \mu_1 = \mu_2 = \ldots = \mu_M$, and the alternative hypothesis H_1 : at least one of the population means is different from the others.

(3) If $R < F_{m_1, m_2, 1-\alpha}$, accept H_0. If $F_{m_1, m_2, 1-\alpha} < R$, accept H_1,

where

$F_{m_1, m_2, 1-\alpha}$ = abscissa value of the F distribution with m_1 and m_2 degrees of freedom and cumulative area of $1-\alpha$

$$m_1 = M-1, \quad m_2 = \left(\sum_{i=0}^{M-1} N_i\right) - M$$

N_i = sample size of treatment i, i=0, 1, 2, ..., M
α = level of significance

Example 11.5: ANOVA Test for the Effect of Sugar on Density

For the experiment results summarized in Tables 11.2 and 11.4 perform an ANOVA test at the 5% level of significance.

Solution

α=0.05. From the said tables M=3, N_0=N_1=N_2=8 samples. We follow the ANOVA procedure described above. From equation (11.11),

$$S_w^2 = \frac{\sum_{i=0}^{M-1}(N_i-1)S_{X_i}^2}{\left(\sum_{i=0}^{M-1}N_i\right)-M} = \frac{(8-1)\times1.2429+(8-1)\times32.2457+(8-1)\times2.2570}{8+8+8-3}$$

$$= 11.9152\left(\frac{gr}{cm^3}\right)^2$$

From equation (11.14),

$$S_B^2 = \frac{\sum_{i=0}^{M-1}N_i\overline{X}_i^2 - \frac{\left(\sum_{i=0}^{M-1}N_i\overline{X}_i\right)^2}{\sum_{i=0}^{M-1}N_i}}{M-1} = \frac{8\times1010.45^2+8\times1038.55^2+8\times1060.26^2}{3-1}$$

$$- \frac{\frac{(8\times1010.45+8\times1038.55+8\times1060.26)^2}{8+8+8}}{3-1}$$

$$= 4989.2936\left(\frac{gr}{cm^3}\right)^2$$

From equation (11.15),

$$R = \frac{S_B^2}{S_w^2} = \frac{4989.7650}{11.9152} = 418.7738$$

$H_0: \mu_0 = \mu_1 = \mu_2$. H_1: at least one of the three populations is different from the others. $m_1 = M - 1 = 3 - 1 = 2$. $m_2 = N_0 + N_1 + N_2 - M = 8 + 8 + 8 - 3 = 21$. From Table A.4, $F_{2,21,0.95} = 3.467$. Since $F_{2,21,0.95} < R$, we accept H_1. We say, at the 5% level of significance, there is reason to believe that at least one of the treatment means is different from the others. Although we do not know exactly where these differences lie, it is rather obvious from the data in Table 11.2, and from the side-by-side box plots of Figure 11.2, that sugar has a marked effect on liquid density and the resulting ethanol concentration.

Example 11.6: Computer Program of an ANOVA Test

Write a Maple program that implements an ANOVA test for Example 11.5.

Solution

The following Maple worksheet reads the data, programs equations (11.11) and (11.15), and executes steps (1) through (3) for an ANOVA test described in this section.

```
[ Initialize and enter data as a series of arrays and estimate sample size.
[ > restart: with(stats): M:=3: alpha:=0.05:
    X[1]:=[1011.5,1012.4,1010.3,1009.8,1008.8,1009.7,1010.4,1010.7]:
    X[2]:=[1040.2,1039.8,1024.7,1038.9,1040.5,1041.3,1042.1,1040.9]:
    X[3]:=[1061.0,1063.4,1058.8,1059.6,1060.7,1058.7,1059.9,1060.0]:
    for i from 1 to M do
        N[i]:=describe[count](X[i]);
    end do:
[ Calculate sample means.
[ > for i from 1 to M do
        Xbar[i]:=describe[mean](X[i]):
    end do:
[ Calculate the sample variances.
[ > for i from 1 to M do
        SX2[i]:=describe[variance[1]](X[i]):
    end do:
[ Equation (11.11).
[ > Sw2:=sum((N[j]-1)*SX2[j],j=1..M)/(sum(N[j],j=1..M)-M):
[ Equation (11.14).
[ > SB2:=( sum(N[j]*Xbar[j]^2,j=1..M)
            -sum(N[j]*Xbar[j],j=1..M)^2/sum(N[j],j=1..M))/(M-1):
[ Equation (11.15).
[ > R:=SB2/Sw2:
[ Ho: mu0=mu1=mu2. H1: at least one population mean is different from the others.
[ Conmpare F with R and make a decision.
[ > m1:=M-1: m2:=sum(N[j],j=1..M)-M:
    F:=statevalf[icdf,fratio[m1,m2]](1-alpha):
    if(R<F) then `Accept Ho` else `accept H1` end if;
                        accept H1
```

Example 11.7: Maple Intrinsic Functions for ANOVA

Modify the program in Example 11.6 t to use Maple ANOVA functions, instead of programming equations (11.11) and (11.15).

Solution

```
  Initialize, enter sample data for treatments 0, 1, and 2, and form a data set composed of the 3
  treatments.
> restart: with(stats): alpha:=0.05:
  X0:=[1011.5,1012.4,1010.3,1009.8,1008.8,1009.7,1010.4,1010.7]:
  X1:=[1040.2,1039.8,1024.7,1038.9,1040.5,1041.3,1042.1,1040.9]:
  X2:=[1061.0,1063.4,1058.8,1059.6,1060.7,1058.7,1059.9,1060.0]:
  data:=[X0,X1,X2]:
  Ho: mu0 = mu1 = mu 2. H1: at least one of the means is different.
  Use the anova[ ] ( ) [2] function to calculate the m1 and m2 degrees of freedom, and the R statistic.
> Ratio:=anova[oneway](data)[2];
                        Ratio := [ 2, 21, 418.77387609, 1. ]
  Define m1, m2, and R from the ANOVA output vector Ratio.
> m1:=Ratio[1]: m2:=Ratio[2]: R:=Ratio[3]:
  Find the value of the abscissa of the F distribution with m1, m2 degrees of freedom and 1-alpha
  cumulative area.
> F:=statevalf[icdf,fratio[m1,m2]](1-alpha):
  Make a decision.
> if (R<F) then `Accept Ho` else `Accept H1` end if;
                             Accept H1
```

The output of the anova[ ]( )[2] function is a vector that contains much more information than displayed, such as the variance components for each treatment. For consistency with our previous description we only display the second slot, [2], and we name it Ratio. This Ratio vector contains, from left to right, the degrees of freedom m_1 and m_2, the R statistic, and the area under the F density function corresponding to the R abscissa. This area is 1 in this example, indicating that the calculated R is so large that it corresponds to the entire area under the F density function. Since our rejection area is 0.95 (or $1-\alpha=1-0.05=0.95$), we could immediately state that the area corresponding to R, 1, is greater than our critical area and thus reject the null hypothesis. However, for consistency with our description and with Example 11.5 we compare abscissas instead of areas.

Example 11.8: A Test for the Effect of Voltage on Components Life.

The life of an electronic component was to be studied under five different operating voltages. Ten different components were randomly assigned to each of the five operating voltages and the life interval of each component was recorded. The results are summarized in Table 11.5. Using Maple, perform an ANOVA test to investigate if the mean voltages are equal at the 5% level of significance.

Table 11.5: Life Duration Statistics for Experiment in Example 11.8

Statistics Treatment	Sample Size, N_i	Sample Mean, $\bar{L}_i$ (year)	Sample Variance, $S_{L_i}^2$ (year)^2
Voltage V_1	10	3.2	0.46
Voltage V_2	10	3.8	0.51
Voltage V_3	10	4.1	0.39
Voltage V_4	10	4.0	0.20
Voltage V_5	10	3.7	0.28

Solution

Here we have a situation where the data values themselves are not available. We modify the program in Example 11.5 $\alpha = 0.05$, $M = 5$, $N_1 = N_2 = N_3 = N_4 = N_5 = 10$.

```
⌊ Enter data; store sample sizes, sample means and sample variances in appropriate vectors.
⌈ > restart: with(stats): alpha:=0.05: M:=5:
    N:=[10,10,10,10,10]:
    Vbar:=[3.2,3.8,4.1,4.0,3.7]:
⌊   S2:=[0.46,0.51,0.39,0.2,0.28]:
⌈ Ho:=mu1=mu2=mu3=mu4=mu5. H1: at least one of the means is different.
⌊ Program equations (11.14), (11.11), and (11.15). Read F, compare with R, and make a decision.
⌈ > SB2:=(sum(N[i]*Vbar[i]^2,i=1..M)-(sum(N[i]*Vbar[i],i=1..M)^2)
                                     /sum(N[i],i=1..M))/(M-1):
    Sw2:=sum((N[i]-1)*S2[i],i=1..M)/(sum(N[i],i=1..M)-M):
    R:=SB2/Sw2;
    m1:=M-1: m2:=sum(N[i],i=1..M)-M:
    F:=statevalf[icdf,fratio[m1,m2]](1-alpha);
                    R := 3.342391304
                    F := 2.578739184
⌈ > if(R<F) then `Accept Ho` else `Accept H1` end if;
                       Accept H1
```

At the 5% level of significance, there is reason to believe that at least one of the population means is different from the others. The test cannot identify which one may be, but it suggests a way to further study and experimentation to pinpoint the treatment with a different mean.

PROBLEMS

11.1 A Method for determining the percentage of iron in a mixed fertilizer is available and long experience with this method shows that its determinations have a Normal probability density function with a standard deviation of 0.12%. A company producing a certain type of fertilizer wishes to compare the findings of its laboratory with that of a government laboratory. The results are shown in Table 11.6. Test at the 5% level of significance the hypothesis that both labs do equivalent analysis against the hypothesis that the government lab has a downward bias relative to the company lab.

Table 11.6: Percentages of Iron in Fertilizer

Company Lab	8.84	8.86	9.16
Government Lab	8.78	8.96	8.62

11.2 Two methods were used in a study of the latent heat of fusion of ice. Both Method 1 (an electrical Method) and Method 2 (a method of mixtures) were conducted with the specimens cooled to $-0.72^\circ C$. Table 11.7 shows the change in total heat from $-0.72^\circ C$ to water at $0^\circ C$ in *cal/gr*. Assuming the determinations are Normal, test at the 5% level of significance the hypothesis $\mu_{X_1} = \mu_{X_2}$ versus $\mu_{X_1} \neq \mu_{X_2}$, where X_1 and X_2 represent the heat of fusion determined respectively by methods 1 and 2. Assume the variances are equal.

Table 11.7: Latent Heat of Fusion of Water in (*cal/gr*) for Problem 11.2

Method 1	79.98	80.03	80.02	80.04	80.04	80.00	80.02	79.97
	80.02	80.04	80.05	80.03	80.03			
Method 2	80.02	79.97	79.94	80.03	79.98	79.95	79.97	79.97

11.3 In Problem 11.2, assess whether or not the variances are different at the 5% level of significance.

11.4 For the experiment results summarized in Table 11.3 perform an ANOVA test at the 5% level of significance. Indicate the equations applied and use a hand calculator.

11.5 Solve Problem 11.4 using Maple.

Table 11.8: Corrosion Readings for Experiment of Problem 11.6

Paint 1	10.1	11.4	12.1	10.8	11.1	10.7	11.5
Paint 2	13.4	12.9	13.3	13.1	13.2	13.6	12.8
Paint 3	12.7	11.9	12.5	12.3	12.4	12.2	12.6

11.6 An experiment was conducted to determine the effect of different paints on corrosion pipes. A long pipe was cut into 21 equal segments, which were randomly assigned to one of the three paints so that each paint type (treatment) would be used on 7 segments. The segments were painted and allowed to weather for a period of one year. Table 11.8 shows the resulting corrosion readings. (1) Produce side-by-side box plots of the three treatments and interpret them. (2) At the 1% level of significance, is there any evidence to indicate a difference in the mean levels of corrosion of the paints?

11.7 Dissolved oxygen samples were taken at four locations of a lake: north edge, south edge, east side and center of the lake respectively. Table 11.9 shows the results of the experiment. At the 5% level of significance, is there sufficient evidence to indicate a difference in the mean dissolved oxygen at the four locations?

Table 11.9: Dissolved Oxygen Statistics for Problem 11.7

Location Statistics	Sample Size, N_i	Sample Mean, $\bar{O}_i$ (mg/L)	Sample Variance, $S_{O_i}^2$ (mg/L)^2
North	6	4.66	0.05
South	8	6.42	0.04
East	4	6.50	0.03
Center	10	5.40	0.12

12 STOCHASTIC PROCESSES

12.1   THE CONCEPT OF A STOCHASTIC PROCESS

Until now, we have studied engineering systems affected by one or more independent random variables. Recall that in Chapters 2 and 3 we characterized a random variable as a set of outcomes from an experiment that assigns each outcome to a real value according to certain rule. If after repeating the experiment many times the resulting values are independent from each other, we described the ensemble of values a random variable characterized by its probability density function. This sequential independence among the values sampled from a random variable may be more precisely stated by a zero serial correlation function. In Chapter 4, when we generated long sequences of random numbers of a specified distribution for a Monte Carlo simulation model, we required the serial correlation function of the numbers to be as close as possible to zero. This restriction assured that the random numbers came from a single random variable. In Chapter 5 we studied cases of systems affected by more than one random variable. The concept of dependence among the different random variables was mathematically expressed as the cross correlation, the covariance, and the correlation coefficient.

In many situations of practical importance engineering systems may be subject to an input, a parameter, or an output which behaves as a sinuous function of time, or an erratic function of space. Intuitively, we may be tempted to characterize this signal as a random variable. However, its seemingly erratic (i.e., random) nature is accompanied by a strong dependence with respect to time or with respect to space. In other words, in the underlying experiment that generates the signal, successive outcomes at different times of observation are *not independent*. On the contrary, successive observation in time (or in space) of the signal indicates a strong "memory" or dependence on time (or space). This dependence or serial correlation on time or space suggests that a family of random variables may describe the signal. For example, consider the time evolution of the price of a company stock as reported by the Standard & Poor Stock Report over a decade. Depending on the company, one may note an erratic fluctuation in price. This fluctuation may be observed in the sinuous shape of the curve and the unpredicted fast jumps up or down in the price. However, there seems to be a strong time dependence. This may be characterized by a continuous growth over a long term, or even a seasonal periodic variability in the mean price. If we observe the price signal for a period of one day, at intervals of 1 minute, the value of the share fluctuates in an erratic manner. However, the price at a given minute may partially depend on the price one minute before plus other unpredictable (random) factors. The dependency of the price on prior realizations may be measured by the serial correlation. This in theory suggests that a stock price may be represented by a family of random variables, one for every minute. The density function of each random variable may or may not be the same for each minute of the day and its parameters are correlated with those at other times.

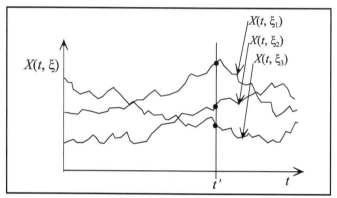

Figure 12.1: Samples of a Stochastic Process

With the preceding notions, let us revisit the concept of a probabilistic experiment described in Chapter 2. We are given an experiment, **E**, specified by its possible outcomes, ξ. The outcomes constitute the sample space, S, with certain subsets of S called events, and with the probabilities assigned to each event. Now let us assume that to each experimental outcome we assign a time function, $X(t, \xi)$, according to a certain rule. In this manner, we have created a family of functions called a *stochastic process* (or a *random process*). When we defined a random variable in Chapter 2, we wrote it as $X(\xi)$ to emphasize that the random variable depends on the random outcomes, ξ, and later we simply denoted it as X. Similarly, a stochastic process is a function of two variables, t and ξ. The domain of ξ is the set S and the domain of t is the set of real numbers (i.e., the time axis).

For a specific outcome of the experiment, ξ_1, $X(t, \xi_1)$ is a single time function (see Figure 12.1). For a specific time, t', $X(t', \xi)$ is a random variable whose outcomes are the intersections between $X(t, \xi)$ and the line $t = t'$. For fixed time, t', and a fixed outcome ξ_1, $X(t', \xi_1)$ is a number. For simplicity, we will represent a stochastic process as $X(t)$, omitting ξ.

Some fundamental properties of random processes may be summarized as follows:

1. For any value of $t=t'$, $X(t')$ is a random variable satisfying the rules of probability described in Chapter 3 (see properties 1 and 2 below equation (3.26)).

2. Two stochastic processes $X(t)$ and $Y(t)$ are said to be equal if their respective time functions are identical for any outcome ξ_i.

3. Operations with stochastic processes such as $X(t)+Y(t)$, $X(t)Y(t)$, integration, differentiation, etc. involving one or more stochastic processes are defined as operations on their time functions.

4. Some stochastic processes are composed of complex curves. For example, if $X(t)$ represents the irregular motion of a sediment particle on a river bed (Figure 12.1), a specific outcome is the trajectory of a particle, such as $X(t, \xi_1)$. This is an irregular curve that cannot be described by a formula. Furthermore, if a specific outcome of $X(t)$ is known at $t<t'$, one cannot predict its future values.

5. Some stochastic processes are composed of very smooth regular curves. For example if seasonal recharge to an aquifer is characterized as $X(t)=A+B\cos(C\pi t)$, where A, B, and C are random variables, then individual outcomes of this process are very regular curves. For a given year (i.e., a given outcome of A, B, and C) the time function, $X(t)$, is a smooth periodic curve. In this case if a specific outcome, $X(t)$, is known at $t<t'$, one may predict its future values.

We invite the reader to reread Example 4.14, Chapter 4. Many of the concepts on stochastic processes were hinted then and we believe that the preceding discussion will clarify some points. Next, we examine a variation of that example.

Example 12.1

A stochastic process is given by $U(t)=V+We^{-Kt}$, where $0 \leq t < \infty$ is time, K is a decay parameter expressed as a random variable such that $K \sim N(0.2, 0.05^2)$, W is a scale parameter such that $W \sim N(10.0, 1.0^2)$, and V is a minimum random variable such that $V \sim N(1.0, 0.1^2)$. V, W, and K are independent random variables. (1) Using 20 samples of V, W, and K plot 20 corresponding samples of $U(t)$. (2) Generate 1000 realizations of

$U(t=20)$. Plot the frequency histogram (i.e., the empirical density function) of the samples.

Solution

(1) Start the statistics module and enter data.

```
> restart: with(stats): with(stats[statplots]):
  N:=20: mu_v:=1: mu_k:=0.2: mu_w:=10:
  sigma_v:=0.1: sigma_k:=0.05: sigma_w:=1.:
```
Generate 20 samples of V, K, and W.
```
> v_data:=[random[normald[mu_v,sigma_v]](N)]:
  k_data:=[random[normald[mu_k,sigma_k]](N)]:
  w_data:=[random[normald[mu_w,sigma_w]](N)]:
```
Generate 20 time curves of U, store them in the vector $g[\ ]$, and display them on a graph.
```
> for i from 1 to N do
      g[i]:=plot(v_data[i]+w_data[i]*exp(-k_data[i]*t),
          t=0..30,color=black,thickness=2),
  end do:
  plots[display](seq( g[i],i=1..N),labels=['t','U(t)'],
      labelfont=[TIMES,ROMAN,16] );
```

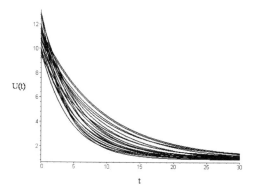

(2) $U(t=20)$ is a random variable. Generate 1000 realizations of U at $t=20$ and plot the histogram.
```
> N:=1000: v_data:=[random[normald[mu_v,sigma_v]](N)]:
  k_data:=[random[normald[mu_k,sigma_k]](N)]:
  w_data:=[random[normald[mu_w,sigma_w]](N)]:
  UT:=[seq(v_data[i]+w_data[i]*exp(-k_data[i]*20),i=1..N)]:
  histogram(UT,color=gray,labels=['U(t=20)','fU'],
      labelfont=[TIMES,ROMAN,16],thickness=2);
```

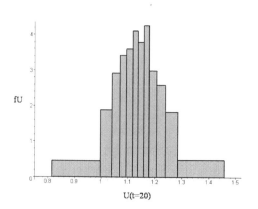

For each outcome of *V, W,* and *K, U(t)* is a smooth exponentially decaying curve. The frequency histogram is obtained from the intersection of 1000 curves of *U* with the line *t*=20. Further analysis would be required to determine which probability density function the random variable *U(t=20)* follows. Furthermore, *U(t=10)* or at any other time might follow a different density function.

Example 12.2

A stochastic process is given by $g_i = g_{i-1} + Z_i$, where *i*=1, 2, 3,..., *N* is time at fixed discrete intervals, $g_0 = 0$, and $Z_i \sim N(0, 1)$. Generate and plot three samples of the process for a period of *N*=1000 units of time.

Solution

```
[ > restart: with(stats): N:=1000:
[ Define a Maple procedure to generate one sample (time curve) of g(t).
[ > sample:=proc(k,n)
    global G: local Z,g,i,data:
    Z:=[random[normald](n)]:
    g[0]:=0:
    for i from 1 to n do g[i]:=g[i-1]+Z[i] end do:
    data:=[seq([i,g[i]],i=0..n)]:
    G[k]:=plot(data,color=black,thickness=2):
    end:
[ Generate 3 sample curves and display them.
[ > Ns:=3: plots[display](seq(sample(i,N),i=1..Ns),
    labels=[`t`,`g(t)`],labelfont=[TIMES,ROMAN,16]);
```

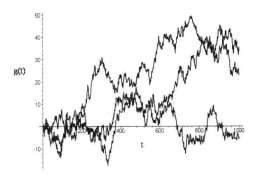

In this case the value of the process at any given time equals that at the previous time step plus a Standard Normal random number. This is an example of the so-called *auto-regressive* processes, which take on values at fixed time intervals that depend on the previous time steps. These processes commonly appear in series of engineering signals sampled at fixed discrete times. Each sample of the process is a complex erratic function of time. To observe a sample of the process we need $N=1000$ Standard Gaussian random numbers. The program defines a Maple procedure proc( ), called "sample," which generates each time curve and stores in coordinates $[t, g(t)]$ for plotting.

12.2 FIRST AND SECOND-ORDER STATISTICS

Density Function and Cumulative Distribution Function

Consider a real stochastic process $X(t)$. For a specific t, $X(t)$ is a random variable. The cumulative distribution function of this random variable will depend on t, in general. As a generalization of equation (3.28),

$$F_X(x, t)=P(X(t){\leq}x) \tag{12.1}$$

where

$F_X(x, t)$=cumulative distribution function of the stochastic process $X(t)$, also called the *first-order distribution* of $X(t)$

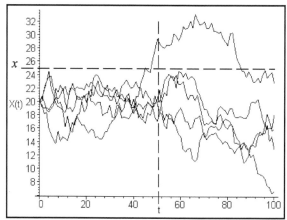

Figure 12.2: Cumulative Distribution, $F_X(x, t)$

Referring to Figure 12.2, for a given set of two real numbers x and t, the function $F_X(x, t)$ is the probability of the event $(X(t){\leq}x)$ consisting of the outcomes ξ such that, at a specified t, the various time curves of the process $X(t)$ do not exceed the given number x. From equation (3.29), the probability density function of the process $X(t)$ is obtained by differentiating the cumulative distribution function:

$$f_X(x, t) = \frac{\partial F_X(x, t)}{\partial x} \qquad (12.2)$$

where
   $f_X(x, t)$=probability density function of the process $X(t)$, also called
      the *first order density* of $X(t)$

Hence, we see that the density function and the cumulative distribution function of a stochastic process are defined in similar terms as those of random variables. A fundamental difference between random variables and random processes is that the latter have the density and the cumulative distribution defined as functions of time.

If N repetitive samples of a stochastic process are obtained via Monte Carlo simulations (see Examples 12.1, and 12.2, and Problems 12.1, and 12.2), then at a given time t and a given x, we may count the number of intercepts of $X(t)$, n, such that the number x is not exceeded (see Figure 12.2). Thus, a point on the *empirical* cumulative distribution function is approximated as $F_X(x, t) \approx n/N$.

Now, given two fixed times t_1 and t_2, then $X(t_1)$ and $X(t_2)$ are two random variables. Their joint cumulative distribution function depends on t_1 and t_2, in general. As a generalization of equation (5.11), we may write

$$F_X(x_1, x_2; t_1, t_2) = P(X(t_1) \le x_1, X(t_2) \le x_2) \qquad (12.3)$$

where
   $F_X(x_1, x_2; t_1, t_2)$=join cumulative distribution function of $X(t_1)$ and
      $X(t_2)$, called the *second-order distribution* of the process $X(t)$

The corresponding join density function (called the *second order density*) may be obtained by differentiating the join cumulative with respect to x_1 and x_2. In other words,

$$f_X(x_1, x_2; t_1, t_2) = \frac{\partial^2 F_X(x_1, x_2; t_1, t_2)}{\partial x_1 \partial x_2} \qquad (12.4)$$

Conditional density functions and higher-order densities may be defined similarly. We see that the complete definition of a stochastic process with the i-th order distribution, i=1, 2, ..., n, is difficult to obtain in many applications. In practical situations, the best one can obtain are the mean and the autocorrelation function.

Mean, and Correlation Functions

The mean of a stochastic process $X(t)$ is defined as a generalization of equation (3.30) as

$$E\{X(t)\}=\mu_X(t)=\int_{-\infty}^{\infty}xf_X(x,\ t)dx \qquad (12.5)$$

where

$\mu_X(t)$=time-dependent mean of the process $X(t)$

The autocovariance of the random variables $X(t_1)$ and $X(t_2)$ may be conceived as a generalization of equation (5.18) as

$$C_X(t_1,\ t_2)=E\{[X(t_1)-\mu_X(t_1)][X(t_2)-\mu_X(t_2)]\} \qquad (12.6)$$

where

$C_X(t_1,\ t_2)$=*autocovariance* of the process $X(t)$ at times t_1 and t_2

Solving the product and using the properties of expectation (see properties 1 through 4 below equation (3.5)), equation (12.6) reduces to,

$$C_X(t_1,\ t_2)=R_X(t_1,\ t_2)-\mu_X(t_1)\mu_X(t_2) \qquad (12.7)$$

where

$R_X(t_1,\ t_2)=E\{X(t_1)X(t_2)\}$=the *autocorrelation function* of $X(t)$ at times t_1 and t_2

The autocorrelation function is the joint moment of the random variables $X(t_1)$ and $X(t_2)$ (see equation (5.19):

$$R_X(t_1,\ t_2)=E\{X(t_1)X(t_2)\}=\int_{-\infty}^{\infty}\int_{-\infty}^{\infty}x_1x_2\ f_X(x_1,\ x_2;\ t_1,\ t_2)dx_1dx_2 \qquad (12.8)$$

If $t_1=t_2$, then equation (12.7) reduces to

$$\sigma_X^2(t)=C_X(t,\ t)=R_X(t,\ t)-\mu_X^2(t) \qquad (12.9)$$

where

$\sigma_X^2(t)$=time-dependent variance of the process $X(t)$

The *autocorrelation coefficient* of the random variables $X(t_1)$ and $X(t_2)$ is defined as their covariance divided by the product of their standard deviations at t_1 and t_2 (see equation (5.20):

$$\rho_X(t_1,\ t_2)=\frac{C_X(t_1,\ t_2)}{\sigma_X(t_1)\sigma_X(t_2)} \tag{12.10}$$

where

$\rho_X(t_1,\ t_2)$=*autocorrelation coefficient* of $X(t)$ at times t_1 and t_2

If $t_1=t_2=t$, and using equation (12.9), then equation (12.10) becomes $\rho_X(t,\ t)=1$.

We see that the first and second-order statistics of a stochastic process are now functions of the observation times. The autocorrelation, autocovariance, and autocorrelation coefficient functions are measures of the serial linear relationship in the random process with time. The autocorrelation coefficient has the same meaning discussed in Chapter 5 (see properties 1 through 5 below equation (5.20)), except that it is now a function of the observation times.

Example 12.3: Analytical Derivation of First and Second-Order Statistics

Consider a simple stochastic process given as $X(t)=At+b$, where A and B are two independent random variables. Derive the following expressions of $X(t)$: (1) the mean, (2) the autocorrelation function, (3) the autocovariance function, (4) the variance, and (5) the autocorrelation coefficient.

Solution

This process consists of a family of straight lines. We apply the properties of expectation (see below equation (3.5), Chapter 3), and the concepts of analytical derivation of second-order statistics introduced in Chapter 4, section 4.4.

(1) Taking expectations on both sides of the equation, the mean of $X(t)$ is given as

$$\mu_X(t)=E\{X(t)\}=E\{A\}t+E\{B\}=\mu_A t+\mu_B$$

where μ_A and μ_B are the means of A and B, respectively.

(2) From equation (12.8) the autocorrelation function at two times t_1 and t_2 is

$$R_X(t_1,\ t_2)=E\{[At_1+B][At_2+B]\}=E\{A^2\}t_1t_2+E\{AB\}(t_1+t_2)+E\{B^2\}$$

Since A and B are independent, then (see property 4 below equation (3.5))

$$R_X(t_1,\ t_2)=E\{A^2\}t_1t_2+\mu_A\mu_B(t_1+t_2)+E\{B^2\}$$

(3) From equation (12.7), the autocovariance function is given as

$$C_X(t_1,\ t_2)=R_X(t_1,\ t_2)-\mu_X(t_1)\mu_X(t_2)$$

$$=E\{A^2\}\ t_1t_2+\mu_A\mu_B(t_1+t_2)+E\{B^2\}-[\mu_At_1+\mu_B][\mu_At_2+\mu_B]$$

$$=\ E\{A^2\}t_1t_2-\mu_A^2t_1t_2+E\{B^2\}-\mu_B^2$$

From equation (3.8), this could be written as

$$C_X(t_1,\ t_2)=E\{A^2\}t_1t_2-\mu_A^2t_1t_2+\sigma_B^2$$

where σ_B^2 is the variance of B.

(4) From equation (12.9) set $t_1=t_2=t$ in the autocovariance to obtain the variance:

$$\sigma_X^2(t)=C_X(t,\ t)=E\{A^2\}t^2-\mu_A^2t^2+\sigma_B^2=\sigma_A^2t^2+\sigma_B^2$$

where σ_A^2 is the variance of A and equation (3.8) was used.

(5) From equation (12.10) the autocorrelation coefficient is given as

$$\rho_X(t_1,\ t_2)=\frac{C_X(t_1,\ t_2)}{\sigma_X(t_1)\sigma_X(t_2)}=\frac{E\{A^2\}t_1t_2-\mu_A^2t_1t_2+\sigma_B^2}{\sqrt{\sigma_A^2t_1^2+\sigma_B^2}\sqrt{\sigma_A^2t_2^2+\sigma_B^2}}$$

Example 12.4: Analytical Derivation of First and Second-Order Statistics

For the stochastic process in Example 12.1, derive (1) the mean, (2) the autocorrelation function, (3) the autocovariance function, (4) the variance, and (5) the autocorrelation coefficient.

Solution
(1) Taking expectations on both sides of the equation, using the properties of expectation, and noting that the random variables are independent, the mean of $U(t)$ is given as

$$E\{U(t)\}=E\{V+We^{-Kt}\}=E\{V\}\ +\ E\{We^{-Kt}\}=E\{V\}+E\{W\}E\{e^{-Kt}\}$$

The last term on the right side includes the expectation of an exponential of the random variable K times t. Since K is a Normal random variable, this could be approximated as in Example 4.16 by expanding the exponential as a Taylor series, taking expectation of the individual terms, and using the properties of moments of Normal random variables. If we feel confident with integrals, or if we use Maple, we could attempt to find its exact value by applying the rules of expectation of a function of a random variable. Thus, from equation (4.15),

$$E\{e^{-Kt}\}=\int_{-\infty}^{\infty}e^{-Kt}f_K(k)dk=\int_{-\infty}^{\infty}e^{-kt}\frac{e^{-\frac{(k-\mu_K)^2}{2\sigma_K^2}}}{\sqrt{2\pi\sigma_K^2}}dk=e^{[\frac{\sigma_K^2 t^2}{2}-\mu_K t]} \qquad (12.11)$$

where $f_K(k)$ is the density function of K (i.e., Gaussian), and μ_K and σ_K^2 are the mean and the variance of K, respectively. Thus, the mean of $U(t)$ becomes

$$\mu_U(t)=E\{U(t)\}=\mu_V+\mu_W e^{[\frac{\sigma_K^2 t^2}{2}-\mu_K t]}$$

where μ_V and μ_W are the means of V and W, respectively.

(2) From equation (12.8) the autocorrelation function at two times t_1 and t_2 is

$$R_U(t_1,\ t_2)=E\{U(t_1)U(t_2)\}=E\{V+We^{-Kt_1}\}E\{V+We^{-Kt_2}\}$$

$$=E\{V^2\}+E\{V\}E\{W\}E\{e^{-Kt_1}\}+E\{V\}E\{W\}E\{e^{-Kt_2}\}+E\{W^2\}E\{e^{-K(t_1+t_2)}\}$$

Applying equation (12.11), the autocorrelation function becomes

$$R_U(t_1,t_2)=\{V^2\}+\mu_V\mu_W e^{[\frac{\sigma_K^2 t_1^2}{2}-\mu_K t_1]}+\mu_V\mu_W e^{[\frac{\sigma_K^2 t_2^2}{2}-\mu_K t_2]}+E\{W^2\}e^{[\frac{\sigma_K^2(t_1+t_2)^2}{2}-\mu_K(t_1+t_2)]}$$

(3) From equation (12.7), the autocovariance function is given as

$$C_U(t_1,\ t_2)=R_U(t_1,\ t_2)-\mu_U(t_1)\mu_U(t_2)$$

$$C_U(t_1,\ t_2)=E\{V^2\}+\mu_V\mu_W e^{[\frac{\sigma_K^2 t_1^2}{2}-\mu_K t_1]}+\mu_V\mu_W e^{[\frac{\sigma_K^2 t_2^2}{2}-\mu_K t_2]}$$

$$+E\{W^2\}e^{[\frac{\sigma_K^2(t_1+t_2)^2}{2}-\mu_K(t_1+t_2)]}\left[-\mu_V+\mu_W e^{(\frac{\sigma_K^2 t_1^2}{2}-\mu_K t_1)}\right]\left[\mu_V+\mu_W e^{(\frac{\sigma_K^2 t_2^2}{2}-\mu_K t_2)}\right]$$

(4) From equation (12.9) set $t_1=t_2=t$ in the autocovariance and solve to obtain the variance:

$$\sigma_U^2(t)=C_U(t,\ t)=E\{V^2\}+2\mu_V\mu_W e^{[\frac{\sigma_K^2 t^2}{2}-\mu_K t]}+E\{W^2\}e^{[2\sigma_K^2 t^2-2\mu_K t]}$$

$$-\left[\mu_V+\mu_W e^{(\frac{\sigma_K^2 t^2}{2}-\mu_K t)}\right]^2\approx\sigma_V^2+\sigma_W^2 e^{[2\sigma_K^2 t^2-2\mu_K t]}\approx\sigma_V^2+\sigma_W^2 e^{-2\mu_K t}$$

where σ_V^2 and σ_W^2 are the variances of V and W, respectively.

(5) From equation (12.10) the autocorrelation coefficient is given as

$$\rho_U(t_1,\ t_2)=\frac{C_U(t_1,\ t_2)}{\sigma_U(t_1)\sigma_U(t_2)}$$

where the autocovariance and the standard deviation are given in items (3) and (4), respectively. Although an analytical derivation of these statistics may seem tedious, these expressions constitute the exact first and second-order moments of a stochastic process. Once they are derived as continuous functions of time, their observation and plotting at any time are straight forward. Alternatively, one may wish to derive estimates of these statistics via the application of a Monte Carlo simulation, as described in Chapter 4. The numerical calculation of each of these statistics is needed for each fixed time t. What we saved in mathematical derivation is now balanced by the increased computer time and memory requirements, and the numerical computational errors.

Example 12.5: Analytical Simulation of First and Second-Order Statistics.

Using the data from Example 12.1, and the expressions derived in Example 12.4, in a single graph plot the following curves of the process $U(t)$ as a function of time: the mean, the mean plus one standard deviation, and the mean minus one standard deviation.

Solution

The following Maple worksheet illustrates the procedure.

```
[ Enter data and define the mean and the standard deviation as functions of time.
[ > mu_k:=0.2: mu_v:=1: mu_w:=10.:
    sigma_k:=0.05: sigma_v:=0.1: sigma_w:=1.:
    mu_u:=t->mu_v+mu_w*exp((sigma_k*t)^2/2-mu_k*t):
    sigma_u2:=t->sigma_v^2+(sigma_w^2)*exp(-2*mu_k*t):
    sigma_u:=t->sqrt(sigma_u2(t)):
[ Plot the mean plus and minus one standard deviation as a function of time.
[ > G1:=plot(mu_u(t),t=0..20,color=black,thickness=2):
    G2:=plot(mu_u(t)+sigma_u(t),t=0..20,linestyle=3,color=black):
    G3:=plot(mu_u(t)-sigma_u(t),t=0..20,linestyle=3,color=black):
    plots[display](G1,G2,G3,labels=[`t`,``],
        labelfont=[TIMES,ROMAN,16]);
```

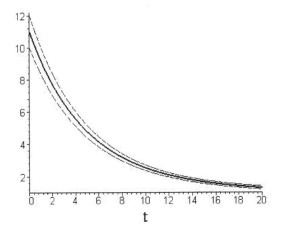

The solid line represents the mean and the dashed lines are the mean plus and the mean minus one standard deviation, respectively. These statistics are now functions of time. The standard deviation, and thus the uncertainty of this process, decreases with time. It is interesting to compare the plot of the statistics with the graph in Example 12.1 that shows sample realizations of the process. The band of the statistics curves cuts through the center of the cloud of sample curves.

Example 12.6: Monte Carlo Simulation of First and Second-Order Statistics

Example 12.5 culminates with a figure that summarizes the statistics of the process $U(t)$ introduced in Example 12.1. This figure was produced after an analytical derivation of the statistics of the process (Example 12.4). Produce a figure of the same process using Monte Carlo simulations. In other words, generate realizations of the random variables U, V, and K as shown in Example 12.1, input them into the formula of $U(t)$, but instead of plotting sample time curves, calculate the mean and the standard deviation of $U(t)$ for various times t. Store these statistics in arrays and plot them.

Solution

We modify the Maple worksheet of Example 12.1 as follows.

```
  Initialize, enter data, and generate N samples of random variables V, K , and U
> restart: with(stats): N:=100:
  mu_k:=0.2: mu_v:=1: mu_w:=10.:
  sigma_v:=0.1: sigma_w:=1.:
  v_data:=[random[normald[mu_v,sigma_v]](N)]:
  k_data:=[random[normald[mu_k,sigma_k]](N)]:
  w_data:=[random[normald[mu_w,sigma_w]](N)]:
  Create a procedure to calculate the sample mean and
  sample standard deviation at a fixed time t-j.
> step:=proc(j)
  global N, v_data, w_data, k_data, U_bar, sigma_U:
  local U_t, i:
  U_t:=[seq(v_data[i]+w_data[i]*exp(-k_data[i]*j),i=1..N)]:
  U_bar[j]:=describe[mean](U_t):
  sigma_U[j]:=describe[standarddeviation](U_t):
  end:
  Execute the procedure for 20 time steps.
> for j from 0 to 20 do step(j) end do:
  Prepare to plot data: create arrays of coordinates [[x1,x2], [x2,y2]....] for the mean.
  the mean plus one standard deviation, and the mean minus one standard deviation.
> Mean:=[seq([j,U_bar[j]],j=0..20)]:
  Dev_plus:=[seq([j,U_bar[j]+sigma_U[j]],j=0..20)]:
  Dev_minus:=[seq([j,U_bar[j]-sigma_U[j]],j=0..20)]:
  Plot the arrays. How do Monte Carlo simulations compare to analytical simulations of Example 12.5?
> G1:=plot(Mean,color=black,thickness=2):
  G2:=plot(Dev_plus,linestyle=3,color=black):
  G3:=plot(Dev_minus,linestyle=3,color=black):
  plots[display](G1,G2,G3,labels=[`t`,``],
    labelfont=[TIMES,ROMAN,16]);
```

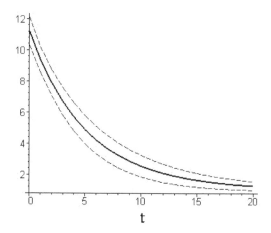

There is a good agreement between the Monte Carlo simulations and the analytical simulation. A good Monte Carlo simulation model must conform to the exact analytical model. An analytical model is usually simpler to implement than a corresponding numerical model. However, in many circumstances a Monte Carlo model constitutes a reasonable avenue because the complexity of the underlying stochastic process makes an analytical derivation difficult. The following example illustrates this point.

Example 12.7: Approximate Analytical derivation of Statistics

An aquifer recharge from rainfall (m) in a region is a stochastic process of the form $R_g(t) = A - B\cos(\pi(t-C)/6)$, where $0 \le t < \infty$ is time in *months*, A is a random variable representing seasonal average that depends on location, B is a random amplitude parameter, and C is a phase random variable. If A, B, and C are independent Normal variables, derive an expression for the mean of $R_g(t)$.

Solution

Taking expectations on both sides of $R_g(t)$ and noting that A, B, and C are independent,

$$\mu_R = E\{R_g\} = \mu_A - \mu_B E\left\{\cos\left(\frac{\pi(t-C)}{6}\right)\right\} \qquad (12.12)$$

Where μ_A and μ_B are the means of A and B, respectively. The last expectation on the right side of equation (12.12) could be solved by applying the formula for the expectation of a function of a random variable, equation (4.15):

$$E\left\{\cos\left(\frac{\pi(t-C)}{6}\right)\right\} = \int_{-\infty}^{\infty} \cos\left(\frac{\pi(t-c)}{6}\right) f_C(c)dc = \int_{-\infty}^{\infty} \cos\left(\frac{\pi(t-c)}{6}\right) \frac{e^{-\frac{(c-\mu_C)^2}{2\sigma_C^2}}}{\sqrt{2\pi\sigma_C^2}} dc$$

where $f_C(c)$ is the (Gaussian) density function of C. This integral is not easy to solve. Thus, we opt for the procedure outlined in Example 4.16, Chapter 4. Using a common trigonometric identity, we write the expectation in question as

$$E\left\{\cos\left(\frac{\pi(t-C)}{6}\right)\right\} = E\left\{\cos\left(\frac{\pi t}{6}\right)\cos\left(\frac{\pi C}{6}\right) + \sin\left(\frac{\pi t}{6}\right)\sin\left(\frac{\pi C}{6}\right)\right\}$$

$$= \cos\left(\frac{\pi t}{6}\right) E\left\{\cos\left(\frac{\pi C}{6}\right)\right\} + \sin\left(\frac{\pi t}{6}\right) E\left\{\sin\left(\frac{\pi C}{6}\right)\right\}$$

From calculus, one may expand the functions inside the expectation operators as infinite Taylor series about $C=0$:

$$E\left\{\cos\left(\frac{\pi(t-C)}{6}\right)\right\}=\cos\left(\frac{\pi t}{6}\right)E\left\{1-\frac{\pi^2 C^2}{6^2 2!}+\frac{\pi^4 C^4}{6^4 4!}+...\right\}$$

$$+\sin\left(\frac{\pi t}{6}\right)E\left\{\frac{\pi C}{6}-\frac{\pi^3 C^3}{6^3 3!}+\frac{\pi^5 C^5}{6^5 5!}-...\right\}$$

At this point, we may proceed as in Example 4.16. That is, express C as a standard Normal variable, and take expectations. Since C follows a Normal distribution, all the odd moments are zero. The even moments are computed by the formula and a large number of terms may be computed. In practical cases of non-Gaussian random variables, or of random variables of unknown densities for which one only knows the mean and the variance, one must truncate the series at the first few terms, take expectations, and hope the series converges. Let us follow this avenue, which is the simplest one.

$$E\left\{\cos\left(\frac{\pi(t-C)}{6}\right)\right\}\approx\cos\left(\frac{\pi t}{6}\right)\left[1-\frac{\pi^2 E\{C^2\}}{6^2 2!}\right]+\sin\left(\frac{\pi t}{6}\right)\left[\frac{\pi E\{C\}}{6}\right]$$

$$=\cos\left(\frac{\pi t}{6}\right)\left[1-\frac{\pi^2(\sigma_C^2+\mu_C^2)}{6^2 2!}\right]+\sin\left(\frac{\pi t}{6}\right)\left[\frac{\pi\mu_C}{6}\right]$$

Where μ_C is the mean of C and equation (3.8) has been used (i.e., $\sigma_C^2=E\{C^2\}-\mu_C^2$). Substituting this expression in equation (12.12),

$$\mu_R(t)=\mu_A-\mu_B\left\{\cos(\frac{\pi t}{6})\left[1-\frac{\pi^2(\sigma_C^2-\mu_C^2)}{6^2 2!}\right]+\sin(\frac{\pi t}{6})\left[\frac{\pi\mu_C}{6}\right]\right\}$$

The analytical derivation of the autocorrelation function and other second-order statistics are lengthy in this example.

Stationarity

A stochastic process is said to be *strictly stationary* if its statistics are not affected by a shift in the origin. This means that two processes $X(t)$ and $X(t+\tau)$ have the same statistics for any $X(t+\tau)$. A consequence of this is that the first order density, $f_X(x,t)=f_X(x)$ is independent of t; $\mu_X(t)=\mu_X$ is a constant; the second order density $f_X(x_1,x_2;t_1,t_2)=f_X(x_1,x_2;\tau)$, $\tau=t_1-t_2$; the autocorrelation function $R_X(t_1,t_2)=R_X(\tau)$; the autocovariance $C_X(t_1,t_2)=C_X(\tau)$; and the autocorrelation coefficient $\rho_X(t_1,t_2)=\rho_X(\tau)$ all depend on the time difference.

Observe the figure in Example 12.5. Since the mean and the variance of the process changes with time, the process is not stationary. A stationary process is simpler to analyze and to model than a non stationary one. In Examples 12.1 through 12.7, non stationarity did not posed much difficulty because we knew the analytical formulae of the processes involved. In many engineering applications, stationarity is assumed out of necessity in order to obtain a simpler and manageable model, or because the data set is insufficient to define a general non stationary process. See Papoulis (1984) for more details and consequences of the above definition of stationarity.

The Correlogram

In terms of a time interval τ, and with the above assumptions, the autocorrelation coefficient equation (12.10) may be written as

$$\rho_X(t,\ t+\tau) = \frac{C_X(t,\ t+\tau)}{\sigma_X^2(t)}\rho_X(t,\ t+\tau) = \frac{C_X(t,\ t+\tau)}{\sigma_X^2(t)} \qquad (12.13)$$

The time difference, τ, is called the *lag time*. Once again, when $\tau=0$, $\rho_X(0)=1$. As the lag increases one expects the value of the autocorrelation coefficient to decrease. Since the autocorrelation coefficient is a measure of linear dependence, a decline in its value as the lag increases suggests a decrease in the "memory" of the process. In other words, samples of the process observed at short intervals of time are expected to be more linearly dependent than those separated by long intervals.

A graph of the autocorrelation coefficient with respect to lag time is called the *correlogram*. An analysis of its shape represents an important tool in diagnosis and modeling of stochastic processes. How fast the value of $\rho_X(\tau)$ decreases as τ increases indicates how serially dependent or independent a process is. A fast decline with lag time indicates a process with short memory, that is a process with sample values independent from each other. Conversely, a slow decline of the autocorrelation function with the lag suggests a process with long memory, that is sample values are serially dependent. The form of the correlogram is very important in the analysis of samples of processes for which an analytical form is unknown. It is customary in engineering modeling to fit a curve to the correlogram resulting from measured data. This in turn may be used in Monte Carlo simulations of the process with the required correlation structure. Sometimes the correlogram of a process does not continuously decline with lag time. Instead, it shows a periodic increase and decrease with the lag. This usually suggests a deterministic component present in the process, such as a seasonal variability in the signal.

Example 12.8: Plotting a Correlogram

Use the results in Example 12.4 to produce a correlogram of the process $U(t)$ and interpret it.

Solution

The following Maple worksheet illustrates the procedure.

```
Define the mean and the variance of U as functions of t.
Neglect the quadratic terms in the exponentials.
> restart:
  mu_u:=t->mu_v+mu_w*exp(-mu_k*t):
  sigma2_u:=t->sigma_v^2+sigma_w^2*exp(-2*mu_k*t):
Define the autocorrelation, autocovariance and autocorrelation coefficient as functions
of two times t1 and t2.
> R_u:=(t1,t2)->sigma_v^2+mu_v^2+mu_v*mu_w*exp(-mu_k*t1)
    +mu_v*mu_w*exp(-mu_k*t2)
    +(sigma_w^2+mu_w^2)*exp(-mu_k*(t1+t2)):
  C_u:=(t1,t2)->R_u(t1,t2)-mu_u(t1)*mu_u(t2):
  rho_u:=(t1,t2)->C_u(t1,t2)/sigma2_u(t1):
Enter statistics data of input variables V, W, and K. Plot the correlogram at times t1=0, t2=tau.
> mu_v:=1: mu_k:=0.2: mu_w:=10:
  sigma_v:=0.1: sigma_k:=0.05: sigma_w:=1.:
  plot(rho_u(0,tau),tau=0..20,color=black,thickness=2,
    labels=[`tau`,`rho (tau)`],labelfont=[TIMES,ROMAN,16],
    labeldirections=[HORIZONTAL,VERTICAL]);
```

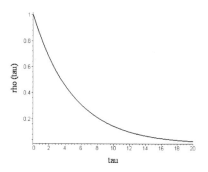

```
Plot the correlogram at times t1=30, t2=30+tau.
> plot(rho_u(30,30+tau),tau=0..20,color=black,thickness=2,
    labels=[`tau`,`rho (tau)`],labelfont=[TIMES,ROMAN,16],
    labeldirections=[HORIZONTAL,VERTICAL],tickmarks=[default,4]);
```

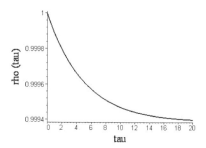

The correlogram exhibits a slowly-decaying correlation coefficient with the lag. This indicates a strong linear dependency, or memory, among subsequent observation times of the process. This is corroborated by the fact that a large portion of the magnitude of the process is controlled by a deterministic function. The correlogram at $t_1=0$ is similar to that at $t_1=20$. However the former decays faster with lag time (by about 0.97 at $\tau=20$) than the latter (by less than 0.1 at $\tau=20$). This indicates once again that U is not stationary and that observations of the process are more independent from one another at earlier times than at prolonged times.

Many signals are not available on a continuous time basis. Instead, the process may be observed and recorded at fixed discrete time intervals Δt. For example, weather and environmental variables are reported daily or at 12-*hour* intervals. In such cases the *sample* autocorrelation coefficient, rather than the *population* one is estimated based on a one-time sample. The sample autocorrelation coefficient measures the linear statistical dependence of subsequent observations at discrete times. The lag time takes on discrete values $\tau=k\Delta t$, $k=0$, 1, 2, ... For a stationary stochastic process observed at fixed time intervals $\Delta t=1$, an estimate of equation (12.13) may be obtained by applying equation (8.2) to a series:

$$\hat{\rho}_X(t, \ t+k\Delta t)\approx r_X(k)=\frac{c_X(k)}{S_X^2}, \qquad k=0, \ 1, \ 2, \ ... \qquad (12.14)$$

where

$\hat{\rho}_x(t, \ t+k\Delta t)$=lag k sample autocorrelation coefficient (the "hat" denotes "sample")
$r_X(k)$=lag k sample autocorrelation coefficient

S_X^2=sample variance of X (see below equation (6.2))
$c_X(k)$=lag k sample autocovariance given from equation (8.1) as

$$c_X(k)=\frac{1}{N-k}\sum_{i=1}^{N-k}\left(X_i-\bar{X}\right)\left(X_{i+k}-\bar{X}\right) \qquad (12.15)$$

where
$\bar{X}$=sample mean ox X, and N=number of data points in the series

For $k=0$ equation (12.14) is equal to 1. For lags k greater than zero there are N-k pairs of data in the summation term. For instance for $k=1$, there are N-1 pairs of data points in equation (12.15), each separated by

$k=1$ unit of time. Hence the lag-1 autocorrelation coefficient measures the linear statistical dependency of data points separated by one unit of time. For $k=2$, there are N-2 pairs of data points in equation (12.15), each separated by $k=2$ units of time. Thus, the lag-2 autocorrelation coefficient measures the linear statistical dependency of data points separated by two units of time. As in the continuous case, the discrete lag-k autocorrelation coefficient is expected to decrease in magnitude as k increases, thus implying that observations separated by longer periods of time are expected to be less correlated than those closely spaced in time. However, this is not always the case. A careful analysis of the serial correlogram may indicate the presence of deterministic components, such as seasonal periodic variability. An important limitation should be kept in mind when applying equations (12.14) and (12.15): as the lag k increases, the number of pairs of data points N-k decreases. This makes the calculated lag k autocorrelation coefficient less accurate for higher lag times. When the sample size N is small, the maximum lag k to compute should be a small fraction of N.

Equations (12.14) and (12.15) introduce the idea that the second-order statistics of a process may be estimated from a single sample curve, rather from the *ensemble* of many samples (i.e., time curves). If the statistics are estimated from a single curve, we obtain the *time average statistics*; if they are estimated from many curves, we obtain the *ensemble* statistics. When the time average statistics are the same as those of the ensemble ones, the process is called *ergodic*. Ergodicity is a much stronger condition than stationarity since it refers to the statistics of all orders. In practice, when only a single sample is available, ergodicity or stationarity is assumed for simplicity, and the the properties of the stochastic process are calculated from a single realization.

Example 12.9: Estimation of the Correlogram from a Single Sample
A stochastic process, Z_i, is produced by generating a Standard Gaussian random number each unit of time, i. Generate a sample of the process for 1000 time intervals, plot, and analyze the correlogram.

Solution

```
Initialize, generate 1000 Standard Gaussian random numbers, store them in an array,
and calculate their mean and variance.
> restart: with(stats): N:=1000:
  z:=[random[normald](N)]:
  Z_bar:=describe[mean](z):
  S_z2:=describe[variance[1]](z):
Store the process sample coordinates [[0, Z(0)],[1, Z(1)],...] in an array and plot.
> sample_z:=[ seq([i,z[i]],i=1..N) ]:
  plot(sample_z,labels=[`Time`,` Z(t) `],color=black,
      labelfont=[TIMES,ROMAN,16]) :
```

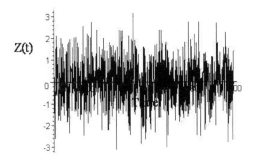

[Define equation (12.14) as a function of lag k.
[> r:=k->sum((z[i]-Z_bar)*(z[i+k]-Z_bar),i=1..N-k)/((N-k)*S_z2):
[Calculate the autocorrelation coefficient for $k=0, 1, ..., 10$, and store the correlogram
  coordinates $[[0, r(0)], [1, r(1)], ...]$ in an array. Then, plot the correlogram.
[> correlogram:=[seq([k,r(k)],k=0..10)]:
    plot(correlogram,color=black,thickness=2,labels=[`Lag k`,`r(k)`],
    labelfont=[TIMES,ROMAN,16]);

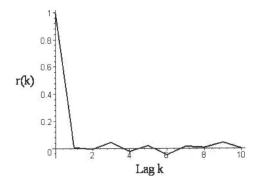

The autocorrelation coefficient becomes almost zero for lags greater or equal to one. This implies that observations of the process are serially independent. It is interesting to observe the shape of a sample of a "purely random," "memoryless," process. Throughout this book, we use a computer to generate random numbers to be used in Monte Carlo simulations. We tested the numbers by plotting their frequency histogram and accepted the numbers if the frequency histogram resembled the theoretical density function the numbers were supposed to follow. As stated in Chapter 4, an additional test is provided by plotting the correlogram of the numbers. Theoretically, random numbers sampled from a probability distribution must be uncorrelated. This implies that acceptable random numbers must exhibit a correlogram that approaches zero for lags greater or equal to one.

Transformations of the Correlation Function: The Spectral Density

From the study of physics, we know that when a beam of white light passes through a prism it is decomposed into an array of colored lights. Each color is deflected in proportion to its harmonic frequency or rate of energy oscillation. The resulting colors are said to be the primary colors of the *spectrum*. The red has the lowest frequency, whereas the violet has the highest frequency. Any particular color may be thought of as a combination of fundamental colors. Similarly, any sound may be decomposed into a set of fundamental sounds. In 1807, French mathematician Jean Baptiste Joseph Fourier showed that virtually any periodic signal or function can be expressed as the sum of a series of sines and/or cosines of increasing frequencies (Fourier, 1878). This discovery revolutionized the fields of mathematics, science, and engineering. Decomposing a signal into its primary harmonics may reveal periodic oscillations not readily observed in the raw signal. This is particularly important in signals subject to random variability such as stochastic processes. Fourier, or harmonic, analysis of samples from stochastic processes constitutes a routine procedure to identify hidden periods of oscillation. For example, Moore (1914, chapter 2), applied Fourier and harmonic analysis to rainfall in the Ohio valley which was found to have a significant period of 8 years. Thus, on the average, a cycle of drought or flooding condition is repeated every 8 years. Larger periods of oscillation may be identified when observing long historical records as in the analysis of global climate change. In principle a sample from a stochastic process may be composed of oscillations of all possible frequencies. Identifying the individual frequencies may help understand periodic patterns of system behavior.

The correlogram described above is a function of time and is therefore said to belong to the time domain. An additional means of studying the serial dependence structure of a stochastic process is achieved via the spectrum or the spectral density of the correlation function. The spectrum displays the frequency domain variability of a correlation function. The term frequency is used here in the harmonic sense and not in the sense of the histogram. Let us define the *spectral density* of a stationary stochastic process as the *Fourier transform* of the autocovariance function (Wiener, 1930):

$$s'(f) = \int_{-\infty}^{\infty} C_X(\tau) e^{-2\pi j f \tau} d\tau = \int_{-\infty}^{\infty} C_X(\tau) \cos(2\pi f \tau) d\tau \tag{12.16}$$

where
$s'(f)$ = spectral density function

$C_X(\tau)$=autocovariance function of a process $X(t)$
f= frequency in cycles per unit time
τ=lag time
$j=\sqrt{-1}$

Since for a stationary stochastic process $C_X(\tau)=C_x(-\tau)$ (i.e., the autocovariance is even), then $s'(f)$ is a real function. From the *Fourier inversion formula*, the autocovariance function, $C_X(\tau)$, may be obtained from the spectral density function:

$$C_X(\tau)=\int_{-\infty}^{\infty}s'(f)e^{2\pi jf\tau}df=\int_{-\infty}^{\infty}s'(f)\cos(2\pi f\tau)df \qquad (12.17)$$

provided that the integrals defined above are finite. For a covariance stationary series, it is convenient to divide $s'(f)$ by the variance σ_X^2. Thus, a normalized spectral density function is defined by replacing the autocovariance in equation (12.16) by the autocorrelation coefficient (see equation (12.10)):

$$s(f)=\int_{-\infty}^{\infty}\rho_X(\tau)e^{-2\pi jf\tau}d\tau=\int_{-\infty}^{\infty}\rho_X(\tau)\cos(2\pi f\tau)d\tau \qquad (12.18)$$

where
   $s(f)$=normalized spectral density function

Similarly, equation (12.17) becomes

$$\rho_X(\tau)=\int_{-\infty}^{\infty}s(f)e^{2\pi jf\tau}df=\int_{-\infty}^{\infty}s(f)\cos(2\pi f\tau)df \qquad (12.19)$$

Since $\rho(0)=\cos(0)=1$, equation (12.19) indicates that the area under the normalized spectral density function is 1. A graph of the spectral density shows the predominant frequencies as compared to those that occur less. As with the probability density function, partial areas under the spectral density function between two frequencies f_1 and f_2 indicate the contribution of the total normalized variance contained in the said frequency band. Thus, the spectral density function is useful in determining those frequencies that dominate the variance of a stochastic process. We may describe the spectral density as a function of frequency in cycles per unit time, frequency in radians per unit time, or period in units of time. The three quantities are related by

$$\omega=\frac{2\pi}{p}=2\pi f \qquad (12.20)$$

where

> ω=angular frequency in radians per unit time
> p=period in units of time
> f = frequency in cycles per unit time

Therefore, $s(f)$ and $s(\omega)$ are related via

$$s(\omega)=2\pi s(f) \tag{12.21}$$

An intuitive understanding of the Fourier transformation of a function, in this case a correlation function, may come after observing the cosine integral of equation (12.16). The function being transformed (i.e., the correlation function) is compared with a periodic function (i.e., cosine). If the correlation function has a cycle similar to that of the cosine function, either positive or negative, the product of the two functions, and thus the integrand, gives a higher positive value. Integrating over all times, τ, produces a function of frequency which will yield a peak at the said cycle (frequency). Such sharply rising values of the spectral density function indicate rhythmic movements in the correlation function. The concept of Fourier transform of signals has now been extended to other integral transforms that include a vast array of weighting functions, not just sines and cosines. *Wavelets* or *wavelet transform* is a generalized process that allows the observation of signals at lower as well as higher scales of resolution. These methods have produced remarkable technological applications in digital image processing and global telecommunications (Burrus et al., 1998; Ogden, 1997; Chan, 1995; Newland, 1993).

Example 12.10: Spectral Density of a Periodic Correlation Function

Consider a periodic (e.g., monthly) autocorrelation coefficient function of the form $\rho(\tau)=\cos(\pi\tau/6)/\pi$. Calculate its spectral density function and interpret it.

Solution

We take advantage of the integral transforms sub-package in Maple, which permits the calculation of the Fourier transform of functions:

```
Call the inverse tranforms package and define the autocorrelation coefficient function.
> restart: with(inttrans):
  rho:=tau->cos(Pi*tau/6)/Pi:
  plot(rho(tau),tau=0..24,color=black,thickness=2,
     labels=[`tau`,`rho(tau)`],
     labeldirections=[HORIZONTAL,VERTICAL],
     labelfont=[TIMES,ROMAN,16]):
```

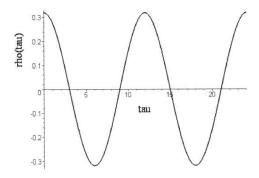

Take the fourier transform of rho(tau) with tau the integration variable and omega the transformation
parameter.

```
> assume(omega>0) :
  fourier(rho(tau),tau,omega) :
  s:=unapply(%,omega) ;
```

$$s := \omega\sim \rightarrow \text{Dirac}\left(\omega\sim -\frac{\pi}{6}\right)$$

The Dirac's delta function equals zero everywhere except at a jump discontinuity at omega=Pi/6. Plot
it as a line.

```
> plottools[line]([Pi/6,0],[Pi/6,1]) :
  plots[display](%,thickness=4,labels=[` omega `,`s(omega)`],
      labeldirections=[HORIZONTAL,VERTICAL],
      labelfont=[TIMES,ROMAN,16]) ;
```

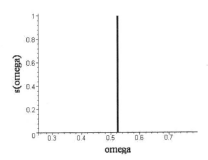

The autocorrelation coefficient function shows a peak repeating every
12 units of lag time, τ. The Fourier transform of the autocorrelation
coefficient yields $s(\omega)=\delta(\omega-\pi/6)$. The Dirac's delta function equals zero,
except when its argument is zero, at which place there is a jump. Thus,
the spectral density function is zero, except at $\omega=\pi/6$, where it has a
jump. From equation (12.20), $f=\omega/2\pi=1/12$ *cycles/month*. Since the
period is the inverse of the frequency, $p=12$*months,* which is a typical
period of oscillation for seasonal series.

In many engineering processes the time scale is discrete rather than continuous. When the process is observed or sampled at equally spaced intervals of time, Δt, the oscillation with the highest frequency from which any information may be obtained has a frequency

$$f_{max} = \Delta t / 2 \qquad (12.22)$$

where
f_{max} = maximum frequency

A sample or estimate of the spectral density function may be computed from the sample autocorrelation coefficient and a numerical integration of equation of equation (12.18) (see Kottegoda, 1980 for details):

$$\hat{s}(f) = \left[r_X(0) + 2 \sum_{k=1}^{M-1} r_X(k)\cos(2\pi k f \Delta t) + r_X(M)\cos(2\pi M f \Delta t) \right] \Delta t \qquad (12.23)$$

where
$\hat{s}(f)$ = sample spectral density function
$r_X(k)$ = lag k sample autocorrelation coefficient
M = maximum lag to compute from a sample of size N
Δt = sample time interval of stochastic process X

The maximum lag to compute, M, should be a small portion of the total number of sample points available, N. Some authors recommend $M \leq 0.25N$. The reason for this restriction is the fact that as the lag k increases the number of data pairs available in the autocorrelation coefficient equation (12.14) decreases to $k/2$ thus rendering $r(k)$ less accurate. Equation (12.23) should be used to compute $\hat{s}(f)$ only for frequencies given by

$$f = \frac{k f_{max}}{M} \qquad (12.24)$$

Some authors recommend smoothing the resulting sample spectral density function to account for increasing errors as the lag k increases in equation (12.23) (see Kottegoda, 1980 for suggestions).

Example 12.11: Spectral Density Function of a Discrete Random Sequence

Modify the program in Example 12.9 in order to calculate and plot the sample spectral density function of the stochastic process.

Solution

The first part of the program is identical to that in Example 12.9. We now incorporate equations (12.22), (12.23) and (12.24).

```
[ Generate a sample of the stochastic process as before ande define the autocorrelation coefficient.
> restart: with(stats): N:=1000: M:=10:
    delta_t:=1: fmax:=delta_t/2:
    z:=[random[normald](N)]:
    Z_bar:=describe[mean](z):
    S_z2:=describe[variance[1]](z):
    r:=k->sum( (z[i]-Z_bar)*(z[i+k]-Z_bar),i=1..N-k )/((N-k)*S_z2):
[ Calculate the autocorrelation coefficient for k=0, 1, ..., M, and store it in the array rr[k].
> rr:=[seq(r(k),k=0..M)]:

[ Define a procedure to calculate the sample spectral density for a given frequency ff (equation (12.23)).
> spectral:=proc(ff)
        global s,delta_t,M,rr:
        local j:
        s:=ff->(rr[1]+
              2*sum(rr[j+1]*cos(2*Pi*j*ff*delta_t),j=1..M-1)
              +rr[M+1]*cos(2*Pi*M*ff*delta_t))*delta_t:
        s(ff):
    end:
[ Calculate the spectral density for discrete frequncies given by equation (12.22),
[ store in array of coordinates [[f, s(f)],...] suitable for plotting.
> f:=k->k*fmax/M:        #equation (12.24)
    s:=[seq([f(k),spectral(f(k))],k=0..M)]:
    plot(s,x=0..fmax,y=0..1.2,color=black,thickness=2,
        labels=[`f`,`s(f)`],
        labelfont=[TIMES,ROMAN,16]);
```

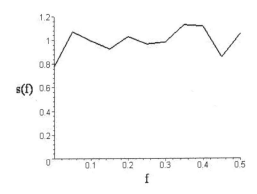

The sample spectral density function is made of values that oscillate around the constant 1 for the different frequencies. This indicates that all frequencies are equally important (i.e., no particular frequency predominate). A process with a constant spectral density implies an uncorrelated stochastic process. This is reemphasized by the shape of the correlogram (see Example 12.9).

Example 12.12: Air Temperature Evolution in Philadelphia

We are interested in the time evolution of mean annual air temperature in Philadelphia recorded for over a century. This data is available through the Franklin Institute of Science and Technology (http://www.fi.edu/weather/data2/wthrmean.txt). Let us assume that you have downloaded this data file; its first two lines contain alphanumeric information describing its contents. Beginning with the third line, the file contains two columns; the left one is the year (from 1874 to 2000) and the right one the average air temperature in degrees Fahrenheit. Assume you named the data file PhillyTemp.txt and stored it in the C drive of your computer at C:\Sergio\ProbabilitySecondEd\Chapter12\PhillyTemp.txt. Write a Maple program that reads the data in this file beginning with the third line; transforms temperature data into degrees Celsius; calculates and plots the sample correlogram; and calculates and plots the sample spectral density of the stochastic process.

Solution

```
Read data file by indicating the path where the text file is located in your disk.
Notice the double "\\" and that the file has 2 columns.
> restart: with(stats): M:=20:
  delta_t:=1: fmax:=delta_t/2:
  TF:=readdata(
  "C:\\Sergio\\ProbabilitySecondEd\\Chapter12\\PhillyTemp.txt",2):
Data is now stored in a vector TF:=[ [year, temperature], [year, temperature], . . .].
Year i is stored in column 1 of TF[i][1].
Temperature in year i is stored in column 2 of TF[i][2].
Now find the number of data points, Np, and transform degrees Fahrenheit into degrees Celsius.
Start in line 3, since the first two lines of data file are just text.
> for i from 3 to 2000 do    #2000 is an arbitrary large number
  Np:=i:
  TC[i-2]:=[TF[i-2][1],(TF[i-2][2]-32)/1.8]:
  end do:
  N:=Np-3:                    #Ignore error when the loop is aborted.
Error, invalid subscript selector

> Ts:=[seq(TC[i][2],i=1..Np-3)]:
  Tsbar:=describe[mean](Ts):
  S_t2:=describe[variance[1]](Ts):
Calculate the autocorrelation coefficient for k=0, 1, ..., M, and store it in the array rr[k].
> r:=k->sum( (Ts[j]-Tsbar)*(Ts[j+k]-Tsbar),j=1..N-k )/((N-k)*S_t2):
  rr:=[seq(r(k),k=0..M)]:
> correlogram:=[ seq([k,r(k)],k=0..M) ]:
  plot(correlogram,color=black,thickness=2,labels=[`Lag k`,`r(k)`],
      labelfont=[TIMES,ROMAN,16]);
```

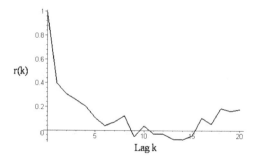

Define a procedure to calculate the sample spectral density for a given frequency ff (equation (12.23)).

```
> spectral:=proc(ff)
    global s,delta_t,M,rr:
    local j:
    s:=ff->(rr[1]+
             2*sum(rr[j+1]*cos(2*Pi*j*ff*delta_t),j=1..M-1)
             +rr[M+1]*cos(2*Pi*M*ff*delta_t))*delta_t:
    s(ff):
  end:
```

Calculate the spectral density for discrete frequncies given by equation (12.22),
store in array of coordinates [[f, s(f)],...] suitable for plotting.

```
> f:=l->l*fmax/M:     #equation (12.24)
  s:=[seq([f(l),spectral(f(l))],l=0..M)]:
  plot(s,x=0..fmax,y=0..5,color=black,thickness=2,
      labels=[`f`,`s(f)`],
      labelfont=[TIMES,ROMAN,16]);
```

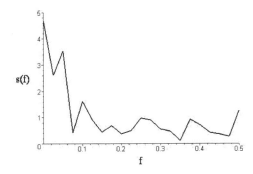

The correlogram suggests some persistence or serial correlation for the first three years in the lag, and then decreases to less than 0.1. Remember that for high lags the accuracy of the correlation coefficient decreases substantially. The spectral density suggests a dominance of cycles of less than 0.08. By playing with various values of M, one may find a consistent peak at frequencies of 0.02 and 0.06, suggesting periods of 50 and 16 years, respectively. This implies that there might be important periodicity in air temperature, which may not be confirmed with certainty with a record of this length.

12.3 SOME THEORETICAL STOCHASTIC PROCESSES

When we studied probability distributions in Chapter 3, we introduced some theoretical densities of common application in engineering uncertainty and risk analyses. Similarly, we now present some theoretical stochastic processes that have been useful in the simulation of engineering systems.

The Random Walk Process

Consider a binomial experiment, $X(t)$, repeated every T units of time for a large number of times N. Each trial has two possible outcomes {0, 1}, such that $P(0)=P(1)=p=1/2$. At each outcome, $X(t)$ experiences an instantaneous step upwards of magnitude s if the outcome is 1, or a step downwards if the outcome is 0. We have thus created a stochastic process known as a *Random Walk*. The magnitude of $X(t=NT)$ depends on the experimental outcome, that is on the particular sequence of zeros and ones. Each sample function of the Random Walk process is of a staircase form with discontinuities at the points $t=NT$.

If after N trials we obtain k outcomes 1, then the process has taken k steps up and $(N-k)$ steps down. Thus, the value of the process is $X(NT)=ks-(N-k)s=(2k-N)s$. In other words, $X(NT)$ is a random variable taking on the values $(2k-N)s$. From equation (3.11) the event {k outcomes 1 in N trials} is given by

$$\left(X(NT)=(2k-N)s\right)=C_k^N\frac{1}{2^N} \tag{12.25}$$

The mean and variance of the Random Walk process are given by

$$E\{X(NT)\}=0, \qquad \sigma_X^2(NT)=Ns^2=E\{X^2(NT)\} \tag{12.26}$$

From the DeMoivre-Laplace theorem, equation (3.25), for large N, and $p=1/2$,

$$P\left(X(NT)=(2k-N)s\right)\approx\frac{e^{-\frac{(k-N/2)^2}{N/2}}}{\sqrt{\pi N/2}} \tag{12.27}$$

Example 12.13: Generation of a Random Walk Sample Function

Write a computer program to generate and plot a Random Walk sample function of specified jump, s, and time interval T.

Solution

The following Maple worksheet illustrates a way to produce a sample function from a sequence of Binomial random numbers.

```
[ Initialize and define step and time interval sizes.
[ > restart: with(stats): with(plottools): s:=1.: T:=1.: N:=100:
[ Generate N Binomial random numbers with 0 and 1 as possible outcomes.
[ > outcomes:=[random[binomiald[1,1/2]](N+1)]:
[ Define the magnitude and direction of each jump.
[ > for i from 1 to N+1 do
      if (outcomes[i]=1.0) then jump[i]:=s
      else jump[i]:=-s end if:
   end do:
[ Calculate the ordinates of X(t).
[ > X[0]:=0:
   for i from 1 to N do
      X[i]:=X[i-1]+jump[i]*s
   end do:
[ Plot X(t) as a sequence of vertical and horizontal lines. Each line has a beginning and and end [x,y]
  coordinates.
[ > vert:=seq( line([i*T,X[i]],[i*T,X[i+1]]) ,i=0..N-1 ):
   horz:=seq( line([i*T,X[i+1]],[(i+1)*T,X[i+1]]) ,i=0..N-1 ):
   plots[display](vert,horz,labels=[`t`,`X(t)`],thickness=2,
   labelfont=[TIMES,ROMAN,16]);
```

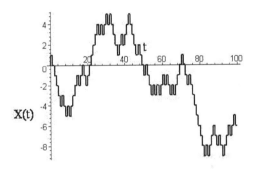

The Brownian Motion Process

The *Brownian Motion* process, also called the *Wiener-Lévy* process may be constructed as a limiting form of the Random Walk process as T tends to zero. If $t=NT$, then equation (12.26) becomes

$$E\{X(t)\}=0, \qquad \sigma_X^2(t)=\frac{ts^2}{T} \qquad (12.28)$$

For a given t, equation(12.28) indicates that in order to keep the variance of $X(t)$ finite and different from zero, s must tend to zero as $\sqrt{T}$. If we assume, in general, that

$$s^2=qT \qquad (12.29)$$

then we define the Brownian Motion process as the limit

$$B(t) = \lim_{T \to 0} X(t) \qquad (12.30)$$

where
$B(t)$=Brownian Motion process

The sample functions of this process are continuous for almost all outcomes. They resemble the sinuous erratic movement of colloidal particles in fluid media. From equations (12.28) the mean and the variance of the Brownian Motion process are given by

$$E\{B(t)\} = 0, \qquad \sigma_B^2(t) = qt = E\{B^2(t)\} \qquad (12.31)$$

where
q=Brownian Motion variance parameter.

For a fixed t, the probability density function of the random variable $B(t)$ is Gaussian with a mean of zero and a variance of qt (see Papoulis, 1984 for proof):

$$f_B(b, \ t) = \frac{e^{-\frac{b^2}{2qt}}}{\sqrt{2\pi qt}} \qquad (12.32)$$

Hence the density function of the Brownian Motion process is a Normal curve with a variance that increases linearly as a function of time, t. The autocorrelation function of the Brownian Motion process is given by

$$R_B(t_1, \ t_2) = E\{B(t_1)B(t_2)\} = q\min(t_1, \ t_2) \qquad (12.33)$$

where
$\min(t_1, \ t_2)$=the minimum between t_1 and t_2

Thus, the autocorrelation function equals qt_1, $t_1 \le t_2$, or qt_2, $t_2 \le t_1$. Since the mean of the Brownian Motion process is zero, then the autocovariance is equal to the autocorrelation function.

Example 12.14: Generation of a Sample Function of the Brownian Motion Process

A particle of sediment released at a stream at t=0min travels longitudinally downstream with a lateral deviation (m) from the origin following a Brownian Motion process with q=0.1 m^2/min. Generate and plot a sample function of the lateral particle trajectory during 1000 min, its mean, and its mean plus and minus one standard deviation.

Solution

In the following program, $Y(t)$ is the lateral displacement in meters, $q = 0.1\,m^2/min$, and $N = 1000\,min$. We set a time step $T = 1\,min$ and calculate the required jump size from equation (12.29). The required Brownian Motion sample is calculated by modifying the program in Example 12.13.

Initialize, define time interval, and calculate step from equation (12.29).

```
> restart: with(stats): with(plottools):
  q:=0.1: T:=1: s:=sqrt(q*T): N:=1000:
```
Generate A Brownian Motion sample as a limiting Random Walk.
```
> outcomes:=[random[binomiald[1,1/2]](N+1)]:
  for i from 1 to N+1 do
      if (outcomes[i]=1.0) then jump[i]:=s
          else jump[i]:=-s end if:
  end do:
  Y[0]:=0:
  for i from 1 to N do Y[i]:=Y[i-1]+jump[i]*s end do:
  vert:=seq( line([i*T,Y[i]],[i*T,Y[i+1]]),i=0..N-1 ):
  horz:=seq( line([i*T,Y[i+1]],[(i+1)*T,Y[i+1]]),i=0..N-1):
  sig1:=line([0,0],[N*T,sqrt(q*N*T)],linestyle=3):
  sig2:=line([0,0],[N*T,-sqrt(q*N*T)],linestyle=3):
  plots[display](vert,horz,sig1,sig2,thickness=2,
      labels=[`t (min)`,`Y (m)`],
      labelfont=[TIMES,ROMAN,16]);
```

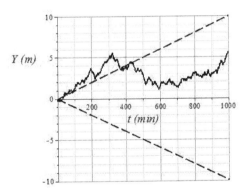

The White Gaussian Noise Process

The *White Gaussian Noise* process results from the formal derivative of the Brownian Motion process. Although the derivative of the Brownian Motion process does not exist for a single sample function, nevertheless it exists in the mean square sense as we will explain in Chapter 13. Thus, formally, we define the process

$$W(t) = \frac{dB(t)}{dt} \tag{12.34}$$

where
> $W(t)$=The White Gaussian Noise process

The mean and variance of the White Noise process are given by (see Jazwinski, 1971 for proof).

$$E\{W(t)\}=0, \quad E\{W(t_1)W(t_2)\}=q\delta(t_1-t_2), \quad \sigma_W^2=q \qquad (12.35)$$

where
> q=White Noise Process variance parameter
> $\delta(t_1-t_2)$=Dirac's delta function.

The first figure in Example 12.9 illustrates a typical sample function of a White Gaussian Noise process, commonly encountered in engineering systems. Notice that the variance is constant over time. The probability density function of the random variable $W(t)$ is Gaussian with a mean of zero and a constant variance q:

$$f_W(w, t)=\frac{e^{-\frac{w^2}{2q}}}{\sqrt{2\pi q}} \qquad (12.36)$$

The White Gaussian Noise process is stationary. Since the mean is zero, the autocovariance equals the autocorrelation function (see equation (12.7)). The autocorrelation coefficient is simply (see equation (12.10))

$$\rho_W(t_1, t_2)=\delta(t_1-t_2) \qquad (12.37)$$

The second figure in Example 12.9 illustrates the sample autocorrelation coefficient of the White Noise process, $\hat{\rho}_W=r_W$. The delta function indicates that there is a jump of magnitude 1 at lag $k=t_1-t_2=0$. At any lag different from zero the autocorrelation coefficient is zero, thus indicating some "memoryless" or serially uncorrelated data.

From equation (12.18) and (12.21) it can be shown that the spectral density function of the White Gaussian Noise process is a constant value of 1. This indicates that the process is composed of all the frequencies in the spectrum with no particular one prevailing. Thus, $W(t)$ is the mathematical analogy of white light, which is composed of all the primary color frequencies. The figure in Example 12.11 shows a sample of the spectral density function of this process. Although the sample spectrum is uncertain, it can be seen that its values oscillate around a constant 1.

The Random Walk, the Brownian Motion, and the White Noises processes, and their associated corollaries, are commonly adopted as theoretical models of engineering systems. Their association with the Gaussian theory and their simple mathematical manipulation make these models popular. A common variation of the White Noise process is the *Colored Noise* process, which has similar sample functions to the White Noise, but its autocorrelation coefficient is an exponentially decayed function. Example 12.8 shows the autocorrelation coefficient of a Colored Noise process. See Problem 12.9 at the end of this chapter for the spectral density function of a Colored Noise process. The spectrum combines several frequencies rendering low frequencies to prevail over high frequencies. To synthesize data from a Colored Noise process we will use the solution of a stochastic differential equation forced by White Noise. See Chapter 13.

12.4 TIME SERIES ANALYSIS

Until now, we have considered stochastic processes for which the *ensemble properties* are available. In the previous sections, we had at our disposal a stochastic process given by a known analytical equation. Because of this we were able to calculate analytically the first and second order statistics of the process and to generate sample functions. In engineering applications the ensemble properties of the process are available after one fits or adopt a specific theoretical model to describe an experimental data. The processes considered in Examples 12.1 and 12.4 illustrate this situation. On the other hand, data from some processes may be available from one sample (i.e., one time curve only). This is the case of the historical record of a variable. In such cases the modeler must infer the properties of a stochastic process from a single realization. A set of observations that measure the variation in time of some aspect of a phenomenon is termed a *time series*. These observations may be made continuously in time, such as the graph from a continuous-recording rain gage; discretely in time, such as the flow rate in a river observed at noon every day; or an average or total value of the variable recorded at discrete times, such as the total daily rain recorded at 6:00 P.M. The resulting time series constitutes *one sample function* from a corresponding stochastic process.

Time Average Versus Ensemble Properties

We may obtain the ensemble properties of a stochastic process from all the sample functions if they are available. For example, if in Figure 12.1 the process $X(t)$ is the time evolution of temperature, the ensemble density function at time t', $f_X(x, t')$, is derived from the intersection of all possible samples with the line $t=t'$. On the other hand, if only one sample is available (e.g., the time curve $X(t, \xi_1)$), the *time average* density function is calculated from all the points included in the single sample (e.g., all points in the curve $X(t, \xi_1)$). Similarly, the ensemble mean of the process at time t', $\mu_X(t')$, is obtained from equation (12.5) based on the intersection of all possible samples with the line $t=t'$. Conversely, the time average mean of the process, $\bar{X}$, is calculated from all the points included in the single sample $X(t, \xi_1)$.

We may calculate the time-average covariance, the time-average variance, the time-average correlation coefficient, etc., based on a single realization of the stochastic process. In general, the time-average properties of stochastic processes differ from those of the ensemble. In the special case when the time-average properties equal the ensemble, true, properties, then the stochastic process is called *ergodic*. Ergodicity is often *assumed* out of necessity in engineering models when the time average properties estimated from a single realization are taken as equal to those of the ensemble. Time series that have been removed from any trend and periodic component approach this ideal condition.

With the above principles in mind, we calculate the time-average properties of stochastic processes with the aid of formulae already introduced for the calculation of sample statistics. Consider a single realization of a stochastic process composed of N values measured at equal time intervals Δt. The time-average mean of the stochastic process $X(t)$, $\bar{X}$, is calculated from equation (6.1), Chapter 6. The time-average autocovariance, $c_X(k)$, is calculated from equation (12.15). The time average autocorrelation coefficient, $r_X(k)$, is calculated from equation (12.14). The time-average variance, S_X^2, is calculated from equation (6.2).

Deterministic Trend

A steady and regular evolution in a time series through which the values are, on average, increasing or decreasing is defined as a long-term trend. Depending on the time scale of observation, trends in a signal could be the result of low frequency oscillations in the observation window. Under small scales of observation, a trend may be mistakenly observed, when in fact a larger scale of observation may correctly reveal a low frequency seasonal or periodic component. On the other hand, long term trends are more appropriate to the steady growth rates in economics. Thus a long-term trend is the part in a stochastic time series that obeys to predictable deterministic laws. These laws may be the result of the application of physical principles in continuum mechanics, such as mass and momentum conservation, or the application of other chemical, biological, and mathematical principles. The deterministic trend usually accounts for the largest portion of the total magnitude of a stochastic process. Let us conceive a common stochastic process as

$$X(t) = T(t) + Z(t, \xi) \qquad (12.38)$$

where
   $T(t)$=deterministic trend
   $Z(t, \xi)$=random component

$T(t)$ is usually the solution of the governing deterministic differential equation of the system, and $Z(t, \xi)$ or simply $Z(t)$ is a stochastic process accounting for the errors in the parameter estimation of the model $T(t)$, inaccuracies incurred upon the adoption of a simplified model and thus the errors in the predictability of the output process $X(t)$. $Z(t)$ is characterized by its statistical properties. The larger the magnitude of its variance, $\sigma_Z^2(t)$ the more important the uncertainty characteristics in $X(t)$. For small variances of Z, the uncertain or statistical analysis of the output variable may be neglected. In practical situations, a governing physical equation is not available to represent $T(t)$. Typical examples include the price of a company stock and the bank prime lending rate. In such cases the trend is represented as a polynomial fitted through the cloud of N historical observations of the process (e.g., a nonlinear regression model), or an exponential function of time. Long-term trends in a stochastic time series may be detected in the spectral density function as low frequencies.

Equation (12.38) is an idealized presentation of a true process. It assumes that the deterministic components may be linearly added to the stochastic ones. Such model may not be realistic since $T(t)$ may include random effects or deterministic effects that may be passed to the random component $Z(t)$. Assuming that equation (12.38) is valid, if a trend is detected and fitted to $T(t)$, then the fitted values of the trend are customarily subtracted from the original signal $X(t)$. The resulting data points constitute a sample of the random component $Z(t)$. The latter are further analyzed in order to identify its serial correlation structure and to detect and remove any periodic component.

Periodicity
Cyclic, periodic, or seasonal fluctuations in time series are deterministic components caused by well-known physical phenomena. As in the study of trends, a periodic component should be represented by the solution to a physically based governing differential equation. When a governing system equation is not available, seasonal components are

observed as cyclic oscillations in the correlogram. If a sample size is large, a reliable sample correlogram may detect the average period of oscillation in the series. Alternatively, the sample spectral density function may show frequencies of oscillations, and thus periods, as jumps or unusually high values.

If a trend and a periodic component are present in a time series, then equation (12.38) becomes

$$X(t) = T(t) + P(t) + Z(t, \xi) \qquad (12.39)$$

where

$P(t)$=periodic component

Again, equation (12.39) is an idealization of a complex natural phenomenon. The trend and periodic components may not realistically add to an assumed independent random component. For example if the periodic component is a solution to a differential equation with a coefficient described as a random variable or a random process, the system output (i.e., the solution of the equation) is not, in general, a linear addition to the random part. Furthermore, if the differential equation is nonlinear, then the separation of $X(t)$ into the summation of deterministic plus stochastic parts is invalid. Nevertheless, equation (12.39) is a useful simplification or approximation in some engineering models. If a periodic component is detected in this manner, a trigonometric function of some sort, is usually employed as a model of periodicity. As in the case of a trend, once a periodic model is fitted, it is usually removed from the original series. The resulting data points constitute a sample of the random component $Z(t)$. The latter are further analyzed to identify its serial correlation structure and to explore auto-regressive models to describe it.

Models for the Random Component

As described in the last two headings, a simple approximation to a stochastic time series $X(t)$ is to assume it is composed of a deterministic component (e.g., trend, periodicity, or both) plus a random component. The deterministic component accounts for the action of fundamental laws in the process. The random component accounts for uncertain errors from measurement, errors arising from simplifications in model development, and unpredictable natural fluctuations. If a deterministic component is detected and a model is fitted to describe it, then it is subsequently subtracted from the original series. The resulting sample of the random component $Z(t)$ is further analyzed in order to study possible models to represent it. If the analysis of trend and periodicity has been done properly, the resulting random component or *residual* should have a mean of zero. In other words, $\bar{Z}$ must be equal to zero.

Several statistical tests are available for the analysis of the residual. Most commonly, a correlogram and a spectral density function of $Z(t)$ are produced in order to detect further hidden periodic components, dominant frequencies, but most importantly to study its serial statistical dependency with time. If the resulting correlogram suggests an uncorrelated series for lags greater than 1, then a delta-correlated stochastic process (i.e., a White Noise process) could be used to represent $Z(t)$ (see Example 12.9). On the other hand, if the correlogram suggests an exponentially decaying curve with lag time, or if the spectral density indicates a smooth combination of several frequencies, then a Colored-Noise process may be a suitable model for $Z(t)$ after adjusting for its variance. For discrete time series, however, if the autocorrelation coefficient function indicates serial dependency up to a given lag time, then a model that reflects this feature should be chosen. These are called *auto-regressive models*. There are literally tens of these models available in the literature, each intending to simulate a particular set of characteristics, such as the reproduction of a finite lag p (e.g., the AR(p) or p-order auto-regressive model), and non stationary properties (e.g., the ARMA or auto-regressive moving average models). The field of time series analysis is so vast that it became a complete specialization in statistics. For a comprehensive introduction see Box and Jenkins (1976).

Example 12.15: An Auto-Regressive Runoff Series
From historical records, annual runoff in a watershed ($mm/year \times 10^2$) has a time-average mean of 4.7, a time-average standard deviation of 0.958, and sample lag-one autocorrelation coefficient of 0.324. If the autocorrelation coefficient is about zero for lags greater than one, and if no significant trend or periodicity is detected, use a first-order autoregressive model to generate runoff data for three years.

Solution
Let X_i be runoff in year i ($mm/year \times 10^2$); $S_X = 0.958(mm/year \times 10^2)$; $r_X(1) = 0.324$; $r_X(k) = 0$, $k > 1$; $T(t) = P(t) = 0$. A stationary stochastic time series X_i with mean $\bar{X}$ and standard deviation S_X may be produced from series Z_i with a mean of zero and a standard deviation of 1 via the transformation

$$Z_i = \frac{X_i - \bar{X}}{S_X} \tag{12.40}$$

A simple first-order auto-regressive model $AR(1)$ for Z_i is given by (see Kottegoda, 1980 for details)

$$Z_i = r_X(1)Z_{i-1} + \sqrt{1 - r_X^2(1)}\ Y_i \qquad (12.41)$$

where $Y_i \sim N(0, 1)$ is a Standard Gaussian random number. Equation (12.41), familiarly known as a *Markov model*, states that the value of the process at time i equals that at the previous time, i-1, scaled by the autocorrelation coefficient plus a Standard Gaussian number with a variance of $1 - r_X^2(1)$. This model neglects any autocorrelation coefficient of lags greater than 1 and assumes that Y_i is independent of Z_{i-1}. From equations (12.40) and (12.41) the runoff series generation equations are:

$$Z_i = 0.324 Z_{i-1} + 0.946 Y_i, \qquad X_i = 4.7 + 0.958 Z_i \qquad (12.42)$$

To start the generation of data, set i=0, Z_0=0, and $X_0 = \bar{X} = 4.7$. Now set i=1, read a random number from a table of Standard Gaussian random numbers or a calculator, Y_1=0.87. From equations (12.42), calculate the value of Z_1=0.823 and then X_1=5.49. Now set i=2, read another random number Y_2=-0.65 and use again equations (12.42) to calculate Z_2=-0.348 and X_2=3.94. Next, i=3, Y_3=1.15, Z_3=-0.975, and X_3=5.63. For more realizations this process is repeated iteratively.

Example 12.16: Fitting a Trend and an Auto-Regressive model
    Table 12.1 lists values of a random process $X(t)$ measured at fixed time intervals t=1, 2, 3, ..., N. Determine if there is a trend in the series, remove it from the original data, and plot the residual as a function of time. Finally, produce and plot a correlogram of the residual.

Table 12.1: Data for Example 12.16

t	1.0	2.0	3.0	4.0	5.0	6.0	7.0	8.0	9.0	1.0
$X(t)$	6.1	5.3	4.2	7.9	9.3	7.2	13.6	11.1	10.5	12.7
t	11.0	12.0	13.0	14.0	15.0	16.0	17.0	18.0	19.0	20.0
$X(t)$	13.9	15.8	14.2	13.8	16.6	18.8	19.5	20.2	19.7	20.9
t	21.0	22.0	23.0	24.0	25.0	26.0	27.0	28.0	29.0	30.0
$X(t)$	22.5	20.7	21.0	23.0	23.3	24.3	23.4	23.4	25.9	23.1
t	31.0	32.0	33.0	34.0	35.0	36.0	37.0	38.0	39.0	40.0
$X(t)$	26.1	26.2	26.8	24.9	27.7	24.7	28.5	27.3	29.7	28.0

Solution
    A plot of the data $X(t)$ versus t suggests a best-fit curve of the form $X = \alpha t^\beta$ where α and β are constants to be determined. This falls into the

category of one-term polynomials that may be log-transformed to a linear regression line (see equation (8.9), Chapter 8). Taking logarithms on both sides of this expression we obtain a straight line of the form $y=ax+b$, where $y=\log(X)$, $a=\beta$, $x=\log(t)$, and $b=\log(\alpha)$.

From equations (8.6) and (8.7) the least-squares estimates of the linear regression coefficients are $\hat{a}=c_{xy}/S_x^2$ and $\hat{b}=\bar{y}-\hat{a}\bar{x}$, where c_{xy} is the covariance between x and y; S_x^2 is the variance of x; $\bar{y}$ is the sample mean of y; and $\bar{x}$ is the sample mean of x. Finally, from $\hat{a}$ and $\hat{b}$ we obtain the best estimates of the trend parameters: $\alpha=10^{\hat{b}}$ and $\beta=\hat{a}$. Thus, by taking logarithms of the data and applying the above equations we obtain a continuous curve of time $T(t)=\alpha t^{\beta}$ describing the trend. The following Maple program shows the calculations.

```
Initialize, enter X(t) data, and store it into an array suitable for plotting,
that is an array [[x1,y1],[x2,y2],...] coordinates. Fit a nonlinear regression trend T(t).
> restart: with(stats):
X:=[6.1,5.3,4.2,7.9,9.3,7.2,13.6,11.1,10.5,12.7,13.9,15.8,
    14.2,13.8,16.6,18.8,19.5,20.2,19.7,20.9,22.5,20.7,21.0,
    23.0,23.3,24.3,23.4,23.4,25.9,23.1,26.1,26.2,26.8,
    24.9,27.7,24.7,28.5,27.3,29.7,28.0]:
N:= describe[count](X):
data_x:=[seq([t,X[t]],t=1..N)]:
y:=[seq(log[10.](X[i]),i=1..N)]:
x:=[seq(log[10.](t),t=1..N)]:
x_bar:=describe[mean](x):
y_bar:=describe[mean](y):
S_x2:=describe[variance[1]](x):
c_xy:=describe[covariance](x,y):
a_hat:=c_xy/S_x2: b_hat:=y_bar-a_hat*x_bar:
alpha:=10^b_hat: beta:=a_hat:
T:=t->alpha*t^beta:
plot(T(t),t=1..N,color=black,thickness=2):
plot(data_x,style=point,color=black,symbol=box):
plots[display](%,%%,labels=[`t`,`X(t), T(t)`],
    labelfont=[TIMES,ROMAN,16],
    labeldirections=[horizontal,vertical]);
```

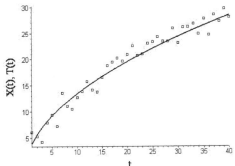

```
De-trend the series: Find the rsidual R(t)=X[t]-T(t). Note that X[ ] is a vector,
whereas T(t) is a continuous function of t.
> R:=[seq(X[t]-T(t),t=1..N)]:
data_R:=[seq([t,R[t]],t=1..N)]:
plot(data_R,color=black,labels=[`t`,`R(t)`],
    labels=[`t`,`X(t), T(t)`],labelfont=[TIMES,ROMAN,16],
    labeldirections=[horizontal,vertical]);
```

```
[ Produce a correlogram for the residual.
 > R_bar:=describe[mean](R) :
   S_R2:=describe[variance[1]](R) :
   r:=k->sum((R[j]-R_bar)*(R[j+k]-R_bar),j=1..N-k)/((N-k)*S_R2) :
   data_R:=[seq([k,r(k)],k=0..6)]:
   plot(data_R,thickness=2,color=black,labels=[`Lag k`,`r(k)`],
     labelfont=[TIMES,ROMAN,16],
     labeldirections=[horizontal,vertical]) ;
```

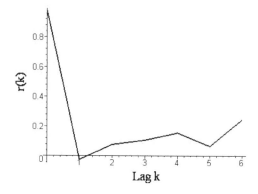

The sample mean of the residual is not zero, indicating less than an ideal trend model. The lag 1 autocorrelation coefficient is -0.0257. It would suggest an uncorrelated series if the autocorrelation coefficient for lags greater than 1 were also very small. On the other hand $N=40$ is a small sample size that produces an unreliable correlogram for large lags.

The above development may be generalized. If the process $Z(t, \xi)=Z_t$ may be expressed as the weighted average of p values of previous observations at times $t-1, t-2, \ldots, t-p$ plus a random number, then

$$Z_t=\varphi_{p,1}Z_{t-1}+\varphi_{p,2}Z_{t-2}+\ldots+\varphi_{p,p}Z_{t-p}+\eta_t=\sum_{i=1}^{p}\varphi_{p,i}Z_{t-i}+\eta_t \qquad (12.43)$$

where

Z_t=value of the residual at time t

$\varphi_{p,i}, i=1, 2,\ldots, p$=weights of prior observations $Z_{t-i}, i=1,\ldots, p$, respectively

η_t=a random number

Equation (12.43) is subject to

$$E\{Z_t\}=E\{\eta_t\}=0 \qquad\qquad Var\{Z_t\}=E\{Z_t^2\}=\sigma_Z^2$$

$$\rho_X(k)=\frac{E\{Z_tZ_{t-k}\}}{\sigma_Z^2}, \ k=1, 2,\ldots \quad E\{\eta_t\eta_{t-k}\}=E\{\eta_tZ_{t-k}\}=0 \qquad (12.44)$$

Without loss of generality, if $\sigma_Z^2=1$, then $X(t)=T(t)+P(t)+\left(S_X Z_t+\overline{X}\right)$. For the first-order auto-regressive model, AR(1), equation (12.43) reduces to

$$Z_t=\varphi_{1,1}Z_{t-1}+\eta_t \qquad (12.45)$$

Now multiply equation (12.45) by Z_{t-1} and take expectations:

$$E\{Z_t Z_{t-1}\}=\varphi_{1,1}E\{Z_{t-1}^2\}+E\{\eta_t Z_{t-1}\} \quad \Rightarrow \quad \rho_X(1)=\varphi_{1,1} \qquad (12.46)$$

Thus, $\varphi_{1,1}$ is simply the lag-one correlation coefficient. It is easy to show that $\sigma_\eta^2=1-\rho_X(1)^2$ or that the estimates $\hat{\sigma}_\eta^2=S_\eta^2=1-r_X(1)^2$. Hence, the AR(1) model for a residual process without trend or periodiciy reduces to

$$X(t)=\overline{X}+S_X Z_t, \quad Z_t=r_X(1)Z_{t-1}+\sqrt{1-r_X(1)^2}\ \eta_t, \quad \eta_t\sim N(0,\ 1) \qquad (12.47)$$

and equation (12.41) in Example 12.15 follows.

For higher-order AR(p) models, we proceed in a similar fashion. Multiply equation (12.43) by Z_{t-1} and take expectations:

$$E\{Z_t Z_{t-1}\}=\varphi_{p,1}E\{Z_{t-1}Z_{t-1}\}+\varphi_{p,2}E\{Z_{t-2}Z_{t-1}\}+...+\varphi_{p,p}E\{Z_{t-p}Z_{t-1}\}+E\{\eta_t Z_{t-1}\}$$

Therefore,

$$\rho_X(1)=\varphi_{p,1}+\varphi_{p,2}\rho_X(1)+\varphi_{p,3}\rho_X(2)+...+\varphi_{p,p}\rho_X(p-1) \qquad (12.48)$$

Repeat the process. Multiplying equation (12.43) by Z_{t-2}, Z_{t-3}, $\ldots$, Z_{t-p}, sequentially, and generalizing, we obtain a matrix equation given by

$$\begin{bmatrix} \rho_X(1) \\ \rho_X(2) \\ \rho_X(3) \\ \vdots \\ \rho_X(p) \end{bmatrix} = \begin{bmatrix} 1 & \rho_X(1) & \rho_X(2) & \cdots & \rho_X(p-1) \\ \rho_X(1) & 1 & \rho_X(1) & \cdots & \rho_X(p-2) \\ \rho_X(2) & \rho_X(1) & 1 & \cdots & \rho_X(3) \\ \vdots & & & & \vdots \\ \rho_X(p-1) & \rho_X(p-2) & \rho_X(p-3) & \cdots & 1 \end{bmatrix} \times \begin{bmatrix} \varphi_{p,1} \\ \varphi_{p,2} \\ \varphi_{p,3} \\ \vdots \\ \varphi_{p,p} \end{bmatrix} \qquad (12.49)$$

which me be written in matrix notation as

$$\rho_p=P_p\varphi_p \quad \Rightarrow \quad \varphi_p=P_p^{-1}\rho_p \qquad (12.50)$$

where

$\boldsymbol{\rho}_p$ =a correlation coefficient column matrix
$\boldsymbol{P}_p$ =a square and symmetric matrix of correlation coefficients
$\boldsymbol{\varphi}_p$ =a column matrix of the p weight coefficients $\varphi_{p,i}$, $i=1,..., p$

In order to use equation (12.50) to estimate the weight coefficients, $\varphi_{p,i}$, $i=1,..., p$, the correlation matrix $\boldsymbol{P}_p$ must possess a positive determinant.

Let us now multiply equation (12.43) by $Z_t + \eta_t$ to obtain

$$Z_t(Z_t + \eta_t) = \left[\sum_{i=1}^{p} \varphi_{p,i} Z_{t-i} + \eta_t \right] (Z_t + \eta_t) \tag{12.51}$$

Taking expectations on equation (12.51) and solving,

$$1 = \sum_{i=1}^{p} \varphi_{p,i} \rho_X(i) + \sigma_\eta^2 \;\Rightarrow\; \sigma_\eta^2 = 1 - \sum_{i=1}^{p} \varphi_{p,i} \rho_X(i) = 1 - R^2 \tag{12.52}$$

which provides the variance of η. For $p=1$, it reduces to $1 - \rho_X^2(1)$ (see equation (12.47)).

PROBLEMS

12.1 For the stochastic process, $R_g(t)$, in Example 12.7, $A \sim N(0.05, 0.01^2)$, $B \sim N(0.02, 0.01^2)$, and $C \sim N(2, 0.5^2)$. (1) Using 20 samples of A, B, and C plot 20 corresponding samples of $R_g(t)$. (2) Generate 1000 realizations of $R_g(t=12)$. Plot the frequency histogram of the samples. *Hint*: use the program in Example 12.1 as a model.

12.2 In Example 12.2, generate 300 samples of the process, each for $N=1000$ units of time. Plot all samples in a single graph.

12.3 For the stochastic process in Problem 12.1, assume that C is not a random variable, but a constant equal to 2. Derive analytically, step by step, the following expressions:(1) the mean, (2) the autocorrelation function, (3) the autocovariance function, (4) the variance, and (5) the autocorrelation coefficient.

12.4 With the results of Problem 12.3, in a single graph plot the mean of the process $R_g(t)$, the mean plus one standard deviation, and the mean minus one standard deviation.

12.5 For the stochastic process of Problem 12.1, design a Monte Carlo simulation model to simulate the mean plus and minus one standard deviation. Plot the results. *Hint*: Modify and adapt the program in Example 12.6 to simulate the process R_g of Problem 12.1. How do the Monte Carlo simulations compare with the analytical results of Problem 12.4?

12.6 Plot the approximate analytical mean of Example 12.7, using the data in Problem 12.1. How does this approximation compare with the Monte Carlo simulations of Problem 12.5 and with those of Problem 12.3?

12.7 With the results of Problem 12.3, and the data of problem 12.1, plot the correlogram of the recharge process and state your observations.

12.8 For the process of Example 12.2, plot the correlogram and interpret it.

12.9 Calculate and plot the spectral density function of a stochastic process with an autocorrelation coefficient function of the form $\rho_X(\tau) = e^{-|\tau|}/2$. Interpret it.

12.10 Modify and combine the programs in Examples 12.9 and 12.11 to plot the correlogram and the spectral density of the tensile strength data of Table 7.4, Chapter 7.

12.11 Modify the program in Example 12.13 to generate and plot a Random Walk sample function composed of 100 steps, each of 5 units in magnitude, and each one spaced by 10 units of time.

12.12 Modify the program in Example 12.14 to produce and plot several sample functions of the Brownian Motion along with the standard deviations. Each sample function simulates the lateral trajectory of an identical particle released at $t=0$.

12.13 Write a computer program of Example 12.14 and generate 100 years of data.

13.14 In Example 12.15, neglect autocorrelation coefficients of lags greater than one and modify the program to fit an $AR(1)$ model for the residual. Generate realizations for another 40 units of time.

". . . *since the basic equations of physics are nonlinear, all of the mathematical physics will have to be done over again. . ."*

Albert Einstein

"The retreat to linearity: On the one hand their analytical methods were not powerful enough to deal with the problem of nonlinearity; on the other hand, purely numerical techniques were not far advanced nor were they feasible frorm the standpoints of economics and engineering. Taking the only possible way out, they changed the problem, tailoring it to fit their modest mathematical means. Specifically, they "linearized" the equations using the perturbation methods . . ."

Philip Duncan Thompson

13 STOCHASTIC DIFFERENTIAL EQUATIONS

13.1 THE ORIGIN OF STOCHASTIC DIFFERENTIAL EQUATIONS

The fundamental building blocks of mathematical models in engineering are composed of solutions to differential equations. These equations result, in general, from the application of fundamental physical, chemical, biological, and other laws governing the system under consideration. Fundamental laws, such as mass conservation, energy conservation, momentum conservation, Fick's law of diffusion, Fourier's law of thermal conductivity, and the three laws of thermodynamics are but a few of the fundamental principles applied in the derivation of governing equations in continuum mechanics.

These laws are manipulated mathematically in congruence with other empirical or fundamental laws intervening in engineering systems undergoing spatial changes (e.g., gradients), or temporal variability (e.g., transient conditions). The resulting set of equations usually contains derivatives. These are functional derivatives of the *dependent variables* or system output with respect to the *independent variables* (i.e., the spatial or temporal coordinates). A system governed by an equation that contains total derivatives with respect to one variable is termed an *ordinary differential equation*. A system that contains partial derivatives with respect to more than one independent variable is termed a *partial differential equation*. The *coefficients* or *parameters* of the differential equation may be constants or functions of the independent variables, in which case the system is governed by a *linear differential equation*. Alternatively, the parameters may be functions of the dependent variable, in which case the system is governed by a *nonlinear differential equation*.

The differential equation may be subject to external excitations, a source, a sink, an input, or a *forcing function*. In addition, the system under consideration may be acting on a physical region with specific geometry. In such cases, the problem is classified as a *boundary-value problem* that requires knowledge of the *boundary conditions* (i.e., knowledge of the system output or dependent variable) at the boundaries of the region. If any of the functions of the equation are time dependent, knowledge of the dependent variable at $t=0$, called the *initial condition*, is necessary.

The solution process of a differential equation is the mathematical procedure followed to obtain an explicit analytical form (i.e., an *analytical solution*) that describes the system output or the dependent variable of the equation. If the procedure followed is numerical in nature, that is one resulting from a discretization of the spatial or temporal domain, followed by the application of linear algebra or finite mathematics techniques, then the result is a *numerical solution*. The solution to the equation is subsequently used to forecast the system output (i.e., *the direct problem*), to test the system behavior, to design an engineering solution for which the magnitude of the system output is required under a specified adverse condition, or to estimate the values of the parameters given some observations of the output (i.e., the *inverse problem*). Many analytical and numerical solutions are preprogramed under various computer languages, coupled with appropriate graphical representations of the output, and added to a friendly end-user interface that facilitates the input of data and the execution of graphical output and other operations. This is the end product of many computer software products available today.

The foregoing description refers to the framework of analysis most engineers learn at a university. The prospective engineer often takes several semesters of calculus and differential equations, followed by major-specific modeling and design courses. In this process the engineer is presented with a fundamental philosophy that the engineering system is deterministic and perfectly predictable. The resulting differential equations and solutions are deterministic, invariable, and without any room for measurement and field uncertainty. From our discussion in Chapter 1, a deterministic system is only a special case of the true and general stochastic system. Problems with a traditional deterministic formulation arise when measuring the values of the system parameters or the forcing function necessary for the application of the model. If repetitive observation of a system parameter under identical conditions yields a different value after each experiment, then the most appropriate description of the parameter is one that defines it as a random variable or a stochastic process.

Another problem with a deterministic formulation results from the measurement errors or uncertain fluctuations in the boundary conditions of a boundary-value problem. A third problem results when the model is compared with independent measurements of the system prototype (i.e., *model calibration* and *verification*). Rarely is the case of a model that perfectly adheres to field-measured values. Indeed, a model is the result of a simplified application of fundamental laws. However, in the process of model simplification some unknown or difficult variables have been neglected. In many circumstances, variables are neglected out of necessity in order to make a mathematical solution possible. In addition the solution process usually involves approximations, instability, and numerical errors. For these reasons, a model is an approximate representation of a more complex prototype. On the other hand, experimental measurements of the dependent variable in the prototype are equally subject to error, depending on the testing procedure, instrumentation, and measurement difficulties. Thus, a comparison between model output (i.e., an approximate representation of a prototype), and prototype measurements (i.e., an uncertain reflection of a prototype) constitutes an approximate means to *calibrate* a model. The correct calibration procedure should induce modifications in the model until the *output statistics* are comparable with those of the prototype. Ideally, a good model is considered acceptable if the statistical properties of the output stochastic process match those of the prototype-measured process. Ordinarily there is not sufficient information to assess this. Instead, if the mean and variance of the model output are similar in magnitude to the corresponding statistics of the prototype measurements, the model is considered acceptable and calibrated.

From this discussion, it is clear that in general differential equations arising in engineering and the sciences are stochastic. If the parameters, the initial condition, the boundary conditions, or the forcing function are stochastic, then the dependent variable, or system output, is also stochastic. The classification of a differential equation as deterministic or stochastic ultimately depends on a sensitivity test of the effect of a random parameter on the system output. If a small variance in the input parameter results in a small or negligible variance of the output, then it is reasonable to assume that the system is deterministic. In this case, disregard of random variability does not affect the ability of the model to predict an accurate output. This is the situation with linear dissipative engineering systems. On the other hand, some nonlinear or unstable systems will tend to amplify a small variance in an input. In such cases, it is appropriate to classify the equation as a stochastic differential equation and to attempt a solution process that produces information on the statistics of the output variable.

Solutions to stochastic differential equations are more difficult that corresponding deterministic ones. This is the reason why their treatment is not introduced in a regular engineering curriculum. However, uncertainty and risk analysis of engineering systems is of utmost importance in today's applications. In the following sections we introduce some fundamental, simple, and systematic techniques for the development of solutions to stochastic differential equations, which could serve as a basis for a general engineering stochastic model.

13.2 STOCHASTIC CONTINUITY, DIFFERENTIATION, AND INTEGRATION

Before considering stochastic differential equations and their solutions, it is important to redefine briefly fundamental calculus concepts occurring in differential equations in general, such as continuity, differentiation and integration. Since the dependent variable(s) in stochastic differential equations is stochastic, these operations are performed on stochastic processes. A stochastic process is a family of functions. What do we mean by continuity or differentiation of a stochastic process? In the following subheadings we define these concepts (see Papoulis, 1984 for proof).

Mean Square Continuity

There are several definitions of continuity of a stochastic process. A definition of strong continuity states that if each sample function of a stochastic process, $X(t)$, is continuous at t, then the process is continuous at t for every experimental outcome. This classical definition appears too restrictive. On the other hand, if $\lim X(t+\tau)=X(t)$, $\tau \to 0$ for almost all outcomes, then the process $X(t)$ is continuous with probability 1. A usual definition of continuity in engineering fields is that of *continuity in the mean square sense*. A process $X(t)$ is continuous in the mean square sense at a point t if

$$E\left\{ \left[X(t+\tau)-X(t)\right]^2 \right\} \to 0 \ as \ \tau \to 0 \qquad (13.1)$$

Expanding and using equation (12.8), then

$$E\left\{ \left[X(t+\tau)-X(t)\right]^2 \right\}=R_X(t+\tau,t+\tau)-R_X(t,t+\tau)-R_X(t+\tau,t)+R_X(t,t) \qquad (13.2)$$

Equation (13.2) tends to zero as $\tau \to 0$. Therefore the process $X(t)$ is continuous at t if its autocorrelation function $R_X(t_1, t_2)$ is continuous at the point $t=t_1=t_2$. A process is continuous for every t if its autocorrelation function $R_X(t_1, t_2)$ is continuous everywhere along the line $t_1=t_2$.

The following properties are the result of the above principle:

1. Mean square continuity does not necessarily imply continuity of each, or any, of the samples of a stochastic process. For example,

from equation (12.28) the autocorrelation function of the Random Walk process is continuous for every t. However, almost all sample functions of a Random Walk process are discontinuous at some points of the t axis.

2. From the mean square continuity of $X(t)$ the expected value must be continuous. Mathematically,

$$E\{X(t+\tau)\} \to E\{X(t)\} \ as \ \tau \to 0 \qquad (13.3)$$

3. We can interchange the order between the limit and the expected value of a function:

$$\lim E\{X(t+\tau)\} = E\{\lim X(t+\tau)\} \ as \ \tau \to 0 \qquad (13.4)$$

where lim denotes the limit.

4. A stationary process is continuous if its autocorrelation function $R(\tau)$ is continuous for $\tau = 0$. A stationary process continuous at $\tau = 0$ is also continuous for all values of τ.

Stochastic Differentiation

The classical definition of a derivative could be extended to the derivative of a stochastic process.

$$X'(t) = \frac{dX(t)}{dt} = \lim_{\tau \to 0} \frac{X(t+\tau) - X(t)}{\tau} \qquad (13.5)$$

If this limit exists for all sample functions of a process $X(t)$, then $X'(t)$ has the traditional meaning of a derivative. If the limit exists in the mean square sense, then $X(t)$ has a derivative in the mean square sense. A process $X(t)$ is said to have a mean square derivative at t if we can find another process $\mathbf{X}'(t)$ such that

$$\lim_{\tau \to 0} E\left\{ \left[\frac{X(t+\tau) - X(t)}{\tau} - X'(t) \right]^2 \right\} = 0 \qquad (13.6)$$

The following properties of mean square derivatives are useful:

1. A stationary process, $X(t)$, is differentiable in the mean square sense if its autocorrelation function, $R(\tau)$, has derivatives of order up to two.

2. A non stationary process, $X(t)$, is differentiable in the mean square sense if the second partial derivative of its autocorrelation function,

$R_X(t_1, t_2)$, with respect to t_1 and t_2 exists for $t_1=t_2$, that is if

$$\frac{\partial^2 R_X(t_1, t_2)}{\partial t_1 \partial t_2}$$

exists for $t_1=t_2$. For example, the autocorrelation function of the Brownian Motion process is given by equation (12.33). $\partial^2 R_B(t_1, t_2)/\partial t_1 \partial t_2$ does not exist; therefore, the Brownian Motion process is not differentiable at any t. The same is true for the Random Walk and the White Noise processes.

3. Property 3 under the heading of Mean Square continuity may be extended to derivatives of stochastic processes:

$$E\{X'(t)\} = \frac{dE\{X(t)\}}{dt} \tag{13.7}$$

In words, the expected value of a stochastic process, $X(t)$, is differentiable and its derivative equals the expected value of $x'(t)$. Furthermore, this property implies that we may exchange expectation and differentiation, a very useful property in the mathematical manipulation of moments of stochastic differential equations.

4. A generalization of property 3 above leads to the concept that the nth derivative of the process $X(t)$, $X^{(n)} = d^n X(t)/dt^n$ exists if

$$\frac{\partial^{2n} R_X(t_1, t_2)}{\partial t_1^n \partial t_2^n}$$

exists.

5. Similarly, a generalization of property 3 above concerning the exchangeability between expectation and differentiation gives

$$E\left\{\frac{d^n X(t)}{dt^n}\right\} = \frac{d^n E\{X(t)\}}{dt^n} \tag{13.8}$$

Example 13.1

A heuristic model of groundwater head, $h(x)$, assumes it to be given as (Serrano, 1995(1)) $h(x) = h_0(x) + h'(x)$, where $h_0(x)$ is a deterministic function of distance x (e.g., the Dupuit parabola), and $h'(x)$ is a Colored Noise process with a mean $E\{h'(x)\} = 0$, a correlation at two points x_1 and

x_2 given as $R_h(x_1, x_2) = C_v^2 e^{-\rho|x_1 - x_2|}$, C_v^2 the coefficient of variability of the transmissivity, and ρ the correlation length parameter. This implies that the correlation between two locations in the aquifer system decreases exponentially with the distance between the two observed points. Derive expressions for the mean, the autocorrelation and the variance functions of the specific discharge, or Darcy's velocity, given by (see equation (1.1), Chapter 1)

$$u(x) = -K \frac{\partial h}{\partial x}$$

where $u(x)$ is the specific discharge, and K is the mean hydraulic conductivity.

Solution

Substituting our model of hydraulic head into the expression of specific discharge,

$$u(x) = -K \frac{\partial}{\partial x} \left(h_0(x) + h'(x) \right)$$

The generation of sample functions of the $q(x)$ process is easily accomplished with the procedures outlined in Chapter 12. Upon taking expectations, using the rules of expectation, and equation (13.7),

$$E\{u(x)\} = \mu_u(x) = E\left\{ -K \frac{\partial}{\partial x} \left(h_0(x) + h'(x) \right) \right\} = -K \frac{\partial h_0}{\partial x} - E\left\{ K \frac{\partial}{\partial x} h'(x) \right\}$$

$$= -K \frac{\partial h_0(x)}{\partial x} - K \frac{\partial}{\partial x} E\{h'(x)\} = -K \frac{\partial h_0(x)}{\partial x}$$

In this case, because of the linearity of the differential equation, the mean of the output coincides with the deterministic component in the system. Now, calculate the correlation of the specific discharge at two points:

$$E\{u(x_1)u(x_2)\} = R_u(x_1, x_2)$$

$$= E\left\{ \left[-K \frac{\partial h_0(x_1)}{\partial x_1} - K \frac{\partial h'(x_1)}{\partial x_1} \right]\left[-K \frac{\partial h_0(x_2)}{\partial x_2} - K \frac{\partial h'(x_2)}{\partial x_2} \right] \right\}$$

$$= K \frac{\partial h_0(x_1)}{\partial x_1} K \frac{\partial h_0(x_2)}{\partial x_2} + K^2 \frac{\partial^2}{\partial x_1 \partial x_2} E\{h'(x_1)h'(x_2)\}$$

$$R_u(x_1,x_2)=K\frac{\partial h_0(x_1)}{\partial x_1}K\frac{\partial h_0(x_2)}{\partial x_2}+K^2\frac{\partial^2}{\partial x_1\partial x_2}\left[C_v^2 e^{-\rho|x_1-x_2|}\right]$$

$$= K\frac{\partial h_0(x_1)}{\partial x_1}K\frac{\partial h_0(x_2)}{\partial x_2}+K^2C_v^2\rho^2 e^{-\rho|x_1-x_2|}$$

Setting $x_1=x_2=x$ in this expression and using equation (12.9), we obtain the variance of the specific discharge:

$$\sigma_u^2=R_u(x,\ x)-\mu_u^2(x)=K^2C_v^2\rho^2=\sigma_K^2\rho^2$$

where $\sigma_K^2=C_v^2K^2$ is the hydraulic conductivity variance.

Example 13.2

Consider again the differential equation in Example 13.1 above, but this time assume that the hydraulic conductivity, K, is a random variable, while the hydraulic head, h, is deterministic. Derive expressions for the mean and the variance of the velocity, u.

Solution

taking expectations on both sides of the differential equation,

$$E\{u(x)\}=\mu_u(x)=-E\left\{K\frac{\partial h}{\partial x}\right\}=-E\{K\}\frac{\partial h}{\partial x}=-\mu_K\frac{\partial h}{\partial x}$$

The two-point correlation is given by

$$E\{uu\}=E\left\{\left[-K\frac{\partial h}{\partial x}\right]\cdot\left[-K\frac{\partial h}{\partial x}\right]\right\}=E\{K^2\}\left(\frac{\partial h}{\partial x}\right)^2$$

from which we may obtain the variance of the output:

$$\sigma_u^2=E\{u(x)^2\}-\mu_u^2=E\{K^2\}\left(\frac{\partial h}{\partial x}\right)^2-\mu_K^2\left(\frac{\partial h}{\partial x}\right)^2=\left[E\{K^2\}-\mu_K^2\right]\left(\frac{\partial h}{\partial x}\right)^2=\sigma_K^2\left(\frac{\partial h}{\partial x}\right)^2$$

Comparing the results from Examples 13.1 and 13.2 we see that the way in which randomness enters the differential equation directly affects the analytical form and the properties of the system output. The following example illustrates the differences between obtaining the moments of the output analytically versus numerically.

Example 13.3

In Example 13.2, assume that the $K \sim N(48.77, 12.29)m/month$, and that the hydraulic gradient is $\partial h/\partial x = -0.03 m/m$. (1) Calculate the mean and the standard deviation of the output, u, using the analytical results of Example 13.2. (2) Develop a Monte Carlo simulation model to plot the density function of the output and estimate the values of it mean and variance.

Solution

(1) Using the expressions derived in Example 13.2,

$$\mu_u = -\mu_K \frac{\partial h}{\partial x} = -48.77 \times (-0.03) = 1.46 m/month$$

$$\sigma_u = \sqrt{\sigma_K^2 \left(\frac{\partial h}{\partial x} \right)^2} = \sqrt{12.29^2 \times (-0.03)^2} = 0.37 m/mes$$

(2) The following Maple program shows the simulations.

```
> restart: with(stats): with(stats[statplots]):
    mu:=48.77: sigma:=12.29: dh_dx:=-0.03:
Generate 1000 samples of K.
> Kdata:=[random[normald[mu,sigma]]](1000):
Apply the differential equation to each sample.
> udata:=-Kdata*dh_dx:
Statistics and frequency histogram of the output.
> mu_u:=describe[mean](udata);
    sigma_u:=describe[standarddeviation](udata);
    histogram(udata,labels=[`u (m/month)`,`f`],color=gray,
        labelfont=[TIMES,ROMAN,16]);
```

$$mu\_u := 1.447736630$$

$$sigma\_u := 0.3676833212$$

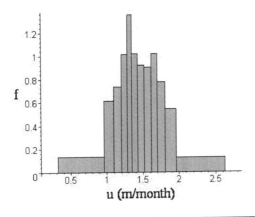

Comparing Examples 13.2 and 13.3, we can draw some conclusions concerning the differences between analytical modeling and Monte Carlo simulations of systems governed by random differential equations:

Analytical method:
1. Does not require knowledge of the probability density function of the random components in the system.

2 Numerical calculations are simple.

3. Requires mathematical analysis and derivations.

Monte Carlo simulation:
1. Does not require mathematical analysis and derivations.

2. Numerical calculations, computer programming and storage, and numerical instability may be important.

3. The probability density function of the random components in the system must be known.

Stochastic Integrals

A stochastic process $X(t)$ is mean square integrable if the limit

$$\int_a^b X(t)dt = \lim_{\Delta t_i \to 0} \sum_i X(t_i)\Delta t_i \tag{13.9}$$

exists in the mean square sense. The following properties of stochastic integration are useful (see Papoulis, 1984 for proof):

1. A process $X(t)$ is integrable over the interval (a, b) if the double integral of the autocorrelation function is bounded. Specifically,

$$\int_a^b \int_a^b |R_X(t_1, t_2)| dt_1 dt_2 < \infty \tag{13.10}$$

2. From property 1 it follows that

$$E\left\{ |\int_a^b X(t)dt|^2 \right\} = \int_a^b \int_a^b R_X(t_1, t_2)dt_1 dt_2 \tag{13.11}$$

3. Property 3 under the heading of Stochastic Differentiation (i.e., equation (13.7)) may be extended to stochastic integrals. In other words, the order of integration and expectation operators may be exchanged. Specifically, if the process $Y(t)$ is the integral of another process $X(t)$, that is if $Y(t) = \int_0^t X(\tau)d\tau$, then

$$E\{Y(t)\} = \mu_Y(t) = E\left\{\int_0^t X(\tau)d\tau\right\} = \int_0^t E\{X(\tau)\}d\tau = \int_0^t \mu_X(\tau)d\tau \qquad (13.12)$$

4. By applying property 3 we may derive expressions of the correlation function of a stochastic integral equation. Specifically, if $Y(t) = \int_0^t X(\tau)d\tau$, then

$$E\{Y(t_1)Y(t_2)\} = R_Y(t_1,\ t_2) = E\left\{\left[\int_0^{t_1} X(t)dt\right]\left[\int_0^{t_2} X(\tau)d\tau\right]\right\}$$

$$= E\left\{\int_0^{t_1}\int_0^{t_2} X(t)X(\tau)dtd\tau\right\} = \int_0^{t_1}\int_0^{t_2} E\{X(t)X(\tau)\}dtd\tau \qquad (13.13)$$

$$= \int_0^{t_1}\int_0^{t_2} R_X(t,\ \tau)dtd\tau$$

Equation (13.13) illustrates the procedure to obtain the autocorrelation function of a stochastic integral, usually resulting from the solution of a stochastic differential equation.

5. Having derived the autocorrelation function, the autocovariance of a stochastic integral may be obtained. From equations (12.7), (13.12), and (13.13),

$$C_Y(t_1,\ t_2) = R_Y(t_1,\ t_2) - \mu_Y(t_1)\mu_Y(t_2)$$

$$= \int_0^{t_1}\int_0^{t_2} R_X(t,\ \tau)dtd\tau - \int_0^{t_1}\int_0^{t_2} \mu_X(t)\mu_X(\tau)dtd\tau \qquad (13.14)$$

6. From equation 12.9, the variance of a stochastic integral is found by setting $t_1 = t_2 = t$ in equation (13.14):

$$\sigma_Y^2(t) = C_Y(t, t) = R_Y(t, t) - \mu_Y^2(t)$$

$$= \int_0^t \int_0^t R_X(\tau, \xi) d\tau d\xi - \int_0^t \int_0^t \mu_X(\tau) \mu_X(\xi) d\tau d\xi \tag{13.15}$$

7. From equations (12.10), (1314), and (13.15), the autocorrelation coefficient may be found.

Example 13.4

A system receives an input, $X(t)$, characterized as a White Gaussian Noise process with variance parameter σ_X^2. Derive expressions for the mean, the correlation and the variance of the output, $Y(t)$, if it is given as $Y(t) = \int_0^t X(\tau) d\tau$.

Solution

From equations (12.35) and (13.12), the mean of $Y(t)$ is given as

$$\mu_Y = \int_0^t \mu_X(\tau) d\tau = 0$$

From equations (12.35) and (13.13), the autocorrelation function of the output is

$$R_Y(t_1, t_2) = \int_0^{t_1} \int_0^{t_2} R_X(t, \tau) dt d\tau = \sigma_X^2 \int_0^{t_1} \int_0^{t_2} \delta(t-\tau) dt d\tau = \sigma_X^2 \int_0^{t_1} d\tau = \sigma_X^2 t_1, \quad t_1 < t_2$$

Note we have used the sampling properties of integrals of the delta function:

$$\int_a^b \delta(x-c) dx = 1, \qquad \int_a^b f(x) \delta(x-c) dx = f(c), \qquad a < c < b \tag{13.16}$$

As expected, the correlation function of the integral of a White Gaussian noise process is the same as that of a Brownian Motion process (see equations (12.33) and (12.34)). Now from equation (13.15), the variance of the output is given by

$$\sigma_Y^2(t) = R_Y(t, t) - \mu_Y^2(t) = \sigma_X^2 t$$

Example 13.5

Consider the stochastic integral with respect to *Brownian Motion increment* given by

$$Y(L) = \int_0^L \sin(ax)dB(x)$$

where $B(x)$ is a Brownian Motion process such that (see equation (12.31), Chapter 12)

$$E\{B(x)\} = 0, \qquad E\{B(x_1)B(x_2)\} = \sigma_B^2 x_1, \quad x_1 < x_2$$

Find the mean and correlation function of Y.

Solution

From Chapter 12, recall that $dB(x)/dx = W(x)$, a White Gaussian Noise in space. Thus, the integral equation may be written as

$$Y(L) = \int_0^L \sin(ax)\frac{dB(x)}{dx}dx$$

Taking expectations,

$$E\{Y(L)\} = E\left\{\int_0^L \sin(ax)\frac{dB(x)}{dx}dx\right\} = \int_0^L \sin(ax)E\left\{\frac{dB(x)}{dx}\right\}dx$$

$$= \int_0^L \sin(ax)\frac{dE\{B(x)\}}{dx}dx = 0$$

For the correlation function at two locations, L_1 and L_2,

$$E\{Y(L_1)Y(L_2)\} = R_Y(L_1, L_2) = E\left\{\left[\int_0^{L_1}\sin(ax)\frac{dB(x)}{dx}dx\right]\left[\int_0^{L_2}\sin(as)\frac{dB(s)}{ds}ds\right]\right\}$$

$$= \int_0^{L_2}\int_0^{L_1}\sin(ax)\sin(as)\frac{\partial^2 E\{B(x)B(s)\}}{\partial x \partial s}dxds$$

$$= \sigma_B^2\int_0^{L_2}\int_0^{L_1}\sin(ax)\sin(as)\frac{\partial^2\min(x,\ s)}{\partial x \partial s}dxds$$

$$E\{Y(L_1)Y(L_2)\} = \sigma_B^2 \int_0^{L_2}\int_0^{L_1} \sin(ax)\sin(as)\delta(x-s)dxds = \sigma_B^2 \int_0^{L_1} \sin^2(ax)dx, \quad L_1 < L_2$$

$$= \sigma_B^2 \left[\frac{x^2}{2} - \frac{\sin(2ax)}{4a} \right]_0^{L_1} = \sigma_B^2 \left[\frac{L_1^2}{2} - \frac{\sin(2aL_1)}{4a} \right]$$

where equation (12.33), Chapter 12, has been used.

13.3 SOLVING APPLIED STOCHASTIC DIFFERENTIAL EQUATIONS

Stochastic differential equations might be viewed as regular differential equations in which one or more of its components are functions of the probability variable. The forcing function (or system input), the parameters, or the coefficients may be defined in probabilistic terms, and the dependent variable (or system output) depends on probability, in addition to space, or time.

Let us briefly review a classification of differential equations. A differential equation may have one or more *independent variables*. These are the spatial coordinates x, y, or z, and the temporal coordinate t, in general. A differential equation may have one or more *dependent variables* or *output variables*. These are the variables with derivatives customarily placed on the left side of the equality. They represent the variables of interest. A differential equation may have one or more *input variables* or *forcing functions*. These are additional functions normally placed on the right side of the equality in systems engineering. A differential equation may contain one or more *coefficients* or *parameters*. These are constants or functions multiplying the derivatives of the differential equation. A *solution* to a differential equation is a function of the independent variables and satisfies the differential equation upon direct substitution. A solution implies manipulating the differential equation analytically or numerically in order to find a description on the form of the dependent variable. This solution no longer contains derivatives of the dependent variable.

Other classification schemes of differential equations are useful. According to the number of independent variables and the type of derivatives the equation contains, we may classify the differential equations as

1. Ordinary differential equations, which have total derivatives with respect to only one independent variable.

2. Partial differential equations, which contain partial derivatives with respect to two or more independent variables.

According to the highest order of the derivative, differential equations may be classified as

1. First-order differential equations, if the highest order derivative is 1.

2. Second order, if the highest order derivative is 2, etc.

According to the functional relation between the coefficients and the dependent variable and the degree of the derivatives, differential equations are classified as

1. Linear, if the coefficients are not functions of the dependent variable or if the degree of all of the derivatives is 1 (i.e., if none of the derivatives is elevated to a power different from 1).

2. Nonlinear if the coefficients of the equation are functions of the dependent variable or if any of the derivatives are elevated to a power different from 1.

Whether or not an equation has random components, it leads to a classification of interest here:

1. Deterministic, if all of its components are deterministic.

2. Stochastic, if the input, initial condition, or the coefficients are defined as probabilistic quantities.

There are other classifications of interest in engineering. For instance an equation may be judged as physically based if it resulted from the application of fundamental physical laws, or empirically based otherwise. A differential equation that describes a causal, or time dependent, system may be classified as a transient or non stationary equation. On the other hand, a time independent equation may be called a steady or stationary equation.

A stochastic differential equation may be further classified according to the way in which the random components enter the equation:

1. Differential equations with random initial conditions.

2. Differential equations with random forcing functions.

3. Differential equations with random boundary conditions.

4. Differential equations with random coefficients.

5. Differential equations with random geometrical domains.

6. Hybrid equations that combine two or more of the above conditions.

A solution to a random differential equation is a stochastic process function of the independent variables that satisfies, in the mean square sense, the differential equation. Finding the probability density function, of all orders, of the dependent variable is usually a hopelessly complicated problem. Most of the time it is simpler to derive the first order moments of the solution. Furthermore, in many applications the only information available on the uncertain properties of the input variables is precisely their first order moments. Consequently, the modeler finds natural to attempt a description of the first order moments of the output variable.

Before embarking on the problem of solving random differential equations, the reader is strongly encouraged to review the basis of the existing theory of applied differential equations. There are many excellent textbooks available. Some of the classical Schaum's Outline Series are extremely attractive since they present hundreds of solved problems of application in engineering (e.g., Ayres, 1952; Spiegel, 1980). Because of its clarity and wide range of applications, we recommend Spiegel (1967), Braun (1983), and Cakmak et al. (1987). For applications of partial differential equation we recommend Powers (1979), Myint-U (1987), and Zauderer, 1983. For the theoretically-minded engineer interested in rigorous treatment including functional analysis, see Reinhard (1987); Hutson and Pym, 1980; and Oden, 1979. On the other hand, if the interest lies in numerical solutions, any textbook on numerical methods describes solutions of differential equations (e.g., Hornbeck, 1975; Chapra and

Canale, 1988). Regarding stochastic differential equations, the field still remains with a heavy emphasis on theory (e.g., Ikeda and Watanabe, 1989; Gardiner, 1985; Jazwinski, 1970, Soong, 1973). A class of stochastic equations with random input is well covered in textbooks on signal and systems analysis (e.g., Cooper and McGillem, 1999). We hope that the following sections will provide a practical presentation of a framework of stochastic differential equations that complement naturally the theory of applied differential equations most engineers cover during the first years at an engineering school. We also present new simple methodologies for the systematic solution of boundary-value initial-value problems in engineering.

Differential Equations with Random Initial Conditions

From the mathematical point of view, the simplest type of random differential equation is one in which its initial condition is defined as a random variable, for ordinary differential equations, or as a random process for partial differential equations. This usually refers to a causal system being excited at $t=0$ by a variable or a function subject to uncertainty, or a system containing some remnant from past performance. Depending on the system equation (i.e., its "memory"), the initial condition will be propagated or diffused, if the system is dissipative or thermodynamical, or it will be amplified, if the system is nonlinear or "cybernetic."

The solution to the differential equation will take the random initial condition and update it into a future condition at time t. The solution thus obtained, the dependent variable, is a stochastic process in general. In theory the probability density function of the output could be obtained if the corresponding density function of the input is available. In practice this problem is difficult due to the analytical form of the solution. If an analytical derivation of the output density function is not feasible, the alternative numerical procedure using Monte Carlo simulation provides a practical means of generating a large sample from the output in order to estimate the output density approximately. In this regard the methods described in Chapters 4 and 12 are useful.

On the other hand, many engineering applications subject to budgetary or technical constraints do not provide information of the complete density function of the input. Instead, only limited information on the first order moments of the input is available (e.g., the mean and correlation functions). As a consequence, the best one can reasonably hope to obtain is the corresponding moments of the system output.

Example 13.6: First-Order Decay Subject to Random Initial Concentration

Many engineering systems, and systems components, are modeled via first-order rates of decay. Consider for example the equation of a (small) lake-wide contaminant decay given by

$$\frac{dC}{dt} + \left(\frac{1}{t_d} + K \right) C = 0, \qquad 0 < t, \ C(0) = C_0 \qquad (13.17)$$

where

C=lake-wide contaminant concentration (kg/m^3)
t_d=lake detention time (*year*), assumed constant
K=contaminant decay coefficient ($year^{-1}$), assumed constant
C_0=initial concentration (kg/m^3), a random variable

If the initial condition, C_0, is a random variable with a mean of $\mu_0 = 300\,kg/m^3$ and a standard deviation of $\sigma_0 = 50\,kg/m^3$, $K = 1/t_d = 1\ year^{-1}$, find the solution to the differential equation, the mean, and the variance of the concentration.

Solution

Let us set $a = 1/t_d + K$. Equation (13.17) becomes

$$\frac{dC}{dt} + aC = 0 \qquad (13.18)$$

This is a first order linear ordinary differential equation with constant coefficients. The integrating factor (consult any book on differential equations) is given by $e^{\int a dt} = e^{at}$. Multiplying the differential equation by the integrating factor,

$$e^{at}\frac{dC}{dt} + ae^{at}C = \frac{d}{dt}\left(Ce^{at}\right) = 0$$

Upon integration,

$$C \frac{d}{dt}e^{at} + e^{at}\frac{dC}{dt}$$

$$C = k_0 e^{-at}$$

At $t=0$, $C=C_0$, and the integration constant is $k_0 = C_0$. Thus, the solution is

$$C = C_0 e^{-at} \qquad (13.19)$$

Since C_0 is a random variable, then the system output, C, is a stochastic process consisting of a family of exponential curves with random initial value. Once the solution is obtained, the analysis of the system output is similar to that of any other stochastic process. The derivation of the mean, the correlation, and the calculation of sample functions is easily accomplished by following the methods described in Chapter 12. Hence, the mean concentration is obtained by taking expectations on both sides of the solution:

$$E\{C(t)\}=\mu_C(t)=E\{C_0\}e^{-at}=\mu_0 e^{-at}$$

The correlation function of the output is given by the expectation of the solution evaluated at $t=t_1$ times the solution evaluated at $t=t_2$:

$$E\{C(t_1)C(t_2)\}=R_C(t_1,\ t_2)=E\left\{\left[C_0 e^{-at_1}\right]\left[C_0 e^{-at_2}\right]\right\}=E\{C_0^2\}e^{-a(t_1+t_2)}$$

$$=\left(\mu_0^2+\sigma_0^2\right)e^{-a(t_1+t_2)}$$

where equation (3.8) was used. Now set $t_1=t_2=t$ in the correlation and subtract the square of the mean to obtain the variance:

$$\sigma_C^2(t)=R_C(t,\ t)-\mu_C^2(t)=\sigma_0^2 e^{-2at}$$

Again, these moments were derived analytically. Analytical derivation is the preferred avenue since it yields the exact moments. However, if the solution is too complicated, one can use the expression of the solution in conjunction with generated random numbers of the initial condition to obtain sample functions of the output process. After a sufficiently large number of sample functions are produced in this manner, one may estimate the moments of the output numerically. This is the Monte Carlo simulation approach. It is simpler mathematically, but it yields the approximate moments, and the mathematical simplicity is replaced by the numerical and computer programming difficulty (see Chapter 12 for more examples).

Example 13.7
Use the results and data from Example 13.6 to produce the following graphs: (1) the mean concentration plus and minus one standard deviation; and (2) 20 sample concentration versus time curves assuming that the probability density function of the initial condition C_0, $f_{C_0}(c_0) \sim N(50, 300^2)$.

Solution

The following program produces the required graphs.

[(1)

```
> restart: with[stats]: with(stats[statplots]):
  mu0:=300.: sigma0:=50.: a:=2.:
  muC:=t->mu0*exp(-a*t):
  sigmaC:=t->sigma0*exp(-a*t):
  Mean:=plot(muC(t),t=0..3,color=black,thickness=2):
  MeanPlus:=plot(muC(t)+sigmaC(t),t=0..3,color=black):
  MeanMinus:=plot(muC(t)-sigmaC(t),t=0..3,color=black):
  plots[display](Mean,MeanPlus,MeanMinus,
      labelfont=[TIMES,ROMAN,16],
      labels=[`t (years)`,`C (kg/m^3)`],
      labeldirections=[horizontal,vertical]);
```

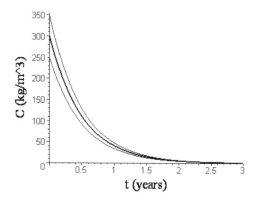

[(2)

```
> N:=20: C0:=[random[normald[mu0,sigma0]](N)]:
  for i from 1 to N do
      C[i]:=plot(C0[i]*exp(-a*t),t=0..3,color=black)
  end do:
  plots[display](seq(C[i],i=1..N),thickness=2,
      labelfont=[TIMES,ROMAN,16],
      labels=[`t (years)`,`C (kg/m^3)`],
      labeldirections=[horizontal,vertical]);
```

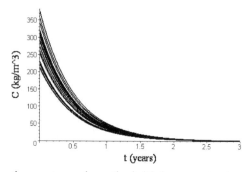

This is a dissipative system where the initial concentration rapidly decays, depending on the value of a. Similarly, the variance of the concentration also decays with time and the system randomness disappears after some time. Notice that in this case each sample function is a smooth curve of time and that at a fixed time the concentration is given by a random variable.

Example 13.8: Probability Density Function of Lake Concentration

In Examples 13.6 and 13.7, derive the analytical for of the probability density function of lake concentration.

Solution

From equation (4.2), Chapter 4, and equation (13.19), the probability density function of a system's output given its input density is given as

$$f_{C(t)}(c) = f_{C_0}(c_0) \left| \frac{dc_0}{dc} \right|_{c_0 = ce^{at}}$$

From the solution equation (13.19),

$$c_0 = ce^{at} \quad \Rightarrow \quad \left| \frac{dc_0}{dc} \right| = e^{at}$$

Hence the output density is given by

$$f_{C(t)}(c) = f_{C_0}(c_0)e^{at} = f_{C_0}(ce^{at})e^{at}$$

Since $f_{C_0}(c_0) \sim N(50, 300^2)$, then

$$f_{C(t)}(c) = \frac{e^{-\frac{(ce^{at}-50)^2}{2 \times 300^2}}}{\sqrt{2\pi \times 300^2}} e^{at}$$

Example 13.9: Damped Vibrations After a Random Initial Excitation.

Consider the second order linear differential equation governing the vibration of a spring with damping given by

$$\frac{d^2Y}{dt^2} + \frac{\beta}{M}\frac{dY}{dt} + \frac{k}{M}Y = 0, \qquad 0 < t, \ Y(0) = Y_0, \ \frac{dY}{dt}(0) = 0 \qquad (13.20)$$

where

$Y(t)$=vertical position of a mass $M(kg)$ with respect to an equilibrium point (m), assumed positively downward.

k=spring constant (kg/s^2)

β=damping constant (kg/s)

Y_0=initial random stretch (m), a random variable with a mean μ_{Y_0} and a variance $\sigma_{Y_0}^2$

The differential equation describes the vibratory movement after the mass M is stretched a random distance Y_0 at t=0 and released. The initial velocity $dY(0)/dt$ is zero. Find a solution to the differential equation.

Solution

Assuming a solution of the form $Y=e^{mt}$, substituting into the differential equation and eliminating the exponential, we obtain the *auxiliary equation*

$$m^2+\frac{\beta}{M}m+\frac{k}{M}=0 \tag{13.21}$$

The roots of this quadratic equation are

$$m_1=\frac{-\beta+\sqrt{\beta^2-4Mk}}{2M}, \qquad m_2=\frac{-\beta-\sqrt{\beta^2-4Mk}}{2M} \tag{12.22}$$

If $\beta^2-4Mk>0$, the roots of the auxiliary equation are real and the solution to the differential equation is given by

$$Y(t)=c_1e^{m_1t}+c_2e^{m_2t}$$

where c_1 and c_2 are constants. From the two initial conditions to equation (13.20), we have two equations with two unknowns from which the constants of integration are found. The solution to the differential equation is then

$$Y(t)=Y_0\left[\frac{m_2}{m_1-m_2}e^{m_1t}+\left(1-\frac{m_2}{m_1-m_2}\right)e^{m_2t}\right] \tag{13.23}$$

From this expression we may derive moments of the output variable $Y(t)$ as well as sample functions, if sample functions of the initial stretch are available (i.e., if the density function of Y_0 is known). Now, if the roots of the auxiliary equation are imaginary (i.e., if $\beta^2-4Mk<0$), then we use Euler's formulae to obtain a general solution of the form

$$Y(t)=e^{at}\big(c_1\sin(bt)+c_2\cos(bt)\big)$$

Once again, from the two boundary conditions substituted into the above general solution we obtain

$$Y(t) = Y_0 e^{at}\left(-\frac{a}{b}\sin(bt) + \cos(bt)\right), \qquad a = -\frac{\beta}{2M}, \quad b = \frac{\sqrt{|\beta^2 - 4Mk|}}{2M} \quad (13.24)$$

Example 13.10: Damped Oscillations Mean and Variance.
In Example 13.9, derive the mean, correlation, and variance of the output.

Solution
If $\beta^2 - 4Mk > 0$, the solution to the differential equations is given by equation (13.23). Taking expectations, we obtain the mean of the output:

$$\mu_Y(t) = \mu_{Y_0}\left[\frac{m_2}{m_1 - m_2}e^{m_1 t} + \left(1 - \frac{m_2}{m_1 - m_2}\right)e^{m_2 t}\right]$$

For the correlation, we take expectations of the product of the output evaluated at t_1 and t_2:

$$R_Y(t_1,t_2) = E\{Y_0^2\}\left[\frac{m_2 e^{m_1 t_1}}{m_1 - m_2} + \left(1 - \frac{m_2}{m_1 - m_2}\right)e^{m_2 t_1}\right]\left[\frac{m_2 e^{m_1 t_2}}{m_1 - m_2} + \left(1 - \frac{m_2}{m_1 - m_2}\right)e^{m_2 t_2}\right]$$

Now set $t_1 = t_2 = t$ in the correlation and subtract the square of the mean to obtain the variance:

$$\sigma_Y^2(t) = \sigma_{Y_0}^2\left[\frac{m_2}{(m_1 - m_2)}e^{m_1 t} + \left(1 - \frac{m_2}{(m_1 - m_2)}\right)e^{m_2 t}\right]^2$$

where the identity $E\{Y_0^2\} = \mu_{Y_0}^2 + \sigma_{Y_0}^2$ has been used. Now, if $\beta^2 - 4Mk < 0$, the solution is given by equation (13.24). Taking expectations, we obtain the mean of the output as

$$\mu_Y(t) = \mu_{Y_0}e^{at}\left(-\frac{a}{b}\sin(bt) + \cos(bt)\right)$$

For the correlation, we take expectations of the product of the output evaluated at t_1 and t_2:

$$R_Y(t_1, t_2) = E\{Y_0^2\}e^{a(t_1 + t_2)}\left(-\frac{a}{b}\sin(bt_1) + \cos(bt_1)\right)\left(-\frac{a}{b}\sin(bt_2) + \cos(bt_2)\right)$$

Now set $t_1 = t_2 = t$ in the correlation and subtract the square of the mean to obtain the variance:

$$\sigma_Y^2(t) = \sigma_{Y_0}^2 e^{2at} \left(-\frac{a}{b} \sin(bt) + \cos(bt) \right)^2$$

Example 13.11: Computer Solution of Differential Equations

Use Maple to solve the following ordinary differential equation:

$$2\frac{d^2z}{dt^2} - 4\frac{dz}{dt} + z = 0, \quad 0 < t, \ z(0) = 0, \ \frac{dz}{dt}(0) = V$$

Solution

[Define the differential equation and the initial conditions.
```
> eq:=2*diff(z(t),t$2)-4*diff(z(t),t)+z(t)=0;
  ini:=z(0)=0, D(z)(0)=V;
```

$$eq := 2\left(\frac{d^2}{dt^2} z(t)\right) - 4\left(\frac{d}{dt} z(t)\right) + z(t) = 0$$

$$ini := z(0) = 0, \ D(z)(0) = V$$

[Solve the differential equation with the dsolve({ }, { }) command. The first argument
of this command specifies a set of the equation with its initial conditions.
The second argument of the command specifies the variable for which to solve.
```
> dsolve({eq,ini},{z(t)});
```

$$z(t) = \frac{1}{2}\sqrt{2}\, V e^{\left[\frac{(2+\sqrt{2})t}{2}\right]} - \frac{1}{2}\sqrt{2}\, V e^{\left[\frac{(-2+\sqrt{2})t}{2}\right]}$$

Example 13.12: Contaminant Dispersion Following an Uncertain Initial Spill

Consider the dispersion of a contaminant in a long aquifer after a point source of uncertain initial mass given by a random variable. The governing equation is the advective dispersive partial differential equation given by (see Serrano, 2010)

$$\frac{\partial C}{\partial t} - D\frac{\partial^2 C}{\partial x^2} + v\frac{\partial C}{\partial x} = 0, \quad -\infty < x < \infty, \ 0 < t, \tag{13.25}$$

$$C(\pm\infty, t) = 0, \ C(x,0) = C_0 = M\delta(x)$$

where

$C(x, t)$ = contaminant concentration (kg/m^3)
D = aquifer longitudinal dispersion coefficient ($m^2/month$)

v=groundwater seepage velocity ($m/month$)
x=distance (m)
t=time after a chemical spill ($month$)
M=initial mass of a spill per unit cross-sectional area perpendicular
  to x (kg/m^2), a random variable with mean μ_M and variance σ_M^2
$\delta(x)$=Dirac's delta function indicating a spill at x=0 of strength M

Equation (13.25) has a first order derivative in t, which requires us to provide a known initial condition (i.e., the initial spill), and a second order derivative in x, which requires us to provide two boundary conditions, one on the left and one on the right far away from the origin. To solve this equation, we could use an analogy with equation (13.18) and write it as

$$\frac{\partial C}{\partial t} + AC = 0, \qquad A = -D\frac{\partial^2}{\partial x^2} + v\frac{\partial}{\partial x} \qquad (13.26)$$

where the operator A now contains the spatial derivatives of the equation. By analogy with equation (13.19), the solution to equation (13.26) could be written as an evolution operator associated with A acting on an initial condition (see Serrano, 1995(2), 1992(1), and 1992(2) for details):

$$C = J_t C_0 = \frac{1}{\sqrt{4\pi Dt}} \int_{-\infty}^{\infty} e^{-\frac{(x-vt-x')^2}{4Dt}} C_0(x')dx'$$

$$= \frac{1}{\sqrt{4\pi Dt}} \int_{-\infty}^{\infty} e^{-\frac{(x-vt-x')^2}{4Dt}} M\delta(x')dx' = \frac{Me^{-\frac{(x-vt)^2}{4Dt}}}{\sqrt{4\pi Dt}} \qquad (13.27)$$

where the properties of integrals of delta functions (equation (13.16)) have been used. The operator J_t is called the analytic *semigroup*. The semigroup transforms the state or system condition at one time (e.g., t=0) into that at a future time, t. In this case the semigroup associated with A is a spatial integral of the system impulse response or Green's function (Roach, 1982). A detailed derivation of equation (13.27) is obtained by applying traditional Fourier transformation and inversion of equation (13.25) and can be found in any book on partial differential equations or heat conduction (e.g., Zauderer, 1983; Carslaw and Jaeger, 1971).

Notice that in equation (13.27) M is a random variable that depends on probability, not a random process that also depends on distance. Hence, it goes out of the spatial integral. Had we defined $M(x)$ as a random process in space, then it would have remained as part of the integral (see Example 13.13). Now the mean concentration is given by

$$E\{C\}=\mu_C=E\left\{\frac{Me^{-\frac{(x-vt)^2}{4Dt}}}{\sqrt{4\pi Dt}}\right\}=\frac{E\{M\}e^{-\frac{(x-vt)^2}{4Dt}}}{\sqrt{4\pi Dt}}=\frac{\mu_M e^{-\frac{(x-vt)^2}{4Dt}}}{\sqrt{4\pi Dt}}$$

The autocorrelation function is given by

$$E\{C(x,t_1)C(x,t_2)\}=R_C(t_1,t_2)=E\left\{\left[\frac{Me^{-\frac{(x-vt_1)^2}{4Dt_1}}}{\sqrt{4\pi Dt_1}}\right]\left[\frac{Me^{-\frac{(x-vt_2)^2}{4Dt_2}}}{\sqrt{4\pi Dt_2}}\right]\right\}$$

$$=E\{M^2\}\left[\frac{e^{-\frac{(x-vt_1)^2}{4Dt_1}}}{\sqrt{4\pi Dt_1}}\right]\left[\frac{e^{-\frac{(x-vt_2)^2}{4Dt_2}}}{\sqrt{4\pi Dt_2}}\right]$$

$$=\left(\mu_M^2+\sigma_M^2(t)\right)\left[\frac{e^{-\frac{(x-vt_1)^2}{4Dt_1}}}{\sqrt{4\pi Dt_1}}\right]\left[\frac{e^{-\frac{(x-vt_2)^2}{4Dt_2}}}{\sqrt{4\pi Dt_2}}\right]$$

Finally, the variance of C,

$$\sigma_C^2(t)=E\{C^2(x,t)\}-\mu_C^2(t)=R_C(t,\ t)-\mu_C^2(t)=\sigma_M^2(t)\frac{e^{-\frac{(x-vt)^2}{2Dt}}}{4\pi Dt}$$

Example 13.13: Contaminant Dispersion Under an Uncertain Spatial Distribution in its Initial Spill

A more realistic model of contaminant dispersion is the case when the initial source is spatially distributed (i.e., a non point source) in an erratic manner. Assume that in Equation (13.25) the initial condition is not a random variable but rather the stochastic process in space given by

$$C_0(x)=C_i+W(x),\ \ C_i=\frac{Me^{-\frac{x^2}{4L^2}}}{\sqrt{4\pi L^2}},\ \ E\{W(x)\}=0,\ \ E\{W(x_1)W(x_2)\}=\sigma_W^2\delta(x_1-x_2)$$

where M is a constant (kg/m^2), $L=1m$, and W(x) is a White Gaussian Noise process in space with variance parameter $\sigma_W^2(kg/m^3)^2$. Find the mean and the variance of the concentration in the aquifer domain.

Solution

In this case, the initial concentration is made of a summation of a deterministic curve, $C_i(x)$, and a random process in distance, $W(x)$. Again, the solution to the differential equation is given by equation (13.27), except that now the initial condition remains inside the spatial integral. Taking advantage of the notational economy of the semigroup operator,

$$C = J_t C_0 = J_t C_i + J_t W(x)$$

$$E\{C(x,\ t)\} = \mu_C(x,\ t) = J_t C_i + J_t E\{W(x)\} = J_t C_i$$

$$= \frac{1}{\sqrt{4\pi Dt}} \int_{-\infty}^{\infty} e^{-\frac{(x-vt-x')^2}{4Dt}} \frac{Me^{-\frac{x'^2}{4L^2}}}{\sqrt{4\pi L^2}} dx' = \frac{Me^{-\frac{(x-vt)^2}{4(L^2+Dt)}}}{\sqrt{4\pi(L^2+Dt)}}$$

See Problem 13.6 for the derivation of the last integral. If one wishes to generate sample functions of the concentration from the first of the above equations, then one must generate a sample function of a White Gaussian Noise process according to the procedure outlined in Chapter 12, and solve the semigroup integral numerically. The autocorrelation function of the concentration at two locations x_1 and x_2 (i.e., $W(x)$ is a spatial random process) is given by

$$E\{C(x_1,\ t)C(x_2,\ t)\} = R_C(x_1,\ x_2)$$

$$= E\{J_t C_0|_{x_1} \cdot J_t C_0|_{x_2}\} = E\left\{ \left[J_t C_i + J_t W \right]_{x_1} \cdot \left[J_t C_i + J_t W \right]_{x_2} \right\}$$

$$= J_t C_i|_{x_1} \cdot J_t C_i|_{x_2} + E\{ J_t W|_{x_1} \cdot J_t W|_{x_2} \}$$

$$= \frac{Me^{-\frac{(x_1-vt)^2}{4(L^2+Dt)}}}{\sqrt{4\pi(L^2+Dt)}} \frac{Me^{-\frac{(x_2-vt)^2}{4(L^2+Dt)}}}{\sqrt{4\pi(L^2+Dt)}}$$

$$+ \frac{\sigma_W^2}{4\pi Dt} \int_{-\infty}^{\infty}\int_{-\infty}^{\infty} e^{-\frac{(x_1-vt-x')^2}{4Dt}} e^{-\frac{(x_2-vt-\xi)^2}{4Dt}} \delta(x'-\xi)dx'd\xi$$

$$R_C(x_1,x_2)=\mu_C(x_1,\ t)\mu_C(x_2,\ t)+\frac{\sigma_W^2}{4\pi Dt}\int_{-\infty}^{\infty}e^{-\frac{(x_1-vt-x')^2}{4Dt}}e^{-\frac{(x_2-vt-x')^2}{4Dt}}dx'$$

$$=\mu_C(x_1,\ t)\mu_C(x_2,\ t)+\sigma_W^2$$

Finally the variance of the concentration is given by

$$\sigma_C^2(x,\ t)=R_C(x,\ x)-\mu_C^2(x,\ t)=\sigma_W^2$$

These simple results derive from the assumption of a delta-correlated noise. Had we assumed a correlated (e.g., a *Colored Noise*) process for $W(x)$, the integrals in the correlation function would reflect this feature.

Example 13.14: Heat Transfer Subject to Random Initial Temperature

Consider the transfer of heat through a metal bar of finite length. The governing equation is given by

$$\frac{\partial T}{\partial t}-k\frac{\partial^2 T}{\partial x^2}=0,\ 0<t,\ 0\le x\le L,\ T(0,t)=0,\ \frac{\partial T}{\partial x}(L,t)=0,$$

$$(13.28)$$

$$T(x,0)=T_0=T_m U(x-a)U(b-x),\ 0<a<b<L,\ E\{T_m\}=\mu_m,\ Var(T_m)=\sigma_i^2$$

where

    $T(x,\ t)$=temperature $(^\circ C)$
    k=thermal diffusivity of the bar (m^2/day)
    x=distance from the left boundary (m)
    t=time (day)
    L=bar length (m)
    $T_0(x)$=initial temperature spatial distribution $(^\circ C)$
    $U(x-a)$=1 for $x>a$ and zero elsewhere (a unit step function)
    T_m=maximum initial temperature, a random variable

The system of equations (13.28) describes a bar of length L subject to a constant zero temperature on the left side, a no flow (i.e., insulated) boundary condition on the right, and an initial temperature distribution represented as a random variable, T_m, between the abscissas $x=a$ and $x=b$. For each realization of T_m, $T_0(x)$ is a rectangle between a and b and zero elsewhere. Find the mean and the variance of the temperature in the bar at any time $t>0$.

Solution

Here we have a partial differential equation in a domain of finite length L. we can use the concept of the semigroup operator to describe the solution to the differential equation. By analogy with equation (13.26), we can write equation (13.28) as

$$\frac{\partial T}{\partial t} + AT = 0, \qquad AT = -k\frac{\partial^2 T}{\partial x^2} \tag{13.29}$$

By analogy with equation (13.27), the solution to equation (13.29) may be written as

$$T(x,t) = J_t T_0(x) = \sum_{n=1}^{\infty} B_n \sin(\lambda_n x) e^{-\lambda_n^2 kt}$$

$$\tag{13.30}$$

$$\lambda_n = \left(\frac{2n-1}{2L}\right)\pi, \ B_n = \frac{2}{L}\int_0^L T_0(x)\sin(\lambda_n x)dx$$

The semigroup associated with the operator A in equation (13.29) is an infinite series of a Fourier coefficient, B_n, and an orthogonal basis function. The detail derivation of the above may be seen in Powers (1979). It is derived from traditional separation of variables in conjunction with Fourier series analysis. Note that the random component appears in the Fourier coefficient, which is a spatial integral over the whole bar domain. Since T_m is a random variable independent of distance (i.e., it is not a stochastic process), it may be taken out of the integral. From the solution equation (13.30) we obtain the mean temperature as

$$E\{T(x,t)\} = \mu_T(x,t) = J_t \left[E\{T_0(x,t)\}\right]$$

$$= J_t \left[E\{T_m\}U(x-a)U(b-x)\right] = J_t \left[\mu_m U(x-a)U(b-x)\right]$$

$$= \sum_{n=1}^{\infty} \left[\frac{2}{L}\int_0^L \mu_m U(x-a)U(b-x)\sin(\lambda_n x)dx\right]\sin(\lambda_n x)e^{-\lambda_n^2 kt}$$

$$= \sum_{n=1}^{\infty} \left[\frac{2\mu_m}{L}\int_a^b \sin(\lambda_n x)dx\right]\sin(\lambda_n x)e^{-\lambda_n^2 kt}$$

$$= \sum_{n=1}^{\infty} \frac{2\mu_m}{\lambda_n L}\left[\cos(\lambda_n a) - \cos(\lambda_n b)\right]\sin(\lambda_n x)e^{-\lambda_n^2 kt}$$

To compute the mean, several terms should be calculated while observing the convergence rate of the series. The number of terms required in the summation should be such that an additional one would add less than a desired resolution (e.g., $0.001^\circ C$). At that stage the series is truncated. It is also possible that the series does not converge. Some parameter values or domain size features are such that convergence is not achieved and this method cannot be used. The autocorrelation function at two times t_1 and t_2 is given by

$$R_T(t_1,t_2) = E\{J_{t_1}T_0 \ J_{t_2}T_0\} = J_{t_1}J_{t_2}E\{T_0^2\}$$

$$= \sum_{n=1}^{\infty}\sum_{j=1}^{\infty}\left[\frac{4}{L^2}\int_a^b\int_a^b E\{T_m^2\}\sin(\lambda_n x)\sin(\lambda_j\xi)dxd\xi\right]$$

$$\cdot\left[\sin(\lambda_n x)\sin(\lambda_j x)e^{-\lambda_n^2 kt_1 - \lambda_j^2 kt_2}\right]$$

$$= \sum_{n=1}^{\infty}\sum_{j=1}^{\infty}\frac{4(\mu_m+\sigma_m^2)}{\lambda_n\lambda_j L^2}(\cos(\lambda_n a)-\cos(\lambda_n b))(\cos(\lambda_j a)-\cos(\lambda_j b))$$

$$\cdot\left[\sin(\lambda_n x)\sin(\lambda_j x)e^{-\lambda_n^2 kt_1 - \lambda_j^2 kt_2}\right]$$

where

$$\lambda_n = \left(\frac{2n-1}{2L}\right)\pi, \qquad \lambda_j = \left(\frac{2j-1}{2L}\right)\pi, \qquad n, j = 1, 2, 3,...$$

Using the mean and the correlation function of the temperature, we finally obtain the variance:

$$\sigma_T^2(x,t) = R_T(t,t) - \mu_T^2(x,t) = \sigma_m^2\left[\sum_{n=1}^{\infty}\frac{2}{\lambda_n L}(\cos(\lambda_n a)-\cos(\lambda_n b))\sin(\lambda_n x)e^{-\lambda_n^2 kt}\right]^2$$

Differential Equations with Random Forcing Functions

In the last heading we studied the solution to differential equations subject to random initial conditions. We covered the cases of first and second-order ordinary differential equations, and that of partial differential equations. Although a random initial condition is rare in engineering applications, the principles involved allowed us to describe the treatment of various kinds of differential equations subject to random components. These principles are common to equations subject to stochastic components appearing anywhere in the system, when attempting to obtain moments, or to generate sample functions.

Consider now the more practical situation of a system governed by a differential equation in which the forcing function is defined in stochastic terms. The system response is no longer limited to an attenuation or dissipation of a stochastic condition present at $t=0$, but rather to the processing of an external excitation over time or distance. In systems analysis notation, the forcing function is written on the right of the equality. As before, we illustrate the solution of various cases with examples.

Example 13.15: First Order Decay Subject to Random Loading

Let us reconsider the model of lake-wide contaminant decay given by equations (13.17) or (13.18) in Example 13.6, when the system is subject to continuous contaminant loading:

$$\frac{dC}{dt}+aC=h+g(t), \quad 0<t, \quad C(0)=C_0=0$$

$$(13.31)$$

$$E\{g(t)\}=0, \quad E\{g(t_1)g(t_2)\}=\sigma_g^2 e^{-\rho|t_1-t_2|}$$

where
   h=a constant average annual load ($kg/year$)
   $g(t)$=random mass input variation ($kg/year$), a Colored Noise process
   ρ=a Colored Noise correlation parameter ($year^{-1}$)
   σ_g^2=the variance of mas input ($kg/year)^2$

The solution to equation (13.31) now has two components: one due to the initial condition C_0 (zero in this case), and one due to the forcing function $h+g(t)$. Since the equation is linear, we can add the two components using the principle of superposition of linear systems. The general solution is then written as

$$C=J_t C_0+\int_0^t J_{t-\tau}(h+g(\tau))d\tau=\frac{h}{a}(1-e^{-at})+\int_0^t J_{t-\tau}g(\tau)d\tau, \quad J_t=e^{-at} \quad (13.32)$$

where the semigroup associated with the constant a in equation (13.31) is simply an exponential of time (see equation 13.19). The component of the solution due to the forcing function is called a *convolution integral* of the system impulse response function (which in this case coincides with the semigroup operator) and the system input (consult any textbook on linear systems theory). It is interesting to note that the semigroup, or a form of it, appears in every component of the solution. Thus, having the semigroup facilitates the solution of a system equation subject to several components. Now, taking expectations in equation (13.32) we obtain the mean of the output concentration:

$$E\{C(t)\}=\mu_C(t)=\frac{h}{a}\left(1-e^{-at}\right)+\int_0^t J_{t-\tau}E\{g(\tau)\}d\tau=\frac{h}{a}\left(1-e^{-at}\right)$$

The autocorrelation function is given by

$$E\{C(t_1)C(t_2)\}=R_C(t_1,t_2)$$

$$=E\left\{\left[\frac{h}{a}\left(1-e^{-at_1}\right)+\int_0^{t_1}J_{t_1-\tau}g(\tau)d\tau\right]\left[\frac{h}{a}\left(1-e^{-at_2}\right)+\int_0^{t_2}J_{t_2-\xi}g(\xi)d\xi\right]\right\}$$

$$=\frac{h^2}{a^2}\left(1-e^{-at_1}\right)\left(1-e^{-at_2}\right)+\int_0^{t_1}\int_0^{t_2}J_{t_1-\tau}J_{t_2-\xi}E\{g(\tau)g(\xi)\}d\tau d\xi$$

$$=\frac{h^2}{a^2}\left(1-e^{-at_1}\right)\left(1-e^{-at_2}\right)+\sigma_g^2\int_0^{t_1}\int_0^{t_2}e^{-a(t_1-\tau)}e^{-a(t_2-\xi)}e^{-\rho|\tau-\xi|}d\tau d\xi$$

$$=\frac{h^2}{a^2}\left(1-e^{-at_1}\right)\left(1-e^{-at_2}\right)$$

$$+\sigma_g^2\left[\frac{e^{-\rho(t_2-t_1)}-\rho e^{-a(t_2-t_1)}+(\rho+a)e^{-a(t_1+t_2)}-ae^{-at_1-\rho t_2}-ae^{-at_2-\rho t_1}}{a(a+\rho)(a-\rho)}\right],\ t_1<t_2$$

See Problem 13.8 for the solution of the double integrals. Now set $t_1=t_2=t$ in the autocorrelation function above and subtract the square of the mean concentration to get the variance of the output:

$$\sigma_C^2(t)=R_C(t,t)-\mu_C^2(t)=\sigma_g^2\left[\frac{(a-\rho)+(a+\rho)e^{-2at}-2ae^{-(a-\rho)t}}{a(a+\rho)(a-\rho)}\right]$$

$$\approx\frac{\sigma_g^2}{a(a+\rho)},\qquad \rho<a,\ t\ large$$

Example 13.16: Damped Vibrations Subject to a Random External Force

Consider again the differential equation governing the vibration of a spring with damping, equation (13.20), but now the system is subject to an external periodic random force:

$$\frac{d^2Y}{dt^2} + \frac{\beta}{M}\frac{dY}{dt} + \frac{k}{M}Y = F(t) = \frac{A}{M}\cos(8t)$$

$$\text{(13.33)}$$

$$0<t, \quad Y(0)=y_0, \quad \frac{dY}{dt}(0)=0, \quad E\{A\}=\mu_A, \quad Var(A)=\sigma_A^2$$

where

$F(t)$=external force per unit mass applied to spring ($m/s^2/kg$)
A=amplitude of force (m/s^2), a random variable

Find the solution to the differential equation, its mean, its autocorrelation function, and its variance.

Solution

The system in equation (13.33) responds to an initial stretch, y_0, assumed constant in this case, and a periodic force with an amplitude represented as a random variable. Since the differential equation is linear, we may use the principle of superposition to obtain the complete solution as a summation of the individual components: one component due to the initial condition, called the *complementary solution*, y_c, and one component due to the external excitation, called the *particular solution*, y_p.

The complementary solution satisfies the differential equation when $F(t)=0$, and thus it is fully described in Example 13.9 (see equation (13.23) or (13.24), depending on the parameter values). The particular solution satisfies the differential equation when $y_0=0$. We seek a function y_p that satisfies the equation. In this example $F(t)$ is a cosine and thus we assume a particular solution $a\sin(8t)+b\cos(8t)$. Upon substitution in equation (13.33), and equating coefficients of like functions, we find two equations with two unknowns, which yield

$$y_p = a\sin(8t)+b\cos(8t), \quad a = \frac{8\beta A}{(k-64M)^2+64\beta^2}, \quad b = \frac{A}{k-64M+\frac{64\beta^2}{k-64M}}$$

This could be written as

$$y_p = A \left[\frac{8\beta\sin(8t)}{(k-64M)^2 + 64\beta^2} + \frac{\cos(8t)}{k-64M + \dfrac{64\beta^2}{k-64M}} \right]$$

The complete solution is $y = y_c + y_p$. The mean of the output is then given by

$$E\{y(t)\} = \mu_y(t) = y_c + E\{y_p\} = y_c + \mu_A \left[\frac{8\beta\sin(8t)}{(k-64M)^2 + 64\beta^2} + \frac{\cos(8t)}{k-64M + \dfrac{64\beta^2}{k-64M}} \right]$$

The autocorrelation function of y is given by

$$E\{y(t_1)y(t_2)\} = R_y(t_1,t_2) = E\left\{ \left[y_c(t_1) + y_p(t_1) \right] \left[y_c(t_2) + y_p(t_2) \right] \right\}$$

$$= y_c(t_1)y_c(t_2) + y_c(t_1)E\{y_p(t_2)\} + y_c(t_2)E\{y_p(t_1)\} + E\{y_p(t_1)y_p(t_2)\}$$

which can be expanded in the usual manner. The variance of y is obtained after setting $t_1 = t_2 = t$ in the correlation above, subtracting the square of the mean of y, and simplifying:

$$\sigma_y^2(t) = R_y(t,t) - \mu_y^2(t) = E\{y_p^2\} - E^2\{y_p\}$$

$$= \sigma_A^2 \left[\frac{8\beta\sin(8t)}{(k-64M)^2 + 64\beta^2} + \frac{\cos(8t)}{k-64M + \dfrac{64\beta^2}{k-64M}} \right]^2$$

Example 13.17: A Model for the Generation of Colored Noise

A common model for random components in engineering is the Colored Noise stochastic process, with a mean of zero and a time autocorrelation given by an exponentially decaying function of lag time. Using a White Noise sample function produced by any computer, it is possible to generate a Colored Noise sample. A linear ordinary differential equation forced by thermal White Noise produces a Colored Noise process as an output (Jazwinski, 1970):

$$\frac{dm}{dt} + \rho m = \sigma\rho W(t), \quad 0 < t, \quad m(0) = m_0 \sim N(0, \frac{\rho\sigma^2}{2})$$

(13.34)

$$E\{W(t)\} = 0, \quad E\{W(t_1)W(t_2)\} = \delta(t_1 - t_2), \quad E\{m_0 W(t)\} = 0$$

where

$m(t)$=system output, a colored noise process (^{o}C) with $E\{m(t)\}=0$,
$E\{m(t_1)m(t_2)\}=\rho\sigma^2 e^{-\rho(t_2-t_1)}/2$, $t_2>t_1$
m_0=initial condition (^{o}C) given as a Normal random variable with a
   mean of zero and a variance $\rho\sigma^2/2$
t=time coordinate (s)
ρ=Colored Noise decay parameter (s^{-1})
σ=constant ($^{o}Cs^{1/2}$)
$W(t)$=White Noise with a unit variance parameter ($s^{-1/2}$)

Show that the output mean and correlation functions have the specified properties of Colored Noise.

Solution
The system equation has a random initial condition, m_0, and a random forcing function, $\sigma\rho W(t)$. The two random components are statistically independent. Since the equation is linear, we may add the solution due to the random initial condition (i.e., the complementary solution) and the solution due to the random forcing function (the particular solution). Equation (13.34) is mathematically identical to Equation (13.18). Thus, the solution due to the random initial condition is given by equation (13.19). The particular solution is given by equation (13.32). Thus, the general solution to equation (13.34) is given by

$$m(t)=m_0 e^{-\rho t}+\rho\sigma\int_0^t e^{-\rho(t-\tau)}W(\tau)d\tau$$

Upon taking expectations we clearly see that $E\{m(t)\}=0$. The correlation of the output is given as

$$E\{m(t_1)m(t_2)\}=R_m(t_1,t_1)$$

$$=E\left\{\left[m_0 e^{-\rho t_1}+\rho\sigma\int_0^{t_1} e^{-\rho(t_1-\tau)}W(\tau)d\tau\right]\left[m_0 e^{-\rho t_2}+\rho\sigma\int_0^{t_2} e^{-\rho(t_2-\xi)}W(\xi)d\xi\right]\right\}$$

$$=\frac{\rho\sigma^2}{2}e^{-\rho(t_1+t_2)}+\rho^2\sigma^2\int_0^{t_1}\int_0^{t_2} e^{-\rho(t_1-\tau)}e^{-\rho(t_2-\xi)}\delta(\tau-\xi)d\xi d\tau$$

$$=\frac{\rho\sigma^2}{2}e^{-\rho(t_1+t_2)}+\rho^2\sigma^2\int_0^{t_2} e^{-\rho(t_1-\xi)}e^{-\rho(t_2-\xi)}d\xi = \frac{\rho\sigma^2}{2}e^{-\rho(t_2-t_1)}, \quad t_2>t_1$$

Example 13.18: Dispersion Under an Uncertain Transient Spill

Consider again the case of a contaminant dispersion into an aquifer of Example 13.12, this time subject to a space-time contaminant loading. Equation (13.25) becomes

$$\frac{\partial C}{\partial t} - D\frac{\partial^2 C}{\partial x^2} + v\frac{\partial C}{\partial x} = F(t), \quad -\infty < x < \infty, \; 0 < t, \; C(\pm\infty, t) = 0, \; C(x, 0) = 0 \tag{13.35}$$

$$F(t) = M(t)U(x-a)U(b-x), \quad E\{M(t)\} = \mu_M, \quad E\{M(t_1)M(t_2)\} = qe^{-\rho(t_2-t_1)}$$

where

$\quad$ $U(\ )$=unit step function

$\quad$ $M(t)$=contaminant mass injected per unit cross-sectional area perpendicular to x, per unit time $(kg/(m^2 month))$, a Colored Noise process

$\quad$ μ_M=mean of $M\,(kg/(m^2 month))$

$\quad$ t=time $(month)$

$\quad$ q=Colored Noise variance parameter $(kg/(m^2 month))^2$

$\quad$ ρ=Colored Noise decay parameter $(month^{-1})$

$\quad$ a, b=constants (m)

The situation depicted is that of a long aquifer receiving a time-dependent contaminant load between the abscissa $x=a$ and $x=b$, that is a strip source. The initial condition is zero. Find the solution to the differential equation and the mean concentration.

Solution

Once again, equation (13.35) can be cast as an evolution equation of the form of equation (13.26):

$$\frac{\partial C}{\partial t} + AC = F(t), \qquad C(x,\ 0) = 0, \qquad A = -D\frac{\partial^2}{\partial x^2} + v\frac{\partial}{\partial x} \tag{13.36}$$

By analogy with equation (13.31), the solution to equation (13.36) is given by

$$C(x,\ t) = \int_0^t J_{t-\tau}F(\tau)d\tau$$

where the semigroup operator, J_t, associated with the partial differential operator A is given by equation (13.27). Thus, the solution becomes

$$C(x,t)=\int_0^t \frac{1}{\sqrt{4\pi D(t-\tau)}} \int_{-\infty}^{\infty} e^{-\frac{(x-v(t-\tau)-x')^2}{4D(t-\tau)}} M(\tau)U(x'-a)U(b-x')dx'$$

$$=\int_0^t \frac{M(\tau)}{\sqrt{4\pi D(t-\tau)}} \int_a^b e^{-\frac{(x-v(t-\tau)-x')^2}{4D(t-\tau)}} dx'$$

$$=\int_0^t \frac{M(t-\tau)}{\sqrt{16\pi D\tau}}\left[erf\left(\frac{x-v\tau-a}{\sqrt{4\pi D\tau}}\right)-erf\left(\frac{x-v\tau-b}{\sqrt{4\pi D\tau}}\right)\right]d\tau$$

where

$erf(\ )=$the "error function" defined as the integral $erf(w)=\dfrac{2}{\sqrt{\pi}}\displaystyle\int_0^w e^{-\frac{\xi^2}{2}}d\xi$

The error functions result after a simple change of variable in the spatial integral (see Serrano, 1988 for details). Values of the error function are obtained from tables or from intrinsic functions in computer programs. Sample functions of $C(x, t)$ may be obtained from a sample function of $M(t)$ (as in Example 13.17 and Problem 13.12), and a numerical integration of the time integral. A six-point Gaussian quadrature provides an accurate approximation (consult any textbook on numerical methods). The mean concentration is given by

$$\mu_C(x,t)=\int_0^t \frac{\mu_M}{\sqrt{16\pi D\tau}}\left[erf\left(\frac{x-v\tau-a}{\sqrt{4\pi D\tau}}\right)-erf\left(\frac{x-v\tau-b}{\sqrt{4\pi D\tau}}\right)\right]d\tau$$

Again the integral must be approximated numerically. The correlation function is obtained in the usual manner.

Example 13.19: Simulation of Mean Contaminant Propagation
Use a numerical integration of the mean contaminant integral in Example 13.18 to simulate the contaminant spatial distribution at t=12, 36, and 120*months*, respectively, assuming that μ_M=10*kg*/(m^2*month*), D=0.1m^2/*month*, v=0.1*m*/*month*, a=10*m*, and b=20*m*.

Solution
Our system is receiving a constant contaminant loading with a mean concentration μ_M=10*kg*/(m^2*month*); the source is located between the abscissas a=10*m*, and b=20*m*. The following Maple program solves numerically the integral of the mean concentration derived in Example 13.17. The command is evalf( ) and its argument is the integral itself. Evalf( ) has at its disposal several numerical integration schemes and the

user may specify a particular one if one so wishes. The figure illustrates the evolution of the contaminant plume distribution as it propagates positively with x and as it disperses in both directions of x.

```
> restart: Dx:=0.1: v:=0.1: a:=10: b:=20: muM:=10:
  f:=(x,t)->muM/sqrt(16*Pi*Dx*t)*
        (erf((x-v*t-a)/sqrt(4*Pi*Dx*t))
                    -erf((x-v*t-b)/sqrt(4*Pi*Dx*t))):
  muC:=(x,t)->evalf(int(f(x,tau),tau=0..t)):
  g1:=plot(muC(x,12),x=0..50,color=black,thickness=2,
        legend=`t=12`):
  g2:=plot(muC(x,36),x=0..50,color=black,thickness=2,
        linestyle=2,legend=`t=36`):
  g3:=plot(muC(x,120),x=0..50,color=black,thickness=2,
        linestyle=3,legend=`t=120`):
  plots[display](g1,g2,g3,labels=[`x (m)`,`kg/m^3`],
        labelfont=[TIMES,ROMAN,16],
        labeldirections=[horizontal,vertical]);
```

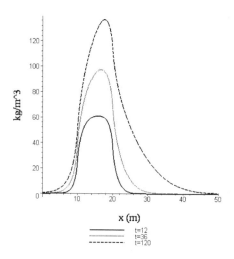

Solution of Random Equations with Decomposition

In Chapter 4 we studied the *method of decomposition* of Adomian (Adomian, 1994, 1991, 1986, 1983) as an important procedure to obtain an analytical series solution of nonlinear algebraic equations. Each term in the series is found from the previous ones and the series usually converges fast. In fact, decomposition has become one of the best *systematic* methods to solve linear and nonlinear differential equations. We illustrate the method with examples of linear ordinary and partial differential equations and extend it to the case of nonlinear differential equations. We encourage the reader to explore Adomian's works, which present a multitude of examples of equations in physics and engineering.

Example 13.20: Convergence of Decomposition to the Exact Solution
Solve equation (13.18) with decomposition and show that it converges to the exact solution.

Solution
To simplify notation, let $L_t = d/dt$. The differential equation becomes

$$L_t C + aC = 0$$

Multiplying by L_t^{-1}, that is the time integral from 0 to t, and rearranging,

$$C = C_0 - L_t^{-1} aC$$

Now expand C in the right side as the infinite series $C = C_0 + C_1 + C_2 + ...$

$$C = C_0 - L_t^{-1} a \left(C_0 + C_1 + C_2 + ... \right)$$

Clearly, the first term is the initial condition, C_0, and subsequent terms are derived recursively:

$$C_1 = -L_t^{-1} aC_0 = -atC_0$$

$$C_2 = -L_t^{-1} aC_1 = \frac{a^2 t^2 C_0}{2}$$

$$\vdots$$

$$C_n = -L_t^{-1} aC_{n-1} = (-1)^n \frac{(at)^n}{n!} C_0$$

Assembling the series we obtain

$$C = \sum_{n=0}^{\infty} C_0 (-1)^n \frac{(at)^n}{n!}$$

This may be written as

$$C = C_0 \left(1 - at + \frac{(at)^2}{2} - ... \right) = C_0 e^{-at}$$

which is identical to equation (13.19)

From Example 13.20 we see that decomposition is a simple procedure that converges to the exact solution. Rigorous convergence of decomposition series has already been established in the mathematical community (Abbaoui and Cherruault, 1994, Cherruault, 1989, and Cherruault et al., 1992). It is also important to mention the rigorous mathematical framework for the convergence of decomposition series developed by Gabet (1994, 1993, 1992). He connected the method of decomposition to well-known formulations where classical theorems (e.g., the fixed-point theorem, substituted series, etc.) could be used. For the class of equations appearing in heat flow and mass transport, theorems on the convergence of decomposition series are shown in Serrano (2003, 1998). In many complex linear and nonlinear problems an exact closed form solution is difficult to obtain. However, in these cases the usual fast convergence rate of decomposition series provides the modeler with a sufficiently accurate approximate solution. A convergent decomposition series made of the first few terms usually provides an effective model in practical applications. With the concepts of partial decomposition and of double decomposition (Adomian, 1994, 1991), the process of obtaining an approximate solution was further simplified. Also recent contributions suggest that the choice of the initial term greatly influences the rate of convergence and the complexity in the calculation of individual terms, especially for nonlinear equations (Wazwaz and Gorguiz, 2004; Wazwaz, 2000). Thus, as long as the initial term in a decomposition series, usually the forcing function or the initial condition, is described in analytic form, a partial decomposition procedure may offer a simplified approximate solution to many modeling problems.

Example 13.21: Two-Dimensional Flow in an Unconfined Aquifer

Consider the partial differential equation of steady flow in an unconfined aquifer subject to mixed boundary conditions (Serrano, 2010):

$$\frac{\partial^2 h}{\partial x^2} + \frac{\partial^2 h}{\partial y^2} = -\frac{R_g}{T}, \quad 0 \le x \le l_x, \ 0 \le y \le l_y$$

$$(13.37)$$

$$h(0, y) = f_1(y), \quad \frac{\partial h}{\partial x}(l_x, y) = 0, \quad h(x, 0) = f_2(x), \quad h(x, l_y) = f_3(x)$$

where

  $h(x, y)$=hydraulic head (m)
  x, y=horizontal (planar) coordinates (m)
  l_x, l_y=aquifer domain dimensions in the x and y directions, respectively (m)
  T=aquifer transmissivity ($m^2/month$)
  R_g=recharge from rainfall ($m/month$)

For boundary conditions, we have a specified head at $x=0$, $y=0$, and $y=l_y$, and a no-flow boundary condition at $x=l_x$, with $f_1(y)=241-0.001y$, $f_2(x)=Cx^2+Ax+f_1(0)$, $f_3(x)=Ex^2+Bx+f_1(l_y)$, $C=-R_g/(2T)$, $A=-2Cl_x$, $B=-2El_x$, and $E=-R_g/T$. Solve the differential equation using decomposition.

Solution

We seek to derive a solution expansion $h=h_0+h_1+h_2+...$ Define the operators $L_x=\partial^2/\partial x^2$ and $L_y=\partial^2/\partial y^2$. The inverse operators L_y^{-1} and L_x^{-1} are the corresponding twofold indefinite integrals with respect to x and y, respectively. Write equation (13.37) as

$$L_x h + L_y h = -\frac{R_g}{T} \tag{13.38}$$

There are two partial decomposition expansions to equation (13.38): The x-partial solution and the y-partial solution. Each partial solution is derived with its own expansion and a combination of the two yields the solution series. The x-partial solution, $h=h_x$, results from operating with L_x^{-1} on equation (13.38) and rearranging. Multiplying by the inverse operator L_x^{-1},

$$h_x = -L_x^{-1}\frac{R_g}{T} - L_x^{-1}L_y h_x \tag{13.39}$$

Expanding h_x in the right side as an infinite series $h_x=h_{x0}+h_{x1}+h_{x2}+...$, equation (13.39) becomes

$$h_x = -L_x^{-1}\frac{R_g}{T} - L_x^{-1}L_y\left(h_{x0}+h_{x1}+h_{x2}+...\right) \tag{13.40}$$

The choice of the initial term, h_{x0}, often determines the level of difficulty in calculating subsequent decomposition terms and the rate of convergence (Adomian, 1994; Wazwaz, 2000). A simple choice is to set h_{x0} equal to the first three terms in the right side of equation (13.40). Thus, the first approximation to the solutions is

$$h_{x0}=k_1(y)+k_2(y)x-L_x^{-1}\frac{R_g}{T}=k_1(y)+k_2(y)x-\frac{R_g x^2}{2T} \tag{13.41}$$

Applying the x boundary conditions from equations (13.37),

$$h(0, y)=f_1(y)=h_{x0}(0, y)=k_1(y)$$

$$\frac{\partial h}{\partial x}(l_x, y)=0=\frac{\partial h_{x0}}{\partial x}(l_x, y)=k_2(y)-\frac{R_g l_x}{T} \quad \Rightarrow \quad k_2(y)=\frac{R_g l_x}{T}$$

Equation (13.41) becomes

$$h_{x0} = f_1(y) + \frac{R_g l_x x}{T} - \frac{R_g x^2}{2T}$$

(13.42)

Equation (13.42) satisfies the governing equation (13.37) and the x boundary conditions, but not necessarily those on the y direction. Now, to obtain the y-partial solution to equation (13.38), $h = h_y$, we operate with L_y^{-1} on equation (13.38) and rearrange:

$$h_y = -L_y^{-1} \frac{R_g}{T} - L_y^{-1} L_x h_y$$

(13.43)

Expanding h_y in the right side as an infinite series $h_y = h_{y0} + h_{y1} + h_{y2} + \dots$, equation (13.43) becomes

$$h_y = -L_y^{-1} \frac{R_g}{T} - L_y^{-1} L_x \left(h_{y0} + h_{y1} + h_{y2} + \dots \right)$$

(13.44)

Again, if we take h_{y0} as the first three terms in the right side of equation (13.44) we obtain the first approximation to the solution, that is

$$h_{y0} = k_3(x) + k_4(x)y - L_y^{-1} \frac{R_g}{T} = k_3(x) + k_4(x)y - \frac{R_g y^2}{2T}$$

(13.45)

where the integration "constants" k_3 and k_4 are to be found from the y boundary conditions in equations (13.37):

$$h_{y0} = f_2(x) + \left(\frac{f_3(x) - f_2(x)}{l_y} + \frac{R_g l_y}{2T} \right) y - \frac{R_g y^2}{2T}$$

(13.46)

This y-partial solution satisfies the differential equation (13.37) and the y boundary conditions, but not necessarily those on the x direction. We now have two partial solutions to equation (13.37): the x-partial solution equation (13.42), and the y-partial solution equation (13.46). Since both are solutions to h, a combination of the two partial solutions yields the first term in the complete series, h_0:

$$h_0(x, y) = \left(\frac{h_{x0}(x, y) + h_{y0}(x, y)}{2} \right)$$

(13.47)

To obtain the second term in the combined series, h_1, we need to re derive a new x-partial solution, a new y-partial solution and combine them as above. Thus, the second term in the x-partial solution, h_{x1}, may be derived from equation (13.40):

$$h_{x1}=k_5(y)+k_6(y)x-L_x^{-1}L_yh_0 \qquad (13.48)$$

where h_0 is given by equation (13.47), and k_5 and k_6 are such that equation (13.48) satisfies homogeneous (i.e., zero) x boundary conditions in equations (13.37). Similarly, the second term in the y-partial solution, h_{y1}, may be derived from equation (13.44):

$$h_{y1}=k_7(x)+k_8(x)y-L_y^{-1}L_xh_0 \qquad (13.49)$$

where h_0 is given by equation (13.47), and k_7 and k_8 are such that equation (13.48) satisfies homogeneous (i.e., zero) y boundary conditions in equations (13.37). Subsequently, h_1 is obtained by combining equations (13.48) and (13.49):

$$h_1(x,\ y)=\left(\frac{h_{x1}(x,\ y)+h_{y1}(x,\ y)}{2}\right) \qquad (13.50)$$

Higher-order terms are derived similarly. The i-th order terms in the x-partial solution, h_{xi}, may be derived from equation (13.40):

$$h_{xi}=k_{4i+1}(y)+k_{4i+2}(y)x-L_x^{-1}L_yh_{i-1} \qquad (13.51)$$

where h_{i-1} is the previous combined term in the decomposition series, and k_{4i+1} and k_{4i+2} are such that equation (13.48) satisfies homogeneous (i.e., zero) x boundary conditions in equations (13.37). Similarly, the i-th order term in the y-partial solution, h_{yi}, may be derived from equation (13.44):

$$h_{yi}=k_{4i+3}(x)+k_{4i+4}(x)y-L_y^{-1}L_xh_0 \qquad (13.52)$$

where h_{i-1} is the previous combined term in the decomposition series, and k_{4i+3} and k_{4i+4} are such that equation (13.48) satisfies homogeneous (i.e., zero) y boundary conditions in equations (13.37). Similarly to equation (13.50), the i-th combined term is given by

$$h_i(x,\ y)=\left(\frac{h_{xi}(x,\ y)+h_{yi}(x,\ y)}{2}\right) \qquad (13.53)$$

Lastly, we approximate the solution with N terms, $h \approx h_0+h_1+...+h_N$.

Example 13.22: Simulation of Two-Dimensional Flow

In Example 13.21, $l_x=100m$, $l_y=200m$, $T=700\,m^2/month$, $R_g=0.03$ $m/month$. Approximate h and plot the steady-state contours.

Solution

```
Initialize and enter the boundary conditions
> restart:
  A:=-2*C*lx:  C:=-Rg/2/T:  B:=-2*E*lx:  E:=-Rg/T:
  f1:=y->241-0.001*y:
  f2:=x->C*x^2+A*x+f1(0):
  f3:=x->E*x^2+B*x+f1(ly):
Derive hx0 from equation (13.41).
> k1(y)+k2(y)*x-int(int(Rg,x),x)/T:
  hx0:=unapply(%,x,y):
  hx0(0,y)=f1(y):
  solve(%,k1(y)):
  k1:=unapply(%,y):
  diff(hx0(x,y),x):
  subs(x=lx,%)=0:
  solve(%,k2(y)):
  k2:=unapply(%,y):
Derive hy0 from equation (13.45).
> k3(x)+k4(x)*y-int(int(Rg,y),y)/T:
  hy0:=unapply(%,x,y):
  hy0(x,0)=f2(x):
  solve(%,k3(x)):
  k3:=unapply(%,x):
  hy0(x,ly)=f3(x):
  solve(%,k4(x)):
  k4:=unapply(%,x):
Derive h0 from equation (13.47)
> h0:=(x,y)->(hx0(x,y)+hy0(x,y))/2:
Derive hx1 from equation (13.48).
> k5(y)+k6(y)*x-int(int(diff(h0(x,y),y$2),x),x):
  hx1:=unapply(%,x,y):
  hx1(0,y)=0:
  solve(%,k5(y)):
  k5:=unapply(%,y):
  diff(hx1(x,y),x):
  subs(x=lx,%)=0:
  solve(%,k6(y)):
  k6:=unapply(%,y):
Derive hy1 from equation (13.49).
> k7(x)+k8(x)*y-int(int(diff(h0(x,y),x$2),y),y):
  hy1:=unapply(%,x,y):
  hy1(x,0)=0:
  solve(%,k7(x)):
  k7:=unapply(%,x):
  hy1(x,ly)=0:
  solve(%,k8(x)):
  k8:=unapply(%,x):
Derive h1 from equation (13.50).
> h1:=(x,y)->(hx1(x,y)+hy1(x,y))/2:
Derive hx2 from equation (13.51).
> k9(y)+k10(y)*x-int(int(diff(h1(x,y),y$2),x),x):
  hx2:=unapply(%,x,y):
  hx2(0,y)=0:
  solve(%,k9(y)):
  k9:=unapply(%,y):
  diff(hx2(x,y),x):
  subs(x=lx,%)=0:
  solve(%,k10(y)):
  k10:=unapply(%,y):
```

```
[ Derive hy2 from equation (13.52).
[ > k11(x)+k12(x)*y-int(int(diff(h1(x,y),x$2),y),y):
    hy2:=unapply(%,x,y):
    hy2(x,0)=0:
    solve(%,k11(x)):
    k11:=unapply(%,x):
    hy2(x,ly)=0:
    solve(%,k12(x)):
    k12:=unapply(%,x):
    h2:=(x,y)->(hx2(x,y)+hy2(x,y))/2:
[ Approximate h=h0+h1+h2.
[ > h:=(x,y)->h0(x,y)+h1(x,y)+h2(x,y):
[ Enter data, and check for convergence at one point in the middle of the domain.
[ > lx:=100: ly:=200.0: T:=700: Rg:=0.03:
    h0(lx/2,ly/2), h1(lx/2,ly/2), h2(lx/2,ly/2);
                            241.2080358, -0.1741071429, 0.07700892851
[ Plot hydraulic head in 3D with suitable contours.
[ > plot3d(h(x,y),x=0..lx,y=0..ly,orientation=[-155,60],
            style=PATCHCONTOUR,color=gray,thickness=2):
    plots[display](%,axes=BOXED,labels=[`x(m)`,`y(m)`,`h(m)`],
            labeldirections=[HORIZONTAL,HORIZONTAL,VERTICAL],
            labelfont=[TIMES,ROMAN,16]);
[ > plot(h(0,y),y=0..ly,color=blue):
    plot(f1(y),y=0..ly):
    plots[display](%,%%);
[ > plot(h(x,0),x=0..lx,color=blue):
    plot(f2(x),x=0..lx):
    plots[display](%,%%);
[ > plot(h(x,ly),x=0..lx,color=blue):
    plot(f3(x),x=0..lx):
    plots[display](%,%%);
```

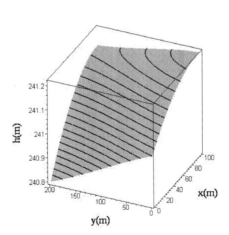

In this program we use contour plotting as well as the symbolic mathematics capabilities of Maple. Since the first term in each coordinate expansion satisfies the given boundary conditions in such an axis, subsequent constants of integration are derived such that they satisfy homogeneous boundary conditions on the sam axis. Notice the fast convergence. In this example only two terms may be necessary to obtain an accurate head prediction. It is important to remark that the convergence rate depends on the magnitude of the parameters, especially T, and on the domain size (l_x, l_y). The usual rules of modeling apply here: the values of the domain size and its parameters must be such that the *Lipschitz condition* is satisfied (see Oden and Demkowicz, 2010; Hutson and Pym, 1980; and Oden, 1979 for details). If this is not possible, it may be necessary to solve the problem in reduced dimensionless coordinates.

From the previous examples some generalizations can be done: (1) Decomposition series converge to the exact solution. (2) Each term in the series is analytic, thus presenting all of the advantages of analytical solutions, such as minimized instability, continuous description of dependent variables, gradients and fluxes, minimal computer programming effort. (3) Each term in the series is calculable based on the previous ones, which is an important difference with traditional small perturbation methods. (4) There is no need of restricting assumptions, such as "smallness" in the parameters (small perturbation methods), or a particular probability density function in the parameters. (5) Decomposition is significantly simpler in its mathematics than traditional analytical producers, such as Fourier series and Laplace transform; it only requires integral calculus. There are other unique advantages of decomposition as we will now describe. One of these is the ability to consider irregular domain shapes. In the past most analytical solutions of *boundary-value initial-value problems* were restricted to rectangles, squares, circles and other regular domain shapes. It was tacitly understood that problems with irregular boundaries had to be solved numerically (i.e., with finite differences or finite elements). However, with analytical decomposition, application of the above procedure to domains of irregular geometry is straight forward, as long as the aquifer boundaries are defined in a functional form.

Example 13.23: Modeling Flow in Irregular Domain Shapes

In Examples 13.21 and 13.22, assume $l_x=860m$, $T=700\,m^2/month$, $R_g= 0.01$ *m/month*, and that the y aquifer dimension, $l_y(x)$, is now a function of x. In other words, $h(x, y)=f_4(x)$, on $y=l_y(x)$, with $l_y(x)=2000+0.4x-0.1\times10^{-2}x^2$. For simplicity, set $h\approx h_0$. Plot the head contours in planar coordinates.

Solution

The solution procedure described in Examples 13.19 and 13.20 remains unchanged, except that in the y-partial solution the integration must now be done over a variable y domain, which is reflected in equation (13.44) with l_y now being a function of x. As described in the following program, the solution is simply approximated by equations (13.41), (13.46), and equation (13.47).

```
> restart:
  A:=-2*C*lx:  C:=-Rg/2/T:  B:=-2*E*lx:  E:=-Rg/T:
  f1:=y->241-0.001*y:
  f2:=x->C*x^2+A*x+f1(0):
  f3:=x->E*x^2+B*x+f1(ly(x)):
  ly:=x->2000+0.4*x-0.1E-2*x^2:
  hx0:=(x,y)->f1(y)+Rg*lx*x/T-Rg*x^2/2/T:
  hy0:=(x,y)->f2(x)+((f3(x)-f2(x))/ly(x)
              +(Rg/2/T)*ly(x))*y-(Rg/2/T)*y^2:
  h0:=(x,y)->(hx0(x,y)+hy0(x,y))/2:
> lx:=860: T:=700: Rg:=0.010:
  g1:=plots[contourplot](h0(x,y),x=0..lx,y=0..ly(x),
      color=black,thickness=2,
      contours=[249,248,247,246,245,244,243,242,241,240]):
  g2:=plot(ly(x),x=0..lx, color=black,thickness=2):
  g3:=plot([[lx,0],[lx,ly(lx)]],color=black,thickness=2):
  plots[display](g1,g2,g3,labels=[`x(m)`,`y(m)`],
      labeldirections=[HORIZONTAL,HORIZONTAL],
      labelfont=[TIMES,ROMAN,16]);
```

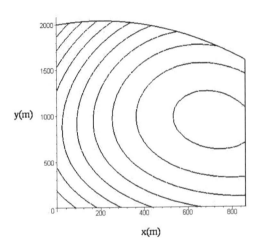

Example 13.24: Transient Groundwater Flow Modeling

Consider the problem of regional groundwater flow through a long, thin, aquifer subject to a seasonal periodic recharge from rainfall. The governing equation is given as (see Serrano, 1997 for details),

$$\frac{\partial h}{\partial t} - \frac{T}{S_y}\frac{\partial^2 h}{\partial x^2} = \frac{R_g(t)}{S_y}, \qquad 0 \le x \le l_x,\ 0 < t$$

$$(13.54)$$

$$h(0,\ t) = V(0), \qquad \frac{\partial h}{\partial x}(l_x,\ t) = 0, \qquad h(x,\ 0) = V(x)$$

where
$h(x, t)$=hydraulic head (m)
x=distance with respect to the left boundary (m)
t=time (*months*)
T=aquifer transmissivity ($m^2/month$)
S_y=aquifer specific yield
$R_g(t)$=recharge from rainfall ($m/month$)
l_x=aquifer length (m)
$V(x)$=initial head (m)

Assume that the recharge term is defined as a random function. Obtain a solution to the differential equation using decomposition.

Solution
We seek to derive a solution expansion $h=h_0+h_1+h_2+....$. There are two possible decomposition solutions to equation (13.53), each with its own independent expansion series: the t-partial solution and the x-partial solution. The t-partial solution, $h=h_t$, results after defining $L_t=\partial/\partial t$ and $L_x=\partial^2/\partial x^2$, multiplying equation (13.54) by L_t^{-1} (i.e., the integral from zero to t) and rearranging:

$$h_t=L_t^{-1}\frac{R_g}{S_y}+\frac{T}{S_y}L_t^{-1}L_x h_t \qquad (13.55)$$

Decompose h_t in the right side of equation (13.55) as $h_t=h_{t0}+h_{t1}+h_{t2}+...$, thus

$$h_t=L_t^{-1}\frac{R_g}{S_y}+\frac{T}{S_y}L_t^{-1}L_x\left(h_{t0}+h_{t1}+h_{t2}+...\right) \qquad (13.56)$$

From equation (13.56), the first term in the series, h_{t0}, is given by

$$h_{t0}=k_0(x)+L_t^{-1}\frac{R_g(t)}{S_y} \qquad (13.57)$$

where $k_0(x)$ must satisfy the initial condition in equations (13.54):

$$h_{t0}=V(x)+L_t^{-1}\frac{R_g(t)}{S_y} \qquad (13.58)$$

Equation (13.58) satisfies the initial condition, but not necessarily the boundary condition in x in equation (13.54). Now the x-partial solution, h_x, results after multiplying equation (13.54) by L_x^{-1} (or the twofold x

indefinite integral), and rearranging:

$$h_x = -L_x^{-1}\frac{R_g(t)}{T} + \frac{S_y}{T}L_x^{-1}L_t h_x \qquad (13.59)$$

Decompose h_x in the right side as $h_x = h_{x0} + h_{x1} + h_{x2} + \ldots$ Then,

$$h_x = -L_x^{-1}\frac{R_g(t)}{T} + \frac{S_y}{T}L_x^{-1}L_t\left(h_{x0} + h_{x1} + h_{x2} + \ldots\right) \qquad (13.60)$$

If we take h_{x0} as the first three terms in the right side of equation (13.60), we obtain

$$h_{x0} = k_1(t) + k_2(t)x - \frac{R_g(t)x^2}{2T} \qquad (13.61)$$

where the constants k_1 and k_2 are found after applying the x boundary conditions in equation (13.54).

$$h_{x0} = V(0) + \frac{R_g(t)l_x x}{T} - \frac{R_g(t)x^2}{2T} \qquad (13.62)$$

Equation (13.62) satisfies the x boundary conditions, but not necessarily the initial condition in equation (13.54). We now have two partial solutions to equation (13.54): The t-partial solution equation (13.58), and the x-partial solution equation (13.62). Since both are approximations to h, a combination of the two yields the first decomposition term, h_0:

$$h_0(x,\ t) = \left(\frac{h_{t0}(x,\ t) + h_{x0}(x,\ t)}{2}\right) \qquad (13.63)$$

Higher-order terms are obtained as described in Example 13.21. In other words, the i-th term in the t-partial expansion is obtained form equation (13.56) as

$$h_{ti} = k_{3i}(x) + \frac{T}{S_y}L_t^{-1}L_x h_{i-1},\quad i>0 \qquad (13.64)$$

where k_{3i} is such that a homogeneous (i.e., zero) initial condition in equation (13.54) is satisfied, and h_{i-1} is the previous combined term in the decomposition series. Then, the i-th term in the x-partial expansion is obtained form equation (13.60) as

$$h_{xi} = k_{3i+1}(t) + k_{3i+2}(t)x + \frac{S_y}{T}L_x^{-1}L_t h_{i-1},\quad i>0 \qquad (13.65)$$

where k_{3i+1} and k_{3i+2} are such that homogeneous (i.e., zero) x-boundary conditions in equation (13.54) are satisfied, and h_{i-1} is the previous combined term in the decomposition series. Next, we combine equations

(13.64) and (13.65) to derive the next term, h_i

$$h_i(x,\ t) = \left(\frac{h_{ti}(x,\ t) + h_{xi}(x,\ t)}{2} \right)$$

(13.66)

It has been found that decomposition series converge rapidly and that only a few terms, $N+1$, are necessary to obtain an accurate solution. In such case, we approximate the solution as

$$h \approx h_0 + h_1 + \ldots + h_N$$

(13.67)

from which the statistics of the output may be computed:

$$E\{h\} = E\{h_0\} + E\{h_1\} + \ldots + E\{h_N\}$$

$$E\{h(x,t_1)h(x,t_2)\} = R_h(t_1,\ t_2)$$

(13.68)

$$= E\{ \left[h_0(x,t_1) + h_1(x,t_1) + \ldots + h_N(x,t_1) \right] \left[h_0(x,t_2) + h_1(x,t_2) + \ldots + h_N(x,t_2 \right.$$

where the correlation between the individual terms is found from the correlation of the input term $R_g(t)$.

Example 13.25: Simulation of Transient Flow

In Example 13.24 $R_g(t) = a - b\sin(c\pi t/6)$; $a = 0.00314 m/month$; $b = 0.00341 m/month$; $c = 1\ month^{-1}$; $l_x = 100 m$; $T = 100\ m^2/month$; $S_y = 0.1$; and $V(x) = 100 + al_x x/T - ax^2/(2T)$. Derive 4 decomposition terms, check for convergence and plot the head temporal variability at $x = 25$, 50, and $100m$, respectively.

Solution

In the following program, the constants of integration in Example 13.24 are derived using the symbolic mathematics features of Maple.

```
> restart: Rg:=t->a-b*sin(Pi*t/6):
  V:=x->100+a*lx*x/T-a*x^2/2/T:
First term in the series
> V(x)+int(Rg(tau),tau=0..t)/Sy:
  ht0:=unapply(%,x,t):
  hx0:=(x,t)->V(0)+Rg(t)*lx*x/T-Rg(t)*x^2/2/T:
  h0:=(x,t)->(ht0(x,t)+hx0(x,t))/2:
Second term.
> k0(x)+T/Sy*int(diff(h0(x,tau),x$2),tau=0..t):
  ht1:=unapply(%,x,t):
  ht1(x,0)=0:
  solve(%,k0(x)):
  k0:=unapply(%,x):
> k1(t)+k2(t)*x+Sy*int(int(diff(h0(x,t),t),x),x)/T:
  hx1:=unapply(%,x,t):
  hx1(0,t)=0:
  solve(%,k1(t)):
  k1:=unapply(%,t):
  diff(hx1(x,t),x):
  subs(x=lx,%)=0:
  solve(%,k2(t)):
  k2:=unapply(%,t):
  h1:=(x,t)->(ht1(x,t)+hx1(x,t))/2:
```

Third term.
```
> k3(x)+T/Sy*int(diff(h1(x,tau),x$2),tau=0..t):
  ht2:=unapply(%,x,t):
  ht2(x,0)=0:
  solve(%,k3(x)):
  k3:=unapply(%,x):
> k4(t)+k5(t)*x+Sy*int(int(diff(h1(x,t),t),x),x)/T:
  hx2:=unapply(%,x,t):
  hx2(0,t)=0:
  solve(%,k4(t)):
  k4:=unapply(%,t):
  diff(hx2(x,t),x):
  subs(x=lx,%)=0:
  solve(%,k5(t)):
  k5:=unapply(%,t):
  h2:=(x,t)->(ht2(x,t)+hx2(x,t))/2:
```
Fourth term.
```
> k6(x)+T/Sy*int(diff(h2(x,tau),x$2),tau=0..t):
  ht3:=unapply(%,x,t):
  ht3(x,0)=0:
  solve(%,k6(x)):
  k6:=unapply(%,x):
> k7(t)+k8(t)*x+Sy*int(int(diff(h2(x,t),t),x),x)/T:
  hx3:=unapply(%,x,t):
  hx3(0,t)=0:
  solve(%,k7(t)):
  k7:=unapply(%,t):
  diff(hx3(x,t),x):
  subs(x=lx,%)=0:
  solve(%,k8(t)):
  k8:=unapply(%,t):
  h3:=(x,t)->(ht3(x,t)+hx3(x,t))/2:
> h:=(x,t)->h0(x,t)+h1(x,t)+h2(x,t)+h3(x,t):
```
Data
```
> a:=0.00314: b:=0.003128: T:=100: Sy:=0.1: lx:=100:
```
Check for convergence at an adverse location x=lx.
```
> h0(lx,36), h1(lx,36), h2(lx,36), h3(lx,36): evalf(%);
          100.7222000, -0.5191470328, 0.1378985164, -0.2048598721
> g1:=plot(h(25,t),t=0..36,color=black,thickness=2,
    legend=`x=25m`):
  g2:=plot(h(50,t),t=0..36,color=black,thickness=2,
    legend=`x=50m`,linestyle=3):
  g3:=plot(h(100,t),t=0..36,color=black,thickness=2,
    legend=`x=100m`,linestyle=4):
  plots[display](g1,g2,g3,labels=[`t`,`h (m)`],
    labelfont=[TIMES,ROMAN,16],
    labeldirections=[horizontal,vertical]);
```

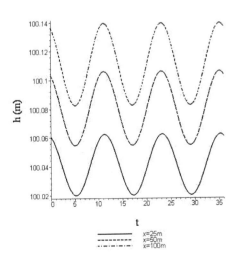

Notice the first three terms suggest a fast convergence, but the fourth one does not decrease accordingly. In any case, two or three terms are reasonably accurate in this case, a resolution commensurate with that of measurement errors in field hydraulic heads ($1cm$). Once again, the domain size and parameter values are critical in determining convergence speed, stability, and accuracy.

Example 13.26: Statistics of Transient Flow

In Example 13.24, $R_g(t)=a-b\sin(\pi t/6)$; $a=0.0314m$; b is a random variable with $\mu_b=0.03128m$, $\sigma_b=0.01m$; $l_x=1100m$; $T=400\,m^2$ /month ; $S_y=0.1$; and $V(x)=100+al_x x/T-ax^2/(2T)$. Assuming that $h\approx h_{t0}$ is an acceptable approximation, calculate the mean of the output head, h, and its variance.

Solution

From equation (13.57) the first term in the series is given by

$$h(x,\ t)\approx h_{t0}=V(x)+L_t^{-1}\frac{R_g(t)}{S_y}=V(x)+\frac{1}{S_y}\left[at+\frac{6b}{\pi}\left(\cos(\frac{\pi t}{6})-1\right)\right]$$

Taking expectation,

$$E\{h(x,t)\}=\mu_h(t)=V(x)+\frac{1}{S_y}\left[at+\frac{6\mu_b}{\pi}\left(\cos(\frac{\pi t}{6})-1\right)\right]$$

Applying the methods derived in Chapter 12, we obtain the variance as

$$\sigma_h^2=E\{h^2(x,\ t)\}-\mu_h^2=\frac{36\sigma_b^2}{S_y^2\pi^2}\left(\cos(\frac{\pi t}{6})-1\right)^2$$

after eliminating common terms. Obviously, if we include h_{x0}, as well as h_{t0}, in the approximation of h, then the variance derivation becomes more involved. Furthermore, including higher-order terms in the approximation of h results in a variance, σ_h^2, possibly requiring information on third and higher-order moments in the random parameter, b. These moments must be known a priori.

Solving Nonlinear Differential Equations

Many engineering systems are governed by nonlinear differential equations, that is equations in which the coefficients are functions of the dependent variable (i.e., the system output), or where the degree of the dependent variable or of any of the derivatives is greater than one. Solution of nonlinear differential equations, even in the deterministic sense is much more difficult than corresponding linear ones. Until recently, it was believed that analytical solutions to nonlinear equations were not possible, in general, except for very specific sets of equations. It was believed that the only way to approximate their solution was by the implementation of a numerical solution (e.g., finite differences, finite elements, boundary elements, etc.). Strictly speaking, however, a numerical solution is in fact a numerical linearization of the true nonlinear equation.

With the advent of decomposition, a systematic procedure for the analytical solution of nonlinear equations is now possible. Even if a close-form solution is not identified from the series, an approximation made of the first few terms may constitute an accurate model. The procedure is a straightforward extension of decomposition for linear equations, except that the nonlinear terms are expanded as a generalized Taylor series about the first term in the series, as described in Chapter 4. There are several decomposition schemes possible, as more theory develops, and we hope that the following examples will stimulate the reader to explore new nonlinear solutions in his/her area of interest. The works of Adomian are required reading in this endeavor. Indeed, it has been said that the universe is nonlinear and stochastic, rather than deterministic and linear, and thus most engineering analysis of complex systems remains to be explored.

Example 13.27: Exact Solution of a Nonlinear Differential Equation

Consider the following nonlinear differential equation subject to an initial condition represented as a random variable:

$$\frac{du}{dt} + au^2 = 0, \quad 0 < t, \ u(0) = u_0, \ E\{u_0\} = \mu_0, \ Var(u_0) = \sigma_0^2 \quad (13.69)$$

Solve the differential equation and approximate the mean of the output.

Solution

Setting $L_t = d/dt$ and using decomposition as before, we write the equation as

$$u = u_0 - aL_t^{-1}Nu, \quad Nu = u^2 \quad (13.70)$$

where L_t^{-1} is the integral from zero to t, and the nonlinear operator, N, is the term u^2. Compare the nonlinear algebraic equation (4.7), Chapter 4, with equation (13.70). The only fundamental difference is that the latter has integral operators. Let us, then, proceed in the manner described in Chapter 4 and expand the nonlinear term as an infinite series

$$u = u_0 - aL_t^{-1} \sum_{n=1}^{\infty} A_n$$

As in equation (4.9), define the A_n as the Adomian polynomials given as a generalized Taylor series expansion about the initial term u_0 (Adomian, 1994):

$$A_0 = Nu_0$$

$$A_1 = u_1 \frac{dNu_0}{du_0}$$

$$A_2 = u_2 \frac{dNu_0}{du_0} + \frac{u_0^2}{2!} \frac{d^2 Nu_0}{du_0^2} \qquad (13.71)$$

$$A_3 = u_3 \frac{dNu_0}{du_0} + u_1 u_2 \frac{d^2 Nu_0}{du_0^2} + \frac{u_1^3}{3!} \frac{d^3 Nu_0}{du_0^3}$$

$$\vdots$$

Alternative application of equations (13.71) and (13.72) produces, one by one, the terms in the series:

$$A_0 = Nu_0 = u_0^2$$

$$u_1 = -L_t^{-1} A_0 = -atu_0^2$$

$$A_1 = u_1 \frac{dNu_0}{du_0} = u_1 u_0 = -2atu_0^3$$

$$u_2 = -aL_t^{-1} A_1 = a^2 t^2 u_0^2$$

$$\vdots$$

Adding the terms,

$$u = u_0 + u_1 + u_2 + \dots$$

$$u = u_0 \left[1 - atu_0 + (atu_0)^2 - (atu_0)^3 + \dots \right]$$

which converges to

$$u = \frac{u_0}{1 + au_0 t}$$

In this particular case, we were able to identify a close-form solution for the series. In many nonlinear equations, however, this is not possible and the modeler has to be content with the series itself. The interesting feature is that one can calculate as many terms as one wishes in order to reach a desired level of accuracy. Thus, modeling accuracy is the desired result, even though mathematical elegance is not always possible. The i-th term requires knowledge of the term $(i-1)$ or less, but not higher than that. This is an important feature of decomposition, as compared to the still prevalent small perturbation schemes that require terms of order higher than i, and the unreasonable requirement of having to neglect the higher-order terms without any physical or mathematical basis.

Another important remark from the nonlinear solution is the fact that sometimes it is not possible to estimate the moments. From the partial series, one may approximate the first two moments of the solution. For example, if in equation (13.74) the mean and variance of the input initial condition are available, the mean of the output is given by

$$E\{u(t)\} \approx E\{u_0[1 - atu_0]\} = \mu_0 - (\mu_0^2 + \sigma_0^2)at$$

When all the moments of the initial condition are known (e.g., if the input density function is known or Gaussian), then in principle the expected value of the output may be calculated from the decomposition series. A final observation is the following: the mean (and in general the moments) of the nonlinear equation does not coincide with that of the corresponding linearized equation.

Example 13.28: Exact Solution of a Nonlinear Partial Differential Equation

Let us now consider the nonlinear advection equation given by (Adomian, 1994)

$$\frac{\partial u}{\partial t} + \frac{\partial u}{\partial x} + u^2 = 0, \quad u(x, 0) = \frac{1}{2x}, \quad u(0, t) = -\frac{1}{t} \qquad (13.72)$$

Derive an exact solution to the equation.

Solution

Let us define the operator $L_t u = -\partial u/\partial x - u^2$, the solution $u = \sum_{n=0}^{\infty} u_n$, and u^2 as $u = \sum A_n$, where A_n are the Adomian polynomials. From equation (13.73) the decomposition series are given by

$$L_t u = -\frac{\partial}{\partial x} \sum_{n=0}^{\infty} u_n - \sum_{n=0}^{\infty} A_n,$$

$$u = u_0 - L_t^{-1} \frac{\partial}{\partial x} \sum_{n=0}^{\infty} u_n - L_t^{-1} A_n,$$

$$u_0 = u(x, \ 0) = \frac{1}{2x},$$

$$u_1 = -L_t^{-1} \frac{\partial}{\partial x} u_0 - L_t^{-1} A_0,$$

$$u_2 = -L_t^{-1} \frac{\partial}{\partial x} u_1 - L_t^{-1} A_1,$$

$$\vdots$$

Substituting the $A_n(u^2)$ from equation (13.71), and summing we have (Adomian, 1994)

$$u = \frac{1}{2x} + \frac{t}{4x^2} + \frac{t^2}{8x^3} + \frac{t^3}{16x^4} + \dots$$

which converges to

$$u = \frac{1}{(2x-t)}$$

Example 13.29: Exact Solution of the Nonlinear Boussinesq Equation

We offer here a more complex nonlinear equation with a slightly different decomposition expansion. Consider the Boussinesq equation subject to a time-dependent boundary condition given by

$$\frac{\partial h}{\partial t} - \frac{1}{2} \frac{\partial^2 h^2}{\partial x^2} = 0, \quad 0 \le x \le 3, \ 0 < t$$

(13.73)

$$h(0, \ t) = h_b(t) = \frac{3}{2} \left(\frac{1}{(t+1)^{1/3}} - \frac{1}{t+1} \right), \quad \frac{\partial h}{\partial x}(3, \ t) = 0, \ h(x, \ 0) = h_i(x) = x$$

where

$h(x, t)$=hydraulic head

x=dimensionless distance

t=dimensionless time

$h_b(t)$=time-dependent left boundary condition

$h_i(x)$=initial condition across the domain

The t-partial decomposition solution of equation (13.73) gives

$$h=h_0+L_t^{-1}N(h) \tag{13.74}$$

where the nonlinear operator $N(h)$ is given by

$$Nh=\frac{1}{2}\frac{\partial}{\partial x^2}\left(h_0+h_1+h_2+\cdots\right)^2=\frac{1}{2}\frac{\partial}{\partial x^2}\left(h_0^2+2h_0h_1+h_1^2+\cdots\right)$$

From equations (13.71) each term in the series is calculated as

$$h_0=h_i$$

$$h_1=(1-h_i)t$$

$$h_2=-(1-h_i)t^2-\frac{t^2}{6}$$

$$\vdots$$

Upon a summation, $h=h_0+h_1+h_2+\cdots$, which converges to

$$h=\frac{h_i(x)}{(t+1)}+h_b(t) \tag{13.75}$$

Equation (13.75) is the exact solution to equation (13.73) (Poluvarinova-Kochina, 1977). It satisfies the differential equation, the boundary and the initial conditions. In this problem the calculation of the moments of the output is a simple manner. For instance, if the boundary condition is represented as a stochastic process in time, the application of expectation would yield the required moments of the output head.

Example 13.30

In Example 13.29 derive the x-partial solution and compare it with the exact solution.

Solution

We define the operator $L_x=\partial^2/\partial x^2$ and operate equation (13.73) with

L_x^{-1}, that is the twofold indefinite integration to obtain

$$h=k_1(t)+k_2(t)x+L_x^{-1}h^{-1}\left\{\frac{\partial h}{\partial t}-\left(\frac{\partial h}{\partial x}\right)^2\right\}$$

The first term in the series is

$$h_0=k_1(t)+k_2(t)x$$

where $k_1(t)$ and $k_2(t)$ must satisfy the boundary conditions of equation (13.73). Thus, $h_0=h_b(t)h_0=h_b(t)$. Higher-order terms are easily obtainable:

$$h_1=L_x^{-1}h_0^{-1}\left\{\frac{\partial h_0}{\partial t}-\left(\frac{\partial h_0}{\partial x}\right)^2\right\}=\frac{1}{h_b}\frac{\partial h_b}{\partial t}x^2+k_3x+k_4$$

$$h_1=\frac{1}{h_b}\frac{\partial h_b}{\partial t}\left(x^2-6x\right)$$

$$h_2=L_x^{-1}\left[h_1(-h_0^{-2})\left\{\frac{\partial h_1}{\partial t}-\left(\frac{\partial h_1}{\partial x}\right)^2\right\}\right]$$

$$\vdots$$

In this case, with the x-partial expansion is difficult to identify a pattern in the series. As an approximation

$$h\approx h_0+h_1=h_b+\frac{1}{h_b}\frac{\partial h_b}{\partial t}\left(x^2-6x\right)$$

$$=\frac{3}{2}\left(\frac{1}{(t+1)^{1/3}}-\frac{1}{(t+1)}\right)-\frac{(t+1)-3(t+1)^{1/3}}{3(t+1)-3(t+1)^{1/3}}(x^2-6x)$$

Compare the above two-term solution with the exact one via the program below:

```
> restart:
  hb:=t->3/2*(1/(t+1)^(1/3)-1/(t+1)):
  hi:=x->x-x^2/6:
  h:=(x,t)->hi(x)/(t+1)+hb(t):
  diff(hb(t),t):
  hb_t:=unapply(%,t):
  hx:=(x,t)->hb(t)+1/hb(t)*hb_t(t)*(x^2-6*x):
  T:=12:
  g1:=plot(h(x,T),x=0..3,Y=0..0.7,color=black,thickness=2,
     legend=`Exact Solution`):
  g2:=plot(hx(x,T),x=0..3,color=black,thickness=2,
     linestyle=3,legend=`x-partial Solution`):
  plots[display](g1,g2,labels=[`x`,`h`],
     labelfont=[TIMES,ROMAN,16]);
```

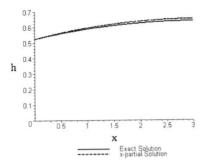

h

Exact Solution
x-partial Solution

In this case a two-term x-partial solution seems reasonably accurate provided that $t>6$, since $h_b \to 0$ as $t \to 0$.

Example 13.31: Horizontal Infiltration with a Sharp Wetting Front

In this example we present a creative combination of decomposition with a successive approximation (Serrano, 2004) that yields a simple approximate solution to a highly nonlinear equation, when traditional numerical solutions present numerous accuracy, complexity, and instability problems. Consider the horizontal infiltration if water equation in a semi-infinite homogeneous soil with a constant boundary condition maintained on one end is

$$\frac{\partial \theta}{\partial t} - \frac{\partial}{\partial x}\left(D(\theta)\frac{\partial \theta}{\partial x}\right) = 0, \quad 0 < x < \infty, \ 0 < t \tag{13.76}$$

$$\theta(0, t) = \theta_b, \qquad \theta(\infty, t) = \theta_i, \qquad \theta(x, 0) = \theta_i$$

where
  θ = soil volumetric water content
  x = horizontal distance (m)
  t = time $(hour)$
  θ_b = 0.458, the water content at the left boundary
  θ_i = 0.086, the initial water content
  $D(\theta)$ = soil-water diffusivity $(m^2/hour)$ given by (Serrano, 1998)

$$D(\theta) = c_1 e^{\lambda \theta^\alpha} - 1 \tag{13.77}$$

where
  $c_1 = 1 m^2/hour$
  $\lambda = 500$
  $\alpha = 11$

The t-partial decomposition expansion of (13.76) is

$$\theta = L_t^{-1} \frac{\partial}{\partial x}\left(D(\theta)\frac{\partial\theta}{\partial x}\right) = L_t^{-1}\frac{\partial}{\partial x}\left(\sum_{j=0}^{\infty}E_j\frac{\partial\theta}{\partial x}\right) \qquad (13.78)$$

where the E_j polynomials are recursively calculated based on successive components of the series $\theta \approx \sum_{j=0}\theta_j$. From equations (13.71), (13.77), and (13.78),

$$\theta_0 = \theta_b, \qquad E_0 = D(\theta_0) = D(\theta_b) = c_1 e^{\lambda\theta_b^\alpha} - 1 \qquad (13.79)$$

If we use the approximation $D(\theta) \approx D(\theta_0) = E_0$, then equation (13.76) reduces to the classical heat flow equation with constant coefficient and whose solution is (Zauderer, 1983)

$$\theta_1 = \theta_i + (\theta_b - \theta_i)erfc\left(\frac{x}{\sqrt{4E_0 t}}\right) \qquad (13.80)$$

where $erfc(\ )$ denotes the "error function complement." Now from equation (13.71) calculate E_1 and obtain an improved diffusivity:

$$D(\theta) \approx E_0 + E_1 = D(\theta_0) + \theta_1 \frac{dD(\theta_0)}{d\theta_0} = D(\theta_0) + \alpha\lambda\theta_1\theta_0^\alpha\left(D(\theta_0) + 1\right) \qquad (13.81)$$

The improved solution to equation (13.76) becomes

$$\theta_2 \approx \theta_i + (\theta_b - \theta_i)erfc\left(\frac{x}{\sqrt{4(E_0 + E_1)t}}\right) \qquad (13.82)$$

This process may be continued. However, we hope the series converges fast and only a few terms are needed. Thus, if θ_2 is a good approximation, we may use it to obtain an improved, final, version of the diffusivity:

$$D(\theta) \approx (E_0 + E_1 + E_2)|_{\theta_2}$$

$$= D(\theta_2) + 2\theta_2 \frac{dD(\theta_2)}{d\theta_2} = D(\theta_2) + 2\alpha\lambda\theta_2^{\alpha+1}\left(D(\theta_2) + 1\right) \qquad (13.83)$$

which we use in equation (13.76) to obtain a final solution

$$\theta \approx \theta_i + (\theta_b - \theta_i)erfc\left(\frac{x}{\sqrt{4(E_0 + E_1 + E_2)t}}\right) \qquad (13.84)$$

Figure 13.1 shows profiles of the water content versus distance profiles at $t = 1 hour$, according to four sources: equation (13.84), the classical numerical solution of Philip (1955), the numerical solution of Parlange (1971), and experimental observations. Figure 13.2 shows the same situation at at $t = 2 hours$. Equation (13.84) appears to be in good agreement with the other solutions and with the observed data. In fact, equation (13.84) appears to better predict the position of the wetting front and the shape of the tail after that than Philip's (1955) or Parlange's (1971) solutions. The decomposition solution is simpler, it provides a continuous spatio-temporal description, and it does not exhibit the stability and discretization restrictions of numerical solutions.

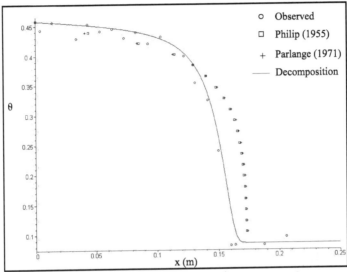

Figure 13.1: Water Content versus Distance at $t=1 hour$

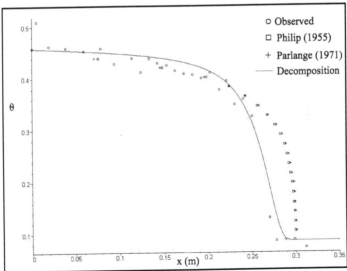

Figure 13.2: Water Content versus Distance at $t=3 hour$

Differential Equations with Random Coefficients

In the previous sections we have studied the solution to differential equations when the random components enter the system in the form of a random initial condition, and in the form of a random forcing function. Another possibility is the case when the coefficients, or the equation parameters, are defined in stochastic terms as random variables or random processes. One of the best procedures to solve these systems is the method of decomposition. By placing the terms that contain the random parameter, and its derivatives, on the right side of the equation, then the equation is cast as one with a random forcing function and subsequently solved in the usual manner. The following examples illustrate the procedure.

Example 13.32: A First-Order Differential Equation

Consider a different version of equation (13.31):

$$\frac{du}{dt} + a(t)u = f(t), \quad u(0) = 0$$

$$(13.85)$$

$$E\{a(t)\} = \bar{a}(t), \quad E\{a(t_1)a(t_2)\} = g(t_1, t_2)$$

Find the mean and the correlation function of the output, $u(t)$.

Solution

From equation (13.32), we already know the general solution to this equation:

$$u = J_t\mu_0 + \int_0^t J_{t-\tau} f(\tau)d\tau = \int_0^t e^{-a(t,\,\xi)(t-\tau)} f(\tau)d\tau$$

However, there is a problem with this direct approach: what is the exponential of a stochastic process? To circumvent this difficulty, let us define the random coefficient as

$$a(t) = \bar{a}(t) + a'(t), \quad E\{a'(t)\} = 0, \quad E\{a'(t_1)a'(t_2)\} = g(t_1, t_2) - \bar{a}(t_1)\bar{a}(t_2) \quad (13.86)$$

The differential equation (13.85) becomes,

$$\frac{du}{dt} + \bar{a}(t)u = f(t) - a'(t, \xi)$$

$$(13.87)$$

We have transformed the original equation (13.85) with a random coefficient into a simpler one, equation (13.87), subject to a random forcing function. Thus, we may use the concept of the semigroup and the techniques derived in the previous sections to deal with differential equations with stochastic forcing functions. The solution to equation (13.87) is now given by

$$u = J_t \mu_0 + \int_0^t J_{t-\tau} f(\tau) d\tau - \int_0^t J_{t-\tau} a'(\tau) d\tau$$

$$u = \int_0^t e^{-\bar{a}(t)(t-\tau)} f(\tau) d\tau - \int_0^t e^{-\bar{a}(t)(t-\tau)} a'(\tau)$$

Upon taking expectation,

$$E\{u\} = \int_0^t e^{-\bar{a}(t)(t-\tau)} f(\tau) d\tau$$

The correlation function, $R_u(t_1, t_2) = E\{u(t_1) u(t_2)\}$, is given by

$$R_u(t_1, t_2) = E\left\{ \left[\int_0^{t_1} J_{t_1-\tau} f(\tau) d\tau - \int_0^{t_1} J_{t_1-s} \bar{a}(s) ds \right] \left[\int_0^{t_2} J_{t_2-w} f(w) dw - \int_0^{t_2} J_{t_2-v} \bar{a}(v) dv \right] \right\}$$

$$= \int_0^{t_2} \int_0^{t_1} J_{t_1+t_2-\tau-w} f(\tau) f(w) d\tau dw - \int_0^{t_2} \int_0^{t_1} J_{t_1+t_2-s-v} E\{\bar{a}(s) \bar{a}(v)\} ds dv$$

$$= \int_0^{t_2} \int_0^{t_1} e^{\bar{a}(t_1+t_2-\tau-w)} f(\tau) f(w) d\tau dw - \int_0^{t_2} \int_0^{t_1} e^{\bar{a}(t_1+t_2-s-v)} [g(s,v) - \bar{a}(s) \bar{a}(v)] ds dv$$

Difficulty in solving the above integrals depends on the form of $\bar{a}(t)$ and $g(t_1, t_2)$.

Example 13.33

Consider the differential equation given by

$$\frac{du}{dt} + au + w'(t)u = f(t) \tag{13.88}$$

$$u(0) = 0, \quad E\{w'(t)\} = 0, \quad E\{w'(t_1) w'(t_2)\} = g(t_1, t_2), \quad g(t,t) = \sigma_w^2(t)$$

The coefficient $w'(t)$ is a colored-noise stochastic process with a known correlation function, $g(t_1, t_2)$. Derive the mean and the correlation function of $u(t)$.

Solution

Define the operator $L_t = d/dt$ and rewrite the differential equation as

$$L_t u + au + w'u = f$$

Operating with L_t^{-1} and rearranging,

$$u = L_t^{-1}f - L_t^{-1}au - L_t^{-1}w'u$$

As before, expand u as $u_0 + u_1 + u_2 + ...$ on the right side to obtain

$$u = L_t^{-1}f - L_t^{-1}a(u_0 + u_1 + u_2 + ...) - L_t^{-1}w'(u_0 + u_1 + u_2 + ...)$$

If we set $u_0 = L_t^{-1}f$, then the first approximation to u is given by

$$u \approx L_t^{-1}f - L_t^{-1}au_0 - L_t^{-1}w'u_0$$

On taking expectations,

$$E\{u\} = L_t^{-1}f - L_t^{-1}au_0$$

The two-times correlation function, $E\{u(t_1)u(t_2)\} = R_u(t_1,t_2)$ is given by

$$R_u(t_1,t_2) = E\left\{\left[L_{t_1}^{-1}f - L_{t_1}^{-1}au_0 - L_{t_1}^{-1}w'u_0\right]\left[L_{t_2}^{-1}f - L_{t_2}^{-1}au_0 - L_{t_2}^{-1}w'u_0\right]\right\}$$

Setting $t_1 = t_2 = t$, the variance of u is given by the last two equations and solving,

$$\sigma_u^2(t) = R_u(t,t) - E^2\{u(t)\} = L_t^{-1}L_t^{-1}\sigma_w^2(t)u_0^2$$

Example 13.34: A Parabolic Partial Differential Equation

Consider a partial differential version of equation (13.88):

$$\frac{\partial u}{\partial t} + Au + w'(t)L_x u = f(x,t)$$

$$(13.89)$$

$$u(x,0) = 0, \quad E\{w'(t)\} = 0, \quad E\{w'(t_1)w'(t_2)\} = g(t_1,t_2), \quad g(t,t) = \sigma_w^2(t)$$

where the operator $L_x = d/dx$, and the operator A has higher-order derivatives in x. Derive the mean and the correlation function of $u(x,t)$.

Solution

As before, define the operator $L_t = \partial/\partial t$ and rewrite equation (13.89) as

$$L_t u + A u + w' L_x u = f$$

Operating with L_t^{-1} and rearranging,

$$u = L_t^{-1} f - L_t^{-1} A u - L_t^{-1} w' L_x u$$

As before, expand u as $u_0 + u_1 + u_2 + ...$ on the right side to obtain

$$u = L_t^{-1} f - L_t^{-1} A \left(u_0 + u_1 + u_2 + ... \right) - L_t^{-1} w' L_x \left(u_0 + u_1 + u_2 + ... \right)$$

If we set $u_0 = L_t^{-1} f$, then the first approximation to u is given by

$$u \approx L_t^{-1} f - L_t^{-1} A u_0 - L_t^{-1} w' L_x u_0$$

On taking expectations,

$$E\{u\} = L_t^{-1} f - L_t^{-1} A u_0$$

The two-times correlation function, $E\{u(t_1)u(t_2)\} = R_u(t_1,t_2)$ is given by

$$R_u(t_1,t_2) = E\left\{ \left[L_{t_1}^{-1} f - L_{t_1}^{-1} A u_0 - L_{t_1}^{-1} w' L_x u_0 \right] \left[L_{t_2}^{-1} f - L_{t_2}^{-1} A u_0 - L_{t_2}^{-1} w' L_x u_0 \right] \right\}$$

Setting $t_1 = t_2 = t$, the variance of u is given by the last two equations and solving,

$$\sigma_u^2(t) = R_u(t,t) - E^2\{u(t)\} = L_t^{-1} L_t^{-1} \sigma_w^2(t) L_x L_x u_0^2$$

Example 13.35: A Second-Order Differential Equation

Consider a second-order differential equation given by

$$\frac{d^2 u}{dt^2} + a \frac{du}{dt} + b u + c w' u = f(t), \quad u(0) = 0, \quad \frac{du}{dt}(0) = 0 \tag{13.90}$$

$$E\{w'(t)\} = 0, \quad E\{w'(t_1)w'(t_2)\} = g(t_1,t_2), \quad E\{w'(t)w'(t)\} = \sigma_w^2(t)$$

Derive the mean and the correlation function of $u(x,t)$.

Solution

Define the operators $L_{tt} = d^2/dt^2$ and $L_t = d/dt$, and rewrite equation (13.90) as

$$L_{tt}u + aL_t u + bu + cw'u = f$$

Operating with L_{tt}^{-1} (i.e., the twofold integral from 0 to t), and rearranging,

$$u = L_{tt}^{-1}f - L_t^{-1}au - bL_{tt}^{-1}u - cL_{tt}^{-1}w'u$$

Now expand u as $u_0 + u_1 + u_2 + ...$ on the right side to obtain

$$u = L_{tt}^{-1}f - L_t^{-1}a\sum_{i=0}^{\infty} u_i - bL_{tt}^{-1}\sum_{i=0}^{\infty} u_i - cL_{tt}^{-1}w'\sum_{i=0}^{\infty} u_i$$

If we set $u_0 = L_{tt}^{-1}f$, then the first approximation to u is given by

$$u = L_{tt}^{-1}f - L_t^{-1}au_0 - bL_{tt}^{-1}u_0 - cL_{tt}^{-1}w'u_0$$

On taking expectations,

$$E\{u(t)\} = L_{tt}^{-1}f - L_t^{-1}au_0 - bL_{tt}^{-1}u_0$$

The two-times correlation function, $E\{u(t_1)u(t_2)\} = R_u(t_1,t_2)$ is given by

$$E\{u(t_1)u(t_2)\} = R_u(t_1,t_2)$$

$$= E\left\{ \left[L_{t_1t_1}^{-1}f - L_{t_1}^{-1}au_0 - bL_{t_1t_1}^{-1}u_0 - cL_{t_1t_1}^{-1}w'u_0 \right]\left[L_{t_2t_2}^{-1}f - L_{t_2}^{-1}au_0 - bL_{t_2t_2}^{-1}u_0 - cL_{t_2t_2}^{-1}w'u_0 \right] \right\}$$

Setting $t_1 = t_2 = t$, the variance of u is given by the last two equations and solving,

$$\sigma_u^2(t) = R_u(t,t) - E^2\{u(t)\} = cL_{tt}^{-1}\sigma_w^2(t)u_0^2$$

Example 13.36: Mechanical Vibrations Subject to Random Damping

Consider again the system governed by equation (13.20) in Example 13.9 when the damping constant, β, is a random variable with $E\{\beta\} = \mu_\beta$, and $Var(\beta) = \sigma_\beta^2$. Solve the differential equation and find the mean and the variance of the output.

Solution

The system equation is given by

$$\frac{d^2Y}{dt^2} + \frac{\beta(t)}{M}\frac{dY}{dt} + \frac{k}{M}Y = 0, \quad 0<t, \quad Y(0)=Y_0, \quad \frac{dY}{dt}(0)=0$$

where the initial condition Y_0 is now a constant. It is easy to see that if the initial condition is statistically independent of the damping constant, a reasonable assumption, the case of a random initial condition along with a random coefficient can be solved without much more complications, since statistical separability would yield the output moments. Let us now define the operator $L_t = d^2/dt^2$. The system equation may be written as

$$Y = Y_0 - \frac{k}{M}L_t^{-1}Y - \frac{\beta}{M}L_t^{-1}\frac{dY}{dt}$$

where L_t^{-1} is the twofold indefinite integral of time. Since β is a random variable, rather than a random process in time, it goes out of the time integrals. Expanding $Y = Y_0 + Y_1 + ... + Y_n$ in the right side, and setting Y_0 as the initial condition, then the second term in the series is given as

$$Y_1 = -\frac{k}{M}L_t^{-1}Y_0 - \frac{\beta}{M}L_t^{-1}\frac{dY_0}{dt} = -\frac{kY_0t^2}{2M}$$

The third term is given by

$$Y_2 = -\frac{k}{M}L_t^{-1}Y_1 - \frac{\beta}{M}L_t^{-1}\frac{dY_1}{dt} = \frac{k^2Y_0t^4}{4!M^2} + \frac{k\beta Y_0t^3}{3!M^2}$$

In general the n-th term is given by

$$Y_n = -\frac{k}{M}L_t^{-1}Y_{n-1} - \frac{\beta}{M}L_t^{-1}\frac{dY_{n-1}}{dt}$$

If we approximate the solution with the first three decomposition terms, then

$$Y \approx Y_0 + Y_1 + Y_2 = Y_0 - \frac{kY_0t^2}{2M} + \frac{k^2Y_0t^4}{4!M^2} + \frac{k\beta Y_0t^3}{3!M^2}$$

$$E\{Y(t)\} \approx Y_0\left[1 - \frac{kt^2}{2M} + \frac{k^2t^4}{4!M^2} + \frac{k\mu_\beta t^3}{3!M^2}\right], \qquad \sigma_Y^2(t) \approx \frac{kY_0^2\sigma_\beta^2 t^6}{6^2M^4}$$

Example 13.37: Transport Equation Subject to General Stochastic Conditions

Consider the general version of the transport equation (13.25) in Example 13.12:

$$\frac{\partial u}{\partial t}=g-Au, \quad A=-D\frac{\partial^2}{\partial x^2}+v\frac{\partial}{\partial x}$$

Derive the solution to the equation for the general case when the initial condition, the parameters, and the forcing function are stochastic.

Solution

Defining $L_t=\partial/\partial t$, the decomposition series are

$$u_0=u(t=0)+L^{-1}g,$$

$$u_1=-L^{-1}\left[-D\frac{\partial^2}{\partial x^2}+v\frac{\partial}{\partial x}\right]u_0=L^{-1}D\frac{\partial^2}{\partial x^2}u_0-L^{-1}v\frac{\partial}{\partial x}u_0$$

$$u_2=-L^{-1}\left[-D\frac{\partial^2}{\partial x^2}+v\frac{\partial}{\partial x}\right]u_1$$

$$\vdots$$

The mean of the output is found by taking expectations on each of the terms:

$$E\{u_0\}=E\{u(t=0)\}+L^{-1}E\{g\},$$

$$E\{u_1\}=L^{-1}\{D\}\frac{\partial^2}{\partial x^2}E\{u_0\}-L^{-1}v\frac{\partial}{\partial x}E\{u_0\}$$

$$E\{u_2\}=L^{-1}\{D\}\frac{\partial^2}{\partial x^2}E\{u_1\}-L^{-1}v\frac{\partial}{\partial x}E\{u_1\}$$

$$\vdots$$

Note that statistical separability between the input functions, the initial condition, and the parameters occurs naturally from the physics of the system. In other words, it is reasonable to assume that the dispersion coefficient is statistically independent of the external source. Thus, no closure approximations, hierarchy, or small perturbation assumptions are needed with decomposition.

Analytical simulations of decomposition schemes maybe easily implemented in Maple programs. By defining the successive integration or differentiation operators as Maple procedures, it is easy to calculate terms in the series recursively. For more details on the simulation of transport equations see Serrano (1996), Serrano and Adomian (1996), and Adomian and Serrano (1998).

PROBLEMS

13.1 Solve Example 13.1 if the hydraulic head correlation function is actually given by $R_h(x_1, x_2) = C_v^2 e^{-\rho(x_1 - x_2)^2}$.

13.2 Repeat Example 13.4 if the input function is of the form $\sin(\omega t)X(t)$.

13.3 Solve Example 13.6 if the coefficient is a linear function of time given by at.

13.4 If in Example 13.9, $M=1000kg$, $\beta=5kg/s$, $k=10kg/s^2$, $\mu_{Y_0}=1m$, $\sigma_{Y_0}=0.5m$, write a program that chooses the correct solution and produces a plot of the mean displacement (m) versus time (s) plus and minus one standard deviation.

13.5 In Example 13.11, assume that the initial gradient, V, is a random variable. In other words, the system equation is given by

$$2\frac{d^2z}{dt^2} - 4\frac{dz}{dt} + z = 0, \quad 0 < t, \ z(0) = 0, \ \frac{dz}{dt}(0) = V, \ \mu_V = 2, \ \sigma_V^2 = 1$$

(1) Manually derive the solution to the differential equation. (2) Find the mean, correlation, and variance functions of the output $z(t)$.

13.6 In Example 13.13, assume that $C_0 = C_i$ (i.e., the deterministic part). (1) Write a computer program to solve the integral. (2) Assume $M=1mg$, $v=1m/month$, and $D=1m^2/month$. In one graph plot the concentration versus distance (i.e., the plume spatial distribution at 1, 6, and 12 months after the spill, respectively).

13.7 In Example 13.14 assume that $L=15$ m, $a=6$ m, $b=8$ m, $k=1$ m^2/day, $\mu_m = 100°C$, and $\sigma_m = 10°C$. Write a computer program that simulates (1) the mean temperature spatial distribution at various times, (2) The mean plus and minus one standard deviation of temperature at a fixed time, and (3) the mean plus and minus one standard deviation of temperature at a fixed distance.

13.8 In Example 13.15, use Maple to verify the double integral in the autocorrelation function.

13.9 In Example 13.15, assume that $a=1year^{-1}$, $h=1\times10^6kg/year$, $\rho=0.1year^{-1}$, and $\sigma_g=0.2\times10^6kg/year$, produce a graph showing of the mean concentration versus time plus and minus its standard deviation.

13.10 If in Example 13.16, $y_0=0m$, $M=1kg$, $\beta=1kg/s$, $k=1kg/s^2$, $\mu_A=1m/s^2$, $\sigma_A=0.5m/s^2$, produce a plot of the mean displacement versus time (s) plus and minus one standard deviation.

13.11 If in Example 13.16, $y_0=0$ and $F(t)=Ae^{-bt}/M$, where A is a random variable and b a constant, derive the solution to the differential equation, its mean and its variance.

13.12 In Example 13.18, use the solution to the differential equation to generate a sample function of 1000 points of a Colored Noise process with parameters $\rho=0.1$, $\sigma^2=1$. *Hint:* Generate a sample function of the White Noise process with a time interval $\Delta t=1$ and solve the stochastic integral numerically.

13.13 Rerun the program in Example 13.22 if the recharge rate increases to $0.1mm/month$.

13.14 Rerun the program in Example 13.25 if the transmissivity increases to $1000m^2/month$.

14 CONCLUSION

As our journey into uncertainty analysis reaches its destination, we wish to recapitulate a method of analysis, this time from the perspective of our experiences gained over the past 13 chapters. Let us repeat the description on the method from Chapter 1, added with some new items more meaningful at this time. The engineer or scientist is faced with the problem of studying a complex engineering system composed of several variables and functions. We wish to *build the simplest model* that will satisfy the project objectives. Our system has deterministic, as well as random, components. We know that a deterministic variable is easier to analyze than a stochastic one and therefore we need to define as random components only those that will truly affect the predictability of the system output. Which variables should be eliminated? Which variables should be deterministic? Which variables should be stochastic? The seven steps of analysis are as follows:

1. Definition of the explicit objectives of the project (e.g., design, analysis, forecast, etc.).

2. Identification of system variables, the knowledge of which will satisfy the project objectives.

3. Model building: application of fundamental physical, chemical, biological, economical, sociological laws, etc. These laws define the relationships between the variables.

4. Identification, measurement, and evaluation of variables.

   4.1. Measurement or data collection of deterministic variables.

   4.2. Repetitive measurement, or experimentation, *under identical conditions*, of uncertain variables.

   4.3. Evaluation of statistical properties.

   4.4. Elimination of variables with a high degree of correlation with others.

   4.5. Sensitivity analysis: variables whose variance effect on a system-output variance is negligible are classified as deterministic.

4.6.   Building of a probabilistic model for uncertain variables, if data permits it.

5. Model solution: application of mathematical, analytical, or numerical methods to solve the system equations. Calculation of output moments and other statistical properties.

6. Model verification and calibration.

6.1.   Verification with theoretical results and existing models.

6.2.   Comparison with limited sets of prototype data.

6.3.   Model is modified until its moments compare well with those of the prototype (i.e., until the statistical properties of the model and those of the prototype are similar).

7. Model application in simulation, data generation, prediction, design or risk assessment.

The above steps will hopefully locate uncertainty analysis under the whole picture of engineering project design or implementation. Clearly, uncertainty analysis is an invaluable tool in engineering practice. It allows the engineer to identify, measure, and evaluate difficult variables. It provides an objective means to design and predict.

Our journey ended where we started. However, our end point is one with a higher perspective than that at the beginning. Thus, our trajectory was *spiral*, rather than circular. We sincerely hope the reader will embark himself or herself into the next cycle of analysis, one that will lead him/her into the analysis and solution of the most interesting nonlinear and stochastic problems still unresolved. This will require the application and the study, more in detail, of some or all of the concepts covered in this book in other more specialized texts. Enjoy the enriching experience of your intellectual journey.

APPENDIX A: TABLES

Table A.1: Cumulative Areas, $F_X(x)$, under the Standard Normal Probability Density Function

$$F_X(x) = \frac{1}{\sqrt{2\pi}} \int_{-\infty}^{x} e^{-\frac{u^2}{2}} du$$

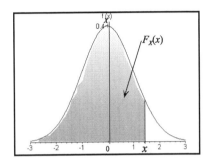

x	0.00	0.01	0.02	0.03	0.04	0.05	0.06	0.07	0.08	0.09
0.00	0.5000	0.5040	0.5080	0.5120	0.5160	0.5199	0.5239	0.5279	0.5319	0.5359
0.10	0.5398	0.5438	0.5478	0.5517	0.5557	0.5596	0.5636	0.5675	0.5714	0.5753
0.20	0.5793	0.5832	0.5871	0.5910	0.5948	0.5987	0.6026	0.6064	0.6103	0.6141
0.30	0.6179	0.6217	0.6255	0.6293	0.6331	0.6368	0.6406	0.6443	0.6480	0.6517
0.40	0.6554	0.6591	0.6628	0.6664	0.6700	0.6736	0.6772	0.6808	0.6844	0.6879
0.50	0.6915	0.6950	0.6985	0.7019	0.7054	0.7088	0.7123	0.7157	0.7190	0.7224
0.60	0.7257	0.7291	0.7324	0.7357	0.7389	0.7422	0.7454	0.7486	0.7517	0.7549
0.70	0.7580	0.7611	0.7642	0.7673	0.7704	0.7734	0.7764	0.7794	0.7823	0.7852
0.80	0.7881	0.7910	0.7939	0.7967	0.7995	0.8023	0.8051	0.8078	0.8106	0.8133
0.90	0.8159	0.8186	0.8212	0.8238	0.8264	0.8289	0.8315	0.8340	0.8365	0.8389
1.00	0.8413	0.8438	0.8461	0.8485	0.8508	0.8531	0.8554	0.8577	0.8599	0.8621
1.10	0.8643	0.8665	0.8686	0.8708	0.8729	0.8749	0.8770	0.8790	0.8810	0.8830
1.20	0.8849	0.8869	0.8888	0.8907	0.8925	0.8944	0.8962	0.8980	0.8997	0.9015
1.30	0.9032	0.9049	0.9066	0.9082	0.9099	0.9115	0.9131	0.9147	0.9162	0.9177
1.40	0.9192	0.9207	0.9222	0.9236	0.9251	0.9265	0.9279	0.9292	0.9306	0.9319
1.50	0.9332	0.9345	0.9357	0.9370	0.9382	0.9394	0.9406	0.9418	0.9429	0.9441
1.60	0.9452	0.9463	0.9474	0.9484	0.9495	0.9505	0.9515	0.9525	0.9535	0.9545
1.70	0.9554	0.9564	0.9573	0.9582	0.9591	0.9599	0.9608	0.9616	0.9625	0.9633
1.80	0.9641	0.9649	0.9656	0.9664	0.9671	0.9678	0.9686	0.9693	0.9699	0.9706
1.90	0.9713	0.9719	0.9726	0.9732	0.9738	0.9744	0.9750	0.9756	0.9761	0.9767
2.00	0.9772	0.9778	0.9783	0.9788	0.9793	0.9798	0.9803	0.9808	0.9812	0.9817
2.10	0.9821	0.9826	0.9830	0.9834	0.9838	0.9842	0.9846	0.9850	0.9854	0.9857
2.20	0.9861	0.9864	0.9868	0.9871	0.9875	0.9878	0.9881	0.9884	0.9887	0.9890
2.30	0.9893	0.9896	0.9898	0.9901	0.9904	0.9906	0.9909	0.9911	0.9913	0.9916
2.40	0.9918	0.9920	0.9922	0.9925	0.9927	0.9929	0.9931	0.9932	0.9934	0.9936
2.50	0.9938	0.9940	0.9941	0.9943	0.9945	0.9946	0.9948	0.9949	0.9951	0.9952
2.60	0.9953	0.9955	0.9956	0.9957	0.9959	0.9960	0.9961	0.9962	0.9963	0.9964
2.70	0.9965	0.9966	0.9967	0.9968	0.9969	0.9970	0.9971	0.9972	0.9973	0.9974
2.80	0.9974	0.9975	0.9976	0.9977	0.9977	0.9978	0.9979	0.9979	0.9980	0.9981
2.90	0.9981	0.9982	0.9982	0.9983	0.9984	0.9984	0.9985	0.9985	0.9986	0.9986
3.00	0.9987	0.9987	0.9987	0.9988	0.9988	0.9989	0.9989	0.9989	0.9990	0.9990

Table A.2: Abscissa Values, $t_{m,1-a}$, Corresponding to Areas 1-a under the Student's t Density Function with m Degrees of Freedom

$$f_T(t) = \frac{\Gamma\left(\dfrac{m+1}{2}\right)}{\sqrt{\pi m} \cdot \Gamma\left(\dfrac{m}{2}\right) \cdot \left(1 + \dfrac{t^2}{m}\right)^{(m+1)/2}}$$

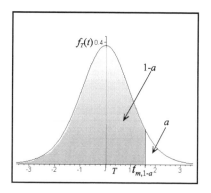

m \ $1-a$	0.600	0.700	0.750	0.800	0.850	0.900	0.950	0.975	0.990	0.995
1	0.3249	0.7265	1.0000	1.3764	1.9626	3.0777	6.3138	12.7062	31.8205	63.6567
2	0.2887	0.6172	0.8165	1.0607	1.3862	1.8856	2.9200	4.3027	6.9646	9.9248
3	0.2767	0.5844	0.7649	0.9785	1.2498	1.6377	2.3534	3.1824	4.5407	5.8409
4	0.2707	0.5686	0.7407	0.9410	1.1896	1.5332	2.1318	2.7764	3.7469	4.6041
5	0.2672	0.5594	0.7267	0.9195	1.1558	1.4759	2.0150	2.5706	3.3649	4.0321
6	0.2648	0.5534	0.7176	0.9057	1.1342	1.4398	1.9432	2.4469	3.1427	3.7074
7	0.2632	0.5491	0.7111	0.8960	1.1192	1.4149	1.8946	2.3646	2.9980	3.4995
8	0.2619	0.5459	0.7064	0.8889	1.1081	1.3968	1.8595	2.3060	2.8965	3.3554
9	0.2610	0.5435	0.7027	0.8834	1.0997	1.3830	1.8331	2.2622	2.8214	3.2498
10	0.2602	0.5415	0.6998	0.8791	1.0931	1.3722	1.8125	2.2281	2.7638	3.1693
11	0.2596	0.5399	0.6974	0.8755	1.0877	1.3634	1.7959	2.2010	2.7181	3.1058
12	0.2590	0.5386	0.6955	0.8726	1.0832	1.3562	1.7823	2.1788	2.6810	3.0545
13	0.2586	0.5375	0.6938	0.8702	1.0795	1.3502	1.7709	2.1604	2.6503	3.0123
14	0.2582	0.5366	0.6924	0.8681	1.0763	1.3450	1.7613	2.1448	2.6245	2.9768
15	0.2579	0.5357	0.6912	0.8662	1.0735	1.3406	1.7531	2.1314	2.6025	2.9467
16	0.2576	0.5350	0.6901	0.8647	1.0711	1.3368	1.7459	2.1199	2.5835	2.9208
17	0.2573	0.5344	0.6892	0.8633	1.0690	1.3334	1.7396	2.1098	2.5669	2.8982
18	0.2571	0.5338	0.6884	0.8620	1.0672	1.3304	1.7341	2.1009	2.5524	2.8784
19	0.2569	0.5333	0.6876	0.8610	1.0655	1.3277	1.7291	2.0930	2.5395	2.8609
20	0.2567	0.5329	0.6870	0.8600	1.0640	1.3253	1.7247	2.0860	2.5280	2.8453
21	0.2566	0.5325	0.6864	0.8591	1.0627	1.3232	1.7207	2.0796	2.5176	2.8314
22	0.2564	0.5321	0.6858	0.8583	1.0614	1.3212	1.7171	2.0739	2.5083	2.8188
23	0.2563	0.5317	0.6853	0.8575	1.0603	1.3195	1.7139	2.0687	2.4999	2.8073
24	0.2562	0.5314	0.6848	0.8569	1.0593	1.3178	1.7109	2.0639	2.4922	2.7969
25	0.2561	0.5312	0.6844	0.8562	1.0584	1.3163	1.7081	2.0595	2.4851	2.7874
26	0.2560	0.5309	0.6840	0.8557	1.0575	1.3150	1.7056	2.0555	2.4786	2.7787
27	0.2559	0.5306	0.6837	0.8551	1.0567	1.3137	1.7033	2.0518	2.4727	2.7707
28	0.2558	0.5304	0.6834	0.8546	1.0560	1.3125	1.7011	2.0484	2.4671	2.7633
29	0.2557	0.5302	0.6830	0.8542	1.0553	1.3114	1.6991	2.0452	2.4620	2.7564
30	0.2556	0.5300	0.6828	0.8538	1.0547	1.3104	1.6973	2.0423	2.4573	2.7500

Table A.2: Abscissa Values of Student's *t* Distribution (Continued)

1-a / m	0.600	0.700	0.750	0.800	0.850	0.900	0.950	0.975	0.990	0.995
31	0.2555	0.5298	0.6825	0.8534	1.0541	1.3095	1.6955	2.0395	2.4528	2.7440
32	0.2555	0.5297	0.6822	0.8530	1.0535	1.3086	1.6939	2.0369	2.4487	2.7385
33	0.2554	0.5295	0.6820	0.8526	1.0530	1.3077	1.6924	2.0345	2.4448	2.7333
34	0.2553	0.5294	0.6818	0.8523	1.0525	1.3070	1.6909	2.0322	2.4411	2.7284
35	0.2553	0.5292	0.6816	0.8520	1.0520	1.3062	1.6896	2.0301	2.4377	2.7238
36	0.2552	0.5291	0.6814	0.8517	1.0516	1.3055	1.6883	2.0281	2.4345	2.7195
37	0.2552	0.5289	0.6812	0.8514	1.0512	1.3049	1.6871	2.0262	2.4314	2.7154
38	0.2551	0.5288	0.6810	0.8512	1.0508	1.3042	1.6860	2.0244	2.4286	2.7116
39	0.2551	0.5287	0.6808	0.8509	1.0504	1.3036	1.6849	2.0227	2.4258	2.7079
40	0.2550	0.5286	0.6807	0.8507	1.0500	1.3031	1.6839	2.0211	2.4233	2.7045
41	0.2550	0.5285	0.6805	0.8505	1.0497	1.3025	1.6829	2.0195	2.4208	2.7012
42	0.2550	0.5284	0.6804	0.8503	1.0494	1.3020	1.6820	2.0181	2.4185	2.6981
43	0.2549	0.5283	0.6802	0.8501	1.0491	1.3016	1.6811	2.0167	2.4163	2.6951
44	0.2549	0.5282	0.6801	0.8499	1.0488	1.3011	1.6802	2.0154	2.4141	2.6923
45	0.2549	0.5281	0.6800	0.8497	1.0485	1.3006	1.6794	2.0141	2.4121	2.6896
46	0.2548	0.5281	0.6799	0.8495	1.0483	1.3002	1.6787	2.0129	2.4102	2.6870
47	0.2548	0.5280	0.6797	0.8493	1.0480	1.2998	1.6779	2.0117	2.4083	2.6846
48	0.2548	0.5279	0.6796	0.8492	1.0478	1.2994	1.6772	2.0106	2.4066	2.6822
49	0.2547	0.5278	0.6795	0.8490	1.0475	1.2991	1.6766	2.0096	2.4049	2.6800
50	0.2547	0.5278	0.6794	0.8489	1.0473	1.2987	1.6759	2.0086	2.4033	2.6778
51	0.2547	0.5277	0.6793	0.8487	1.0471	1.2984	1.6753	2.0076	2.4017	2.6757
52	0.2546	0.5276	0.6792	0.8486	1.0469	1.2980	1.6747	2.0066	2.4002	2.6737
53	0.2546	0.5276	0.6791	0.8485	1.0467	1.2977	1.6741	2.0057	2.3988	2.6718
54	0.2546	0.5275	0.6791	0.8483	1.0465	1.2974	1.6736	2.0049	2.3974	2.6700
55	0.2546	0.5275	0.6790	0.8482	1.0463	1.2971	1.6730	2.0040	2.3961	2.6682
56	0.2546	0.5274	0.6789	0.8481	1.0461	1.2969	1.6725	2.0032	2.3948	2.6665
57	0.2545	0.5273	0.6788	0.8480	1.0459	1.2966	1.6720	2.0025	2.3936	2.6649
58	0.2545	0.5273	0.6787	0.8479	1.0458	1.2963	1.6716	2.0017	2.3924	2.6633
59	0.2545	0.5272	0.6787	0.8478	1.0456	1.2961	1.6711	2.0010	2.3912	2.6618
60	0.2545	0.5272	0.6786	0.8477	1.0455	1.2958	1.6706	2.0003	2.3901	2.6603
61	0.2545	0.5272	0.6785	0.8476	1.0453	1.2956	1.6702	1.9996	2.3890	2.6589
62	0.2544	0.5271	0.6785	0.8475	1.0452	1.2954	1.6698	1.9990	2.3880	2.6575
63	0.2544	0.5271	0.6784	0.8474	1.0450	1.2951	1.6694	1.9983	2.3870	2.6561
64	0.2544	0.5270	0.6783	0.8473	1.0449	1.2949	1.6690	1.9977	2.3860	2.6549
65	0.2544	0.5270	0.6783	0.8472	1.0448	1.2947	1.6686	1.9971	2.3851	2.6536
66	0.2544	0.5269	0.6782	0.8471	1.0446	1.2945	1.6683	1.9966	2.3842	2.6524
67	0.2544	0.5269	0.6782	0.8470	1.0445	1.2943	1.6679	1.9960	2.3833	2.6512
68	0.2543	0.5269	0.6781	0.8469	1.0444	1.2941	1.6676	1.9955	2.3824	2.6501
69	0.2543	0.5268	0.6781	0.8469	1.0443	1.2939	1.6672	1.9949	2.3816	2.6490
70	0.2543	0.5268	0.6780	0.8468	1.0442	1.2938	1.6669	1.9944	2.3808	2.6479
80	0.2542	0.5265	0.6776	0.8461	1.0432	1.2922	1.6641	1.9901	2.3739	2.6387
90	0.2541	0.5263	0.6772	0.8456	1.0424	1.2910	1.6620	1.9867	2.3685	2.6316
100	0.2540	0.5261	0.6770	0.8452	1.0418	1.2901	1.6602	1.9840	2.3642	2.6259
150	0.2538	0.5255	0.6761	0.8440	1.0400	1.2872	1.6551	1.9759	2.3515	2.6090
200	0.2537	0.5252	0.6757	0.8434	1.0391	1.2858	1.6525	1.9719	2.3451	2.6006
300	0.2536	0.5250	0.6753	0.8428	1.0382	1.2844	1.6499	1.9679	2.3388	2.5923
400	0.2535	0.5248	0.6751	0.8425	1.0378	1.2837	1.6487	1.9659	2.3357	2.5882
500	0.2535	0.5247	0.6750	0.8423	1.0375	1.2832	1.6479	1.9647	2.3338	2.5857
1000	0.2534	0.5246	0.6747	0.8420	1.0370	1.2824	1.6464	1.9623	2.3301	2.5808

Table A.3: Abscissa Values, $\chi^2_{m,1-a}$, Corresponding to Areas $1-a$ under the Chi-Squared Density Function with m Degrees of Freedom

$$f_X(\chi) = \frac{\chi^{\frac{m}{2}-1} e^{-\frac{\chi}{2}}}{2^{\frac{m}{2}} \Gamma\left(\frac{m}{2}\right)}$$

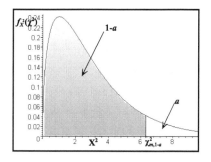

$1-a$ / m	0.005	0.010	0.025	0.050	0.100	0.900	0.950	0.975	0.990	0.995
1	0.0000	0.0002	0.0010	0.0039	0.0158	2.7055	3.8415	5.0239	6.6349	7.8794
2	0.0100	0.0201	0.0506	0.1026	0.2107	4.6052	5.9915	7.3778	9.2103	10.5966
3	0.0717	0.1148	0.2158	0.3518	0.5844	6.2514	7.8147	9.3484	11.3449	12.8382
4	0.2070	0.2971	0.4844	0.7107	1.0636	7.7794	9.4877	11.1433	13.2767	14.8603
5	0.4117	0.5543	0.8312	1.1455	1.6103	9.2364	11.0705	12.8325	15.0863	16.7496
6	0.6757	0.8721	1.2373	1.6354	2.2041	10.6446	12.5916	14.4494	16.8119	18.5476
7	0.9893	1.2390	1.6899	2.1673	2.8331	12.0170	14.0671	16.0128	18.4753	20.2777
8	1.3444	1.6465	2.1797	2.7326	3.4895	13.3616	15.5073	17.5345	20.0902	21.9550
9	1.7349	2.0879	2.7004	3.3251	4.1682	14.6837	16.9190	19.0228	21.6660	23.5894
10	2.1559	2.5582	3.2470	3.9403	4.8652	15.9872	18.3070	20.4832	23.2093	25.1882
11	2.6032	3.0535	3.8157	4.5748	5.5778	17.2750	19.6751	21.9200	24.7250	26.7568
12	3.0738	3.5706	4.4038	5.2260	6.3038	18.5493	21.0261	23.3367	26.2170	28.2995
13	3.5650	4.1069	5.0088	5.8919	7.0415	19.8119	22.3620	24.7356	27.6882	29.8195
14	4.0747	4.6604	5.6287	6.5706	7.7895	21.0641	23.6848	26.1189	29.1412	31.3193
15	4.6009	5.2293	6.2621	7.2609	8.5468	22.3071	24.9958	27.4884	30.5779	32.8013
16	5.1422	5.8122	6.9077	7.9616	9.3122	23.5418	26.2962	28.8454	31.9999	34.2672
17	5.6972	6.4078	7.5642	8.6718	10.0852	24.7690	27.5871	30.1910	33.4087	35.7185
18	6.2648	7.0149	8.2307	9.3905	10.8649	25.9894	28.8693	31.5264	34.8053	37.1565
19	6.8440	7.6327	8.9065	10.1170	11.6509	27.2036	30.1435	32.8523	36.1909	38.5823
20	7.4338	8.2604	9.5908	10.8508	12.4426	28.4120	31.4104	34.1696	37.5662	39.9968
21	8.0337	8.8972	10.2829	11.5913	13.2396	29.6151	32.6706	35.4789	38.9322	41.4011
22	8.6427	9.5425	10.9823	12.3380	14.0415	30.8133	33.9244	36.7807	40.2894	42.7957
23	9.2604	10.1957	11.6886	13.0905	14.8480	32.0069	35.1725	38.0756	41.6384	44.1813
24	9.8862	10.8564	12.4012	13.8484	15.6587	33.1962	36.4150	39.3641	42.9798	45.5585
25	10.5197	11.5240	13.1197	14.6114	16.4734	34.3816	37.6525	40.6465	44.3141	46.9279
26	11.1602	12.1981	13.8439	15.3792	17.2919	35.5632	38.8851	41.9232	45.6417	48.2899
27	11.8076	12.8785	14.5734	16.1514	18.1139	36.7412	40.1133	43.1945	46.9629	49.6449
28	12.4613	13.5647	15.3079	16.9279	18.9392	37.9159	41.3371	44.4608	48.2782	50.9934
29	13.1211	14.2565	16.0471	17.7084	19.7677	39.0875	42.5570	45.7223	49.5879	52.3356
30	13.7867	14.9535	16.7908	18.4927	20.5992	40.2560	43.7730	46.9792	50.8922	53.6720

Table A.3: Abscissa Values, $\chi_{m,1-a}$ of Chi-Squared Distribution (Continued)

$1-a$ / m	0.005	0.010	0.025	0.050	0.100	0.900	0.950	0.975	0.990	0.995
31	14.4578	15.6555	17.5387	19.2806	21.4336	41.4217	44.9853	48.2319	52.1914	55.0027
32	15.1340	16.3622	18.2908	20.0719	22.2706	42.5847	46.1943	49.4804	53.4858	56.3281
33	15.8153	17.0735	19.0467	20.8665	23.1102	43.7452	47.3999	50.7251	54.7755	57.6484
34	16.5013	17.7891	19.8063	21.6643	23.9523	44.9032	48.6024	51.9660	56.0609	58.9639
35	17.1918	18.5089	20.5694	22.4650	24.7967	46.0588	49.8018	53.2033	57.3421	60.2748
36	17.8867	19.2327	21.3359	23.2686	25.6433	47.2122	50.9985	54.4373	58.6192	61.5812
37	18.5858	19.9602	22.1056	24.0749	26.4921	48.3634	52.1923	55.6680	59.8925	62.8833
38	19.2889	20.6914	22.8785	24.8839	27.3430	49.5126	53.3835	56.8955	61.1621	64.1814
39	19.9959	21.4262	23.6543	25.6954	28.1958	50.6598	54.5722	58.1201	62.4281	65.4756
40	20.7065	22.1643	24.4330	26.5093	29.0505	51.8051	55.7585	59.3417	63.6907	66.7660
41	21.4208	22.9056	25.2145	27.3256	29.9071	52.9485	56.9424	60.5606	64.9501	68.0527
42	22.1385	23.6501	25.9987	28.1440	30.7654	54.0902	58.1240	61.7768	66.2062	69.3360
43	22.8595	24.3976	26.7854	28.9647	31.6255	55.2302	59.3035	62.9904	67.4593	70.6159
44	23.5837	25.1480	27.5746	29.7875	32.4871	56.3685	60.4809	64.2015	68.7095	71.8926
45	24.3110	25.9013	28.3662	30.6123	33.3504	57.5053	61.6562	65.4102	69.9568	73.1661
46	25.0413	26.6572	29.1601	31.4390	34.2152	58.6405	62.8296	66.6165	71.2014	74.4365
47	25.7746	27.4158	29.9562	32.2676	35.0814	59.7743	64.0011	67.8206	72.4433	75.7041
48	26.5106	28.1770	30.7545	33.0981	35.9491	60.9066	65.1708	69.0226	73.6826	76.9688
49	27.2493	28.9406	31.5549	33.9303	36.8182	62.0375	66.3386	70.2224	74.9195	78.2307
50	27.9907	29.7067	32.3574	34.7643	37.6886	63.1671	67.5048	71.4202	76.1539	79.4900
51	28.7347	30.4750	33.1618	35.5999	38.5604	64.2954	68.6693	72.6160	77.3860	80.7467
52	29.4812	31.2457	33.9681	36.4371	39.4334	65.4224	69.8322	73.8099	78.6158	82.0008
53	30.2300	32.0185	34.7763	37.2759	40.3076	66.5482	70.9935	75.0019	79.8433	83.2526
54	30.9813	32.7934	35.5863	38.1162	41.1830	67.6728	72.1532	76.1920	81.0688	84.5019
55	31.7348	33.5705	36.3981	38.9580	42.0596	68.7962	73.3115	77.3805	82.2921	85.7490
56	32.4905	34.3495	37.2116	39.8013	42.9373	69.9185	74.4683	78.5672	83.5134	86.9938
57	33.2484	35.1305	38.0267	40.6459	43.8161	71.0397	75.6237	79.7522	84.7328	88.2364
58	34.0084	35.9135	38.8435	41.4920	44.6960	72.1598	76.7778	80.9356	85.9502	89.4769
59	34.7704	36.6982	39.6619	42.3393	45.5770	73.2789	77.9305	82.1174	87.1657	90.7153
60	35.5345	37.4849	40.4817	43.1880	46.4589	74.3970	79.0819	83.2977	88.3794	91.9517
61	36.3005	38.2732	41.3031	44.0379	47.3418	75.5141	80.2321	84.4764	89.5913	93.1861
62	37.0684	39.0633	42.1260	44.8890	48.2257	76.6302	81.3810	85.6537	90.8015	94.4187
63	37.8382	39.8551	42.9503	45.7414	49.1105	77.7454	82.5287	86.8296	92.0100	95.6493
64	38.6098	40.6486	43.7760	46.5949	49.9963	78.8596	83.6753	88.0041	93.2169	96.8781
65	39.3831	41.4436	44.6030	47.4496	50.8829	79.9730	84.8206	89.1771	94.4221	98.1051
66	40.1582	42.2402	45.4314	48.3054	51.7705	81.0855	85.9649	90.3489	95.6257	99.3304
67	40.9350	43.0384	46.2610	49.1623	52.6588	82.1971	87.1081	91.5194	96.8278	100.5540
68	41.7135	43.8380	47.0920	50.0202	53.5481	83.3079	88.2502	92.6885	98.0284	101.7759
69	42.4935	44.6392	47.9242	50.8792	54.4381	84.4179	89.3912	93.8565	99.2275	102.9962
70	43.2752	45.4417	48.7576	51.7393	55.3289	85.5270	90.5312	95.0232	100.4252	104.2149
80	51.1719	53.5401	57.1532	60.3915	64.2778	96.5782	101.8795	106.6286	112.3288	116.3211
90	59.1963	61.7541	65.6466	69.1260	73.2911	107.5650	113.1453	118.1359	124.1163	128.2989
100	67.3276	70.0649	74.2219	77.9295	82.3581	118.4980	124.3421	129.5612	135.8067	140.1695
150	109.1422	112.6676	117.9845	122.6918	128.2751	172.5812	179.5806	185.8004	193.2077	198.3602
200	152.2410	156.4320	162.7280	168.2786	174.8353	226.0210	233.9943	241.0579	249.4451	255.2642
300	240.6634	245.9725	253.9123	260.8781	269.0679	331.7885	341.3951	349.8745	359.9064	366.8444
400	330.9028	337.1553	346.4818	354.6410	364.2074	436.6490	447.6325	457.3055	468.7245	476.6064
500	422.3034	429.3875	439.9360	449.1468	459.9261	540.9303	553.1268	563.8515	576.4928	585.2066
1000	1002.0000	1002.0000	914.2572	927.5944	1002.0000	1057.7239	1074.6794	1089.5309	1106.9690	1118.9481

Table A.4: Abscissa Values, $F_{m_1,m_2,1-a}$, Corresponding to Areas $1-a$, $a=0.05$, under the F Density Function with m_1 and m_2 Degrees of Freedom

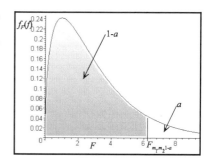

m_2 \ m_1	1	2	3	4	5	6	7	8	9	10
1	161.448	199.500	215.707	224.583	230.162	233.986	236.768	238.883	240.543	241.882
2	18.513	19.000	19.164	19.247	19.296	19.330	19.353	19.371	19.385	19.396
3	10.128	9.552	9.277	9.117	9.013	8.941	8.887	8.845	8.812	8.786
4	7.709	6.944	6.591	6.388	6.256	6.163	6.094	6.041	5.999	5.964
5	6.608	5.786	5.409	5.192	5.050	4.950	4.876	4.818	4.772	4.735
6	5.987	5.143	4.757	4.534	4.387	4.284	4.207	4.147	4.099	4.060
7	5.591	4.737	4.347	4.120	3.972	3.866	3.787	3.726	3.677	3.637
8	5.318	4.459	4.066	3.838	3.687	3.581	3.500	3.438	3.388	3.347
9	5.117	4.256	3.863	3.633	3.482	3.374	3.293	3.230	3.179	3.137
10	4.965	4.103	3.708	3.478	3.326	3.217	3.135	3.072	3.020	2.978
11	4.844	3.982	3.587	3.357	3.204	3.095	3.012	2.948	2.896	2.854
12	4.747	3.885	3.490	3.259	3.106	2.996	2.913	2.849	2.796	2.753
13	4.667	3.806	3.411	3.179	3.025	2.915	2.832	2.767	2.714	2.671
14	4.600	3.739	3.344	3.112	2.958	2.848	2.764	2.699	2.646	2.602
15	4.543	3.682	3.287	3.056	2.901	2.790	2.707	2.641	2.588	2.544
16	4.494	3.634	3.239	3.007	2.852	2.741	2.657	2.591	2.538	2.494
17	4.451	3.592	3.197	2.965	2.810	2.699	2.614	2.548	2.494	2.450
18	4.414	3.555	3.160	2.928	2.773	2.661	2.577	2.510	2.456	2.412
19	4.381	3.522	3.127	2.895	2.740	2.628	2.544	2.477	2.423	2.378
20	4.351	3.493	3.098	2.866	2.711	2.599	2.514	2.447	2.393	2.348
21	4.325	3.467	3.072	2.840	2.685	2.573	2.488	2.420	2.366	2.321
22	4.301	3.443	3.049	2.817	2.661	2.549	2.464	2.397	2.342	2.297
23	4.279	3.422	3.028	2.796	2.640	2.528	2.442	2.375	2.320	2.275
24	4.260	3.403	3.009	2.776	2.621	2.508	2.423	2.355	2.300	2.255
25	4.242	3.385	2.991	2.759	2.603	2.490	2.405	2.337	2.282	2.236
26	4.225	3.369	2.975	2.743	2.587	2.474	2.388	2.321	2.265	2.220
27	4.210	3.354	2.960	2.728	2.572	2.459	2.373	2.305	2.250	2.204
28	4.196	3.340	2.947	2.714	2.558	2.445	2.359	2.291	2.236	2.190
29	4.183	3.328	2.934	2.701	2.545	2.432	2.346	2.278	2.223	2.177
30	4.171	3.316	2.922	2.690	2.534	2.421	2.334	2.266	2.211	2.165
40	4.085	3.232	2.839	2.606	2.449	2.336	2.249	2.180	2.124	2.077
50	4.034	3.183	2.790	2.557	2.400	2.286	2.199	2.130	2.073	2.026
60	4.001	3.150	2.758	2.525	2.368	2.254	2.167	2.097	2.040	1.993
80	3.960	3.111	2.719	2.486	2.329	2.214	2.126	2.056	1.999	1.951
100	3.936	3.087	2.696	2.463	2.305	2.191	2.103	2.032	1.975	1.927
200	3.888	3.041	2.650	2.417	2.259	2.144	2.056	1.985	1.927	1.878
300	3.873	3.026	2.635	2.402	2.244	2.129	2.040	1.969	1.911	1.862
500	3.860	3.014	2.623	2.390	2.232	2.117	2.028	1.957	1.899	1.850
800	3.853	3.007	2.616	2.383	2.225	2.110	2.021	1.950	1.892	1.843
1000	3.851	3.005	2.614	2.381	2.223	2.108	2.019	1.948	1.889	1.840

Table A.4: Abscissa Values, $F_{m_1,m_2,1-a}$, $a=0.05$, under the F Distribution

(Continued)

m_2\m_1	12	15	20	24	30	40	50	60	80	100
1	243.906	245.950	248.013	249.052	250.095	251.143	251.774	252.196	252.724	253.041
2	19.413	19.429	19.446	19.454	19.462	19.471	19.476	19.479	19.483	19.486
3	8.745	8.703	8.660	8.639	8.617	8.594	8.581	8.572	8.561	8.554
4	5.912	5.858	5.803	5.774	5.746	5.717	5.699	5.688	5.673	5.664
5	4.678	4.619	4.558	4.527	4.496	4.464	4.444	4.431	4.415	4.405
6	4.000	3.938	3.874	3.841	3.808	3.774	3.754	3.740	3.722	3.712
7	3.575	3.511	3.445	3.410	3.376	3.340	3.319	3.304	3.286	3.275
8	3.284	3.218	3.150	3.115	3.079	3.043	3.020	3.005	2.986	2.975
9	3.073	3.006	2.936	2.900	2.864	2.826	2.803	2.787	2.768	2.756
10	2.913	2.845	2.774	2.737	2.700	2.661	2.637	2.621	2.601	2.588
11	2.788	2.719	2.646	2.609	2.570	2.531	2.507	2.490	2.469	2.457
12	2.687	2.617	2.544	2.505	2.466	2.426	2.401	2.384	2.363	2.350
13	2.604	2.533	2.459	2.420	2.380	2.339	2.314	2.297	2.275	2.261
14	2.534	2.463	2.388	2.349	2.308	2.266	2.241	2.223	2.201	2.187
15	2.475	2.403	2.328	2.288	2.247	2.204	2.178	2.160	2.137	2.123
16	2.425	2.352	2.276	2.235	2.194	2.151	2.124	2.106	2.083	2.068
17	2.381	2.308	2.230	2.190	2.148	2.104	2.077	2.058	2.035	2.020
18	2.342	2.269	2.191	2.150	2.107	2.063	2.035	2.017	1.993	1.978
19	2.308	2.234	2.155	2.114	2.071	2.026	1.999	1.980	1.955	1.940
20	2.278	2.203	2.124	2.082	2.039	1.994	1.966	1.946	1.922	1.907
21	2.250	2.176	2.096	2.054	2.010	1.965	1.936	1.916	1.891	1.876
22	2.226	2.151	2.071	2.028	1.984	1.938	1.909	1.889	1.864	1.849
23	2.204	2.128	2.048	2.005	1.961	1.914	1.885	1.865	1.839	1.823
24	2.183	2.108	2.027	1.984	1.939	1.892	1.863	1.842	1.816	1.800
25	2.165	2.089	2.007	1.964	1.919	1.872	1.842	1.822	1.796	1.779
26	2.148	2.072	1.990	1.946	1.901	1.853	1.823	1.803	1.776	1.760
27	2.132	2.056	1.974	1.930	1.884	1.836	1.806	1.785	1.758	1.742
28	2.118	2.041	1.959	1.915	1.869	1.820	1.790	1.769	1.742	1.725
29	2.104	2.027	1.945	1.901	1.854	1.806	1.775	1.754	1.726	1.710
30	2.092	2.015	1.932	1.887	1.841	1.792	1.761	1.740	1.712	1.695
40	2.003	1.924	1.839	1.793	1.744	1.693	1.660	1.637	1.608	1.589
50	1.952	1.871	1.784	1.737	1.687	1.634	1.599	1.576	1.544	1.525
60	1.917	1.836	1.748	1.700	1.649	1.594	1.559	1.534	1.502	1.481
80	1.875	1.793	1.703	1.654	1.602	1.545	1.508	1.482	1.448	1.426
100	1.850	1.768	1.676	1.627	1.573	1.515	1.477	1.450	1.415	1.392
200	1.801	1.717	1.623	1.572	1.516	1.455	1.415	1.386	1.346	1.321
300	1.785	1.700	1.606	1.554	1.497	1.435	1.393	1.363	1.323	1.296
500	1.772	1.686	1.592	1.539	1.482	1.419	1.376	1.345	1.303	1.275
800	1.764	1.679	1.584	1.531	1.473	1.409	1.366	1.335	1.292	1.264
1000	1.762	1.676	1.581	1.528	1.471	1.406	1.363	1.332	1.289	1.260

Table A.5: Abscissa Values, $F_{m_1,m_2,1-a}$, Corresponding to Areas 1-a, a=0.025, under the F Density Function with m_1 and m_2 Degrees of Freedom

m_2 \ m_1	1	2	3	4	5	6	7	8	9	10
1	647.789	799.500	864.163	899.583	921.848	937.111	948.217	956.656	963.285	968.627
2	38.506	39.000	39.165	39.248	39.298	39.331	39.355	39.373	39.387	39.398
3	17.443	16.044	15.439	15.101	14.885	14.735	14.624	14.540	14.473	14.419
4	12.218	10.649	9.979	9.605	9.364	9.197	9.074	8.980	8.905	8.844
5	10.007	8.434	7.764	7.388	7.146	6.978	6.853	6.757	6.681	6.619
6	8.813	7.260	6.599	6.227	5.988	5.820	5.695	5.600	5.523	5.461
7	8.073	6.542	5.890	5.523	5.285	5.119	4.995	4.899	4.823	4.761
8	7.571	6.059	5.416	5.053	4.817	4.652	4.529	4.433	4.357	4.295
9	7.209	5.715	5.078	4.718	4.484	4.320	4.197	4.102	4.026	3.964
10	6.937	5.456	4.826	4.468	4.236	4.072	3.950	3.855	3.779	3.717
11	6.724	5.256	4.630	4.275	4.044	3.881	3.759	3.664	3.588	3.526
12	6.554	5.096	4.474	4.121	3.891	3.728	3.607	3.512	3.436	3.374
13	6.414	4.965	4.347	3.996	3.767	3.604	3.483	3.388	3.312	3.250
14	6.298	4.857	4.242	3.892	3.663	3.501	3.380	3.285	3.209	3.147
15	6.200	4.765	4.153	3.804	3.576	3.415	3.293	3.199	3.123	3.060
16	6.115	4.687	4.077	3.729	3.502	3.341	3.219	3.125	3.049	2.986
17	6.042	4.619	4.011	3.665	3.438	3.277	3.156	3.061	2.985	2.922
18	5.978	4.560	3.954	3.608	3.382	3.221	3.100	3.005	2.929	2.866
19	5.922	4.508	3.903	3.559	3.333	3.172	3.051	2.956	2.880	2.817
20	5.871	4.461	3.859	3.515	3.289	3.128	3.007	2.913	2.837	2.774
21	5.827	4.420	3.819	3.475	3.250	3.090	2.969	2.874	2.798	2.735
22	5.786	4.383	3.783	3.440	3.215	3.055	2.934	2.839	2.763	2.700
23	5.750	4.349	3.750	3.408	3.183	3.023	2.902	2.808	2.731	2.668
24	5.717	4.319	3.721	3.379	3.155	2.995	2.874	2.779	2.703	2.640
25	5.686	4.291	3.694	3.353	3.129	2.969	2.848	2.753	2.677	2.613
26	5.659	4.265	3.670	3.329	3.105	2.945	2.824	2.729	2.653	2.590
27	5.633	4.242	3.647	3.307	3.083	2.923	2.802	2.707	2.631	2.568
28	5.610	4.221	3.626	3.286	3.063	2.903	2.782	2.687	2.611	2.547
29	5.588	4.201	3.607	3.267	3.044	2.884	2.763	2.669	2.592	2.529
30	5.568	4.182	3.589	3.250	3.026	2.867	2.746	2.651	2.575	2.511
40	5.424	4.051	3.463	3.126	2.904	2.744	2.624	2.529	2.452	2.388
50	5.340	3.975	3.390	3.054	2.833	2.674	2.553	2.458	2.381	2.317
60	5.286	3.925	3.343	3.008	2.786	2.627	2.507	2.412	2.334	2.270
80	5.218	3.864	3.284	2.950	2.730	2.571	2.450	2.355	2.277	2.213
100	5.179	3.828	3.250	2.917	2.696	2.537	2.417	2.321	2.244	2.179
200	5.100	3.758	3.182	2.850	2.630	2.472	2.351	2.256	2.178	2.113
300	5.075	3.735	3.160	2.829	2.609	2.451	2.330	2.234	2.156	2.091
500	5.054	3.716	3.142	2.811	2.592	2.434	2.313	2.217	2.139	2.074
800	5.043	3.706	3.132	2.802	2.582	2.424	2.303	2.208	2.130	2.064
1000	5.039	3.703	3.129	2.799	2.579	2.421	2.300	2.204	2.126	2.061

Table A.5: Abscissa Values, $F_{m_1,m_2,1-a}$, $a=0.025$, under the F Distribution (Continued)

m_2 \ m_1	12	15	20	24	30	40	50	60	80	100
1	976.708	984.867	993.103	997.249	1001.414	1005.598	1008.117	1009.800	1011.908	1013.175
2	39.415	39.431	39.448	39.456	39.465	39.473	39.478	39.481	39.485	39.488
3	14.337	14.253	14.167	14.124	14.081	14.037	14.010	13.992	13.970	13.956
4	8.751	8.657	8.560	8.511	8.461	8.411	8.381	8.360	8.335	8.319
5	6.525	6.428	6.329	6.278	6.227	6.175	6.144	6.123	6.096	6.080
6	5.366	5.269	5.168	5.117	5.065	5.012	4.980	4.959	4.932	4.915
7	4.666	4.568	4.467	4.415	4.362	4.309	4.276	4.254	4.227	4.210
8	4.200	4.101	3.999	3.947	3.894	3.840	3.807	3.784	3.756	3.739
9	3.868	3.769	3.667	3.614	3.560	3.505	3.472	3.449	3.421	3.403
10	3.621	3.522	3.419	3.365	3.311	3.255	3.221	3.198	3.169	3.152
11	3.430	3.330	3.226	3.173	3.118	3.061	3.027	3.004	2.974	2.956
12	3.277	3.177	3.073	3.019	2.963	2.906	2.871	2.848	2.818	2.800
13	3.153	3.053	2.948	2.893	2.837	2.780	2.744	2.720	2.690	2.671
14	3.050	2.949	2.844	2.789	2.732	2.674	2.638	2.614	2.583	2.565
15	2.963	2.862	2.756	2.701	2.644	2.585	2.549	2.524	2.493	2.474
16	2.889	2.788	2.681	2.625	2.568	2.509	2.472	2.447	2.415	2.396
17	2.825	2.723	2.616	2.560	2.502	2.442	2.405	2.380	2.348	2.329
18	2.769	2.667	2.559	2.503	2.445	2.384	2.347	2.321	2.289	2.269
19	2.720	2.617	2.509	2.452	2.394	2.333	2.295	2.270	2.237	2.217
20	2.676	2.573	2.464	2.408	2.349	2.287	2.249	2.223	2.190	2.170
21	2.637	2.534	2.425	2.368	2.308	2.246	2.208	2.182	2.148	2.128
22	2.602	2.498	2.389	2.331	2.272	2.210	2.171	2.145	2.111	2.090
23	2.570	2.466	2.357	2.299	2.239	2.176	2.137	2.111	2.077	2.056
24	2.541	2.437	2.327	2.269	2.209	2.146	2.107	2.080	2.045	2.024
25	2.515	2.411	2.300	2.242	2.182	2.118	2.079	2.052	2.017	1.996
26	2.491	2.387	2.276	2.217	2.157	2.093	2.053	2.026	1.991	1.969
27	2.469	2.364	2.253	2.195	2.133	2.069	2.029	2.002	1.966	1.945
28	2.448	2.344	2.232	2.174	2.112	2.048	2.007	1.980	1.944	1.922
29	2.430	2.325	2.213	2.154	2.092	2.028	1.987	1.959	1.923	1.901
30	2.412	2.307	2.195	2.136	2.074	2.009	1.968	1.940	1.904	1.882
40	2.288	2.182	2.068	2.007	1.943	1.875	1.832	1.803	1.764	1.741
50	2.216	2.109	1.993	1.931	1.866	1.796	1.752	1.721	1.681	1.656
60	2.169	2.061	1.944	1.882	1.815	1.744	1.699	1.667	1.625	1.599
80	2.111	2.003	1.884	1.820	1.752	1.679	1.632	1.599	1.555	1.527
100	2.077	1.968	1.849	1.784	1.715	1.640	1.592	1.558	1.512	1.483
200	2.010	1.900	1.778	1.712	1.640	1.562	1.511	1.474	1.425	1.393
300	1.988	1.877	1.755	1.688	1.616	1.536	1.484	1.446	1.395	1.361
500	1.971	1.859	1.736	1.669	1.596	1.515	1.462	1.423	1.370	1.336
800	1.961	1.849	1.726	1.658	1.585	1.503	1.449	1.410	1.356	1.321
1000	1.958	1.846	1.722	1.654	1.581	1.499	1.445	1.406	1.352	1.316

Table A.6: Abscissa Values, $F_{m_1,m_2,\,1-a}$, Corresponding to Areas 1-a, a=0.01, under the F Density Function with m_1 and m_2 Degrees of Freedom

m_2 \ m_1	1	2	3	4	5	6	7	8	9	10
1	4052.181	4999.500	5403.352	5624.583	5763.650	5858.986	5928.356	5981.070	6022.473	6055.847
2	98.503	99.000	99.166	99.249	99.299	99.333	99.356	99.374	99.388	99.399
3	34.116	30.817	29.457	28.710	28.237	27.911	27.672	27.489	27.345	27.229
4	21.198	18.000	16.694	15.977	15.522	15.207	14.976	14.799	14.659	14.546
5	16.258	13.274	12.060	11.392	10.967	10.672	10.456	10.289	10.158	10.051
6	13.745	10.925	9.780	9.148	8.746	8.466	8.260	8.102	7.976	7.874
7	12.246	9.547	8.451	7.847	7.460	7.191	6.993	6.840	6.719	6.620
8	11.259	8.649	7.591	7.006	6.632	6.371	6.178	6.029	5.911	5.814
9	10.561	8.022	6.992	6.422	6.057	5.802	5.613	5.467	5.351	5.257
10	10.044	7.559	6.552	5.994	5.636	5.386	5.200	5.057	4.942	4.849
11	9.646	7.206	6.217	5.668	5.316	5.069	4.886	4.744	4.632	4.539
12	9.330	6.927	5.953	5.412	5.064	4.821	4.640	4.499	4.388	4.296
13	9.074	6.701	5.739	5.205	4.862	4.620	4.441	4.302	4.191	4.100
14	8.862	6.515	5.564	5.035	4.695	4.456	4.278	4.140	4.030	3.939
15	8.683	6.359	5.417	4.893	4.556	4.318	4.142	4.004	3.895	3.805
16	8.531	6.226	5.292	4.773	4.437	4.202	4.026	3.890	3.780	3.691
17	8.400	6.112	5.185	4.669	4.336	4.102	3.927	3.791	3.682	3.593
18	8.285	6.013	5.092	4.579	4.248	4.015	3.841	3.705	3.597	3.508
19	8.185	5.926	5.010	4.500	4.171	3.939	3.765	3.631	3.523	3.434
20	8.096	5.849	4.938	4.431	4.103	3.871	3.699	3.564	3.457	3.368
21	8.017	5.780	4.874	4.369	4.042	3.812	3.640	3.506	3.398	3.310
22	7.945	5.719	4.817	4.313	3.988	3.758	3.587	3.453	3.346	3.258
23	7.881	5.664	4.765	4.264	3.939	3.710	3.539	3.406	3.299	3.211
24	7.823	5.614	4.718	4.218	3.895	3.667	3.496	3.363	3.256	3.168
25	7.770	5.568	4.675	4.177	3.855	3.627	3.457	3.324	3.217	3.129
26	7.721	5.526	4.637	4.140	3.818	3.591	3.421	3.288	3.182	3.094
27	7.677	5.488	4.601	4.106	3.785	3.558	3.388	3.256	3.149	3.062
28	7.636	5.453	4.568	4.074	3.754	3.528	3.358	3.226	3.120	3.032
29	7.598	5.420	4.538	4.045	3.725	3.499	3.330	3.198	3.092	3.005
30	7.562	5.390	4.510	4.018	3.699	3.473	3.304	3.173	3.067	2.979
40	7.314	5.179	4.313	3.828	3.514	3.291	3.124	2.993	2.888	2.801
50	7.171	5.057	4.199	3.720	3.408	3.186	3.020	2.890	2.785	2.698
60	7.077	4.977	4.126	3.649	3.339	3.119	2.953	2.823	2.718	2.632
80	6.963	4.881	4.036	3.563	3.255	3.036	2.871	2.742	2.637	2.551
100	6.895	4.824	3.984	3.513	3.206	2.988	2.823	2.694	2.590	2.503
200	6.763	4.713	3.881	3.414	3.110	2.893	2.730	2.601	2.497	2.411
300	6.720	4.677	3.848	3.382	3.079	2.862	2.699	2.571	2.467	2.380
500	6.686	4.648	3.821	3.357	3.054	2.838	2.675	2.547	2.443	2.356
800	6.667	4.632	3.806	3.343	3.040	2.825	2.662	2.533	2.429	2.343
1000	6.660	4.626	3.801	3.338	3.036	2.820	2.657	2.529	2.425	2.339

Table A.6: Abscissa Values, $F_{m_1,m_2,1-a}$, $a=0.01$, under the F Distribution (Continued)

m_2 \ m_1	12	15	20	24	30	40	50	60	80	100
1	6106.321	6157.285	6208.730	6234.631	6260.649	6286.782	6302.517	6313.030	6326.197	6334.110
2	99.416	99.433	99.449	99.458	99.466	99.474	99.479	99.482	99.487	99.489
3	27.052	26.872	26.690	26.598	26.505	26.411	26.354	26.316	26.269	26.240
4	14.374	14.198	14.020	13.929	13.838	13.745	13.690	13.652	13.605	13.577
5	9.888	9.722	9.553	9.466	9.379	9.291	9.238	9.202	9.157	9.130
6	7.718	7.559	7.396	7.313	7.229	7.143	7.091	7.057	7.013	6.987
7	6.469	6.314	6.155	6.074	5.992	5.908	5.858	5.824	5.781	5.755
8	5.667	5.515	5.359	5.279	5.198	5.116	5.065	5.032	4.989	4.963
9	5.111	4.962	4.808	4.729	4.649	4.567	4.517	4.483	4.441	4.415
10	4.706	4.558	4.405	4.327	4.247	4.165	4.115	4.082	4.039	4.014
11	4.397	4.251	4.099	4.021	3.941	3.860	3.810	3.776	3.734	3.708
12	4.155	4.010	3.858	3.780	3.701	3.619	3.569	3.535	3.493	3.467
13	3.960	3.815	3.665	3.587	3.507	3.425	3.375	3.341	3.298	3.272
14	3.800	3.656	3.505	3.427	3.348	3.266	3.215	3.181	3.138	3.112
15	3.666	3.522	3.372	3.294	3.214	3.132	3.081	3.047	3.004	2.977
16	3.553	3.409	3.259	3.181	3.101	3.018	2.967	2.933	2.889	2.863
17	3.455	3.312	3.162	3.084	3.003	2.920	2.869	2.835	2.791	2.764
18	3.371	3.227	3.077	2.999	2.919	2.835	2.784	2.749	2.705	2.678
19	3.297	3.153	3.003	2.925	2.844	2.761	2.709	2.674	2.630	2.602
20	3.231	3.088	2.938	2.859	2.778	2.695	2.643	2.608	2.563	2.535
21	3.173	3.030	2.880	2.801	2.720	2.636	2.584	2.548	2.503	2.475
22	3.121	2.978	2.827	2.749	2.667	2.583	2.531	2.495	2.450	2.422
23	3.074	2.931	2.781	2.702	2.620	2.535	2.483	2.447	2.401	2.373
24	3.032	2.889	2.738	2.659	2.577	2.492	2.440	2.403	2.357	2.329
25	2.993	2.850	2.699	2.620	2.538	2.453	2.400	2.364	2.317	2.289
26	2.958	2.815	2.664	2.585	2.503	2.417	2.364	2.327	2.281	2.252
27	2.926	2.783	2.632	2.552	2.470	2.384	2.330	2.294	2.247	2.218
28	2.896	2.753	2.602	2.522	2.440	2.354	2.300	2.263	2.216	2.187
29	2.868	2.726	2.574	2.495	2.412	2.325	2.271	2.234	2.187	2.158
30	2.843	2.700	2.549	2.469	2.386	2.299	2.245	2.208	2.160	2.131
40	2.665	2.522	2.369	2.288	2.203	2.114	2.058	2.019	1.969	1.938
50	2.562	2.419	2.265	2.183	2.098	2.007	1.949	1.909	1.857	1.825
60	2.496	2.352	2.198	2.115	2.028	1.936	1.877	1.836	1.783	1.749
80	2.415	2.271	2.115	2.032	1.944	1.849	1.788	1.746	1.690	1.655
100	2.368	2.223	2.067	1.983	1.893	1.797	1.735	1.692	1.634	1.598
200	2.275	2.129	1.971	1.886	1.794	1.694	1.629	1.583	1.521	1.481
300	2.244	2.099	1.940	1.854	1.761	1.660	1.594	1.547	1.483	1.441
500	2.220	2.075	1.915	1.829	1.735	1.633	1.566	1.517	1.452	1.408
800	2.207	2.061	1.901	1.814	1.721	1.618	1.550	1.501	1.434	1.390
1000	2.203	2.056	1.897	1.810	1.716	1.613	1.544	1.495	1.428	1.383

Table A.7: Abscissa Values, $F_{m_1,m_2,1-a}$, Corresponding to Areas 1-a, a=0.005, under the F Density Function with m_1 and m_2 Degrees of Freedom

m_2 \ m_1	1	2	3	4	5	6	7	8	9	10
1	16210.723	19999.500	21614.741	22499.583	23055.798	23437.111	23714.566	23925.406	24091.004	24224.487
2	198.501	199.000	199.166	199.250	199.300	199.333	199.357	199.375	199.388	199.400
3	55.552	49.799	47.467	46.195	45.392	44.838	44.434	44.126	43.882	43.686
4	31.333	26.284	24.259	23.155	22.456	21.975	21.622	21.352	21.139	20.967
5	22.785	18.314	16.530	15.556	14.940	14.513	14.200	13.961	13.772	13.618
6	18.635	14.544	12.917	12.028	11.464	11.073	10.786	10.566	10.391	10.250
7	16.236	12.404	10.882	10.050	9.522	9.155	8.885	8.678	8.514	8.380
8	14.688	11.042	9.596	8.805	8.302	7.952	7.694	7.496	7.339	7.211
9	13.614	10.107	8.717	7.956	7.471	7.134	6.885	6.693	6.541	6.417
10	12.826	9.427	8.081	7.343	6.872	6.545	6.302	6.116	5.968	5.847
11	12.226	8.912	7.600	6.881	6.422	6.102	5.865	5.682	5.537	5.418
12	11.754	8.510	7.226	6.521	6.071	5.757	5.525	5.345	5.202	5.085
13	11.374	8.186	6.926	6.233	5.791	5.482	5.253	5.076	4.935	4.820
14	11.060	7.922	6.680	5.998	5.562	5.257	5.031	4.857	4.717	4.603
15	10.798	7.701	6.476	5.803	5.372	5.071	4.847	4.674	4.536	4.424
16	10.575	7.514	6.303	5.638	5.212	4.913	4.692	4.521	4.384	4.272
17	10.384	7.354	6.156	5.497	5.075	4.779	4.559	4.389	4.254	4.142
18	10.218	7.215	6.028	5.375	4.956	4.663	4.445	4.276	4.141	4.030
19	10.073	7.093	5.916	5.268	4.853	4.561	4.345	4.177	4.043	3.933
20	9.944	6.986	5.818	5.174	4.762	4.472	4.257	4.090	3.956	3.847
21	9.830	6.891	5.730	5.091	4.681	4.393	4.179	4.013	3.880	3.771
22	9.727	6.806	5.652	5.017	4.609	4.322	4.109	3.944	3.812	3.703
23	9.635	6.730	5.582	4.950	4.544	4.259	4.047	3.882	3.750	3.642
24	9.551	6.661	5.519	4.890	4.486	4.202	3.991	3.826	3.695	3.587
25	9.475	6.598	5.462	4.835	4.433	4.150	3.939	3.776	3.645	3.537
26	9.406	6.541	5.409	4.785	4.384	4.103	3.893	3.730	3.599	3.492
27	9.342	6.489	5.361	4.740	4.340	4.059	3.850	3.687	3.557	3.450
28	9.284	6.440	5.317	4.698	4.300	4.020	3.811	3.649	3.519	3.412
29	9.230	6.396	5.276	4.659	4.262	3.983	3.775	3.613	3.483	3.377
30	9.180	6.355	5.239	4.623	4.228	3.949	3.742	3.580	3.450	3.344
40	8.828	6.066	4.976	4.374	3.986	3.713	3.509	3.350	3.222	3.117
50	8.626	5.902	4.826	4.232	3.849	3.579	3.376	3.219	3.092	2.988
60	8.495	5.795	4.729	4.140	3.760	3.492	3.291	3.134	3.008	2.904
80	8.335	5.665	4.611	4.029	3.652	3.387	3.188	3.032	2.907	2.803
100	8.241	5.589	4.542	3.963	3.589	3.325	3.127	2.972	2.847	2.744
200	8.057	5.441	4.408	3.837	3.467	3.206	3.010	2.856	2.732	2.629
300	7.997	5.393	4.365	3.796	3.428	3.167	2.972	2.818	2.694	2.592
500	7.950	5.355	4.330	3.763	3.396	3.137	2.941	2.789	2.665	2.562
800	7.923	5.334	4.311	3.745	3.379	3.120	2.925	2.772	2.648	2.546
1000	7.915	5.326	4.305	3.739	3.373	3.114	2.919	2.766	2.643	2.541

Table A.7: Abscissa Values, $F_{m_1,m_2,1-a}$, $a=0.005$, under the F Distribution (Continued)

m_1 / m_2	12	15	20	24	30	40	50	60	80	100
1	24426.366	24630.205	24835.971	24939.565	25043.628	25148.153	25211.089	25253.137	25305.799	25337.450
2	199.416	199.433	199.450	199.458	199.466	199.475	199.480	199.483	199.487	199.490
3	43.387	43.085	42.778	42.622	42.466	42.308	42.213	42.149	42.070	42.022
4	20.705	20.438	20.167	20.030	19.892	19.752	19.667	19.611	19.540	19.497
5	13.384	13.146	12.903	12.780	12.656	12.530	12.454	12.402	12.338	12.300
6	10.034	9.814	9.589	9.474	9.358	9.241	9.170	9.122	9.062	9.026
7	8.176	7.968	7.754	7.645	7.534	7.422	7.354	7.309	7.251	7.217
8	7.015	6.814	6.608	6.503	6.396	6.288	6.222	6.177	6.121	6.088
9	6.227	6.032	5.832	5.729	5.625	5.519	5.454	5.410	5.356	5.322
10	5.661	5.471	5.274	5.173	5.071	4.966	4.902	4.859	4.805	4.772
11	5.236	5.049	4.855	4.756	4.654	4.551	4.488	4.445	4.391	4.359
12	4.906	4.721	4.530	4.431	4.331	4.228	4.165	4.123	4.069	4.037
13	4.643	4.460	4.270	4.173	4.073	3.970	3.908	3.866	3.812	3.780
14	4.428	4.247	4.059	3.961	3.862	3.760	3.698	3.655	3.602	3.569
15	4.250	4.070	3.883	3.786	3.687	3.585	3.523	3.480	3.427	3.394
16	4.099	3.920	3.734	3.638	3.539	3.437	3.375	3.332	3.279	3.246
17	3.971	3.793	3.607	3.511	3.412	3.311	3.248	3.206	3.152	3.119
18	3.860	3.683	3.498	3.402	3.303	3.201	3.139	3.096	3.042	3.009
19	3.763	3.587	3.402	3.306	3.208	3.106	3.043	3.000	2.946	2.913
20	3.678	3.502	3.318	3.222	3.123	3.022	2.959	2.916	2.861	2.828
21	3.602	3.427	3.243	3.147	3.049	2.947	2.884	2.841	2.786	2.753
22	3.535	3.360	3.176	3.081	2.982	2.880	2.817	2.774	2.719	2.685
23	3.475	3.300	3.116	3.021	2.922	2.820	2.756	2.713	2.658	2.624
24	3.420	3.246	3.062	2.967	2.868	2.765	2.702	2.658	2.603	2.569
25	3.370	3.196	3.013	2.918	2.819	2.716	2.652	2.609	2.553	2.519
26	3.325	3.151	2.968	2.873	2.774	2.671	2.607	2.563	2.508	2.473
27	3.284	3.110	2.928	2.832	2.733	2.630	2.565	2.522	2.466	2.431
28	3.246	3.073	2.890	2.794	2.695	2.592	2.527	2.483	2.427	2.392
29	3.211	3.038	2.855	2.759	2.660	2.557	2.492	2.448	2.391	2.357
30	3.179	3.006	2.823	2.727	2.628	2.524	2.459	2.415	2.358	2.323
40	2.953	2.781	2.598	2.502	2.401	2.296	2.230	2.184	2.125	2.088
50	2.825	2.653	2.470	2.373	2.272	2.164	2.097	2.050	1.989	1.951
60	2.742	2.570	2.387	2.290	2.187	2.079	2.010	1.962	1.900	1.861
80	2.641	2.470	2.286	2.188	2.084	1.974	1.903	1.854	1.789	1.748
100	2.583	2.411	2.227	2.128	2.024	1.912	1.840	1.790	1.723	1.681
200	2.468	2.297	2.112	2.012	1.905	1.790	1.715	1.661	1.590	1.544
300	2.431	2.260	2.074	1.973	1.866	1.749	1.673	1.619	1.545	1.498
500	2.402	2.230	2.044	1.943	1.835	1.717	1.640	1.584	1.509	1.460
800	2.385	2.214	2.027	1.926	1.818	1.699	1.621	1.565	1.489	1.438
1000	2.380	2.208	2.022	1.921	1.812	1.693	1.615	1.558	1.482	1.431

APPENDIX B: ANSWERS TO PROBLEMS

Chapter 1
1.4 Velocity mean and standard deviation are now 1.584, and 0.793, respectively.
1.5 For 5000 samples, the mean and standard deviation are 1.503 and 0.870, respectively.
1.6 The coefficient of variability is 301.2%
1.7 The coefficient of variability is 61.2%

Chapter 2
2.2 C_2={ll, lm, lh, ml, mm, mh, hl, hm, hh}
2.3 C_3={lll, llm, llh, lml, lmm, lmh, lhl, lhm, lhh, mll, mlm, mlh, mml, mmm, mmh, mhl, mhm, mhh, hll, hlm, hlh, hml, hmm, hmh, hhl, hhm, hhh}
2.4 2,176,782,336.
2.5 (1) 3,776,965,920. (2) 5,245,786.
2.6 10,272,278,170.
2.7 4,076,357,558,689,587,171,000.
2.8 1/3.
2.9 $P(M)$=4/9.
2.10 $P(H)$=19/27=0.704, or 70.4%
2.11 $P(H_2)$=6/27=2/9=0.222..
2.12 0.0143%
2.13 $P(W)$=0.0258.
2.14 (1) $P(W)$=0.0258. (2) $P(W)$=0.0256.
2.17 1/7.
2.18 $P(S)$=0.028.
2.19 $P(F_B \cap F_A^c | F$=0.136
2.20 Elements in vector W contain the required probabilities:
W:=[.4294310665,.3734183187,.1515320713,
.03806602758,.006620178709,.0008443126472,
.00008157610118,.6080206300×10^{-5},.3524757275×10^{-6},
.1589262539×10^{-7},.5527869701×10^{-9},.1456619158×10^{-10},
.2814723009×10^{-12},.3765515731×10^{-14},.3118439529×10^{-16},
.1205194021×10^{-18}]
2.21 (1) $P(F)$=0.078. (2) $P(L|F)$=0.632, $P(L^c|F^c)$=0.999.

Chapter 3
3.1 $P(X=2)$=0.325,, $P(X=1)$=0.057, and $P(X=0)$=0.004.
3.2 $F_X(x)$=0, $X<0$ $F_X(x)$=0.004, $0 \le X<1$; $F_X(x)$=0.061, $1 \le X<2$;
$F_X(x)$=0.386, $2 \le X<3$; and $F_x(x)$=1, $3 \le X$.
3.3 $E\{X\}$=2.848.
3.4 μ_E=2.45, σ_E=1.023.

3.5 $p_E(1)=0.25$, $p_E(2)=0.2$, $p_E(3)=0.4$, and $p_E(4)=0.15$.
$F_E(0)=0, F_E(1)=0.25$, $F_E(2)=0.45$, $F_E(3)=0.85$, $F_E(4)=1$.

3.6 $\mu_Q=19.92$, $\sigma_Q=7.19$.

3.7 $P_O(3)=0.0036$.

3.8 0.269

3.9 0.0249.

3.10 $E\{X\}=18.32$ days. $\sigma=17.81$ days.

3.11 (1) $P(X=2)=0.018$. (2) $E\{X\}=73.3$ days. $\sigma=35.62$ days.

3.12 0.217.

3.13 (1) $P(X=2)=0.262$. (2) $P(X\le6)=0.765$.

3.14 (1) $a=12$. (2) $P(X\le0.2)=0.0272$. (3) $P(0.4\le X\le0.6)=0.296$.
    (4) $P(X>0.9)=0.0523$.

3.15 $\mu=0.6$, $\sigma=0.2$, $C_v=0.333$, $\gamma=-0.2857$.

3.16 0.810.

3.17 0.414.

3.18 (1) 0.6554. (2) 0.6554. (3) 0.2881. (4) \$510,194.

3.19 (1) 0.7118. (4) 513.9 thousands of dollars.

3.20 (1) 0.712. (2) 0.607. (3) 0.356. (4) \$513,855.

3.21 $F_X(x)=\pi/2+\tan^{-1}[(x-a)/\alpha]/\pi$

Chapter 4

4.1 $f_Y(y)=2e^{-|4/y|}/y^2$.

4.2 $f_Y(y)=1/\sqrt{64y}$, $9\le Y\le49$.

4.3 A symmetric density function, such as the Normal, becomes positively skewed after passing through an exponentially-decaying system.

4.4 $X=1.069$.

4.5 (1) $X-\dfrac{X^4}{4}-\dfrac{3U}{4}=0$,

(2) $X=\dfrac{3U}{4}+\dfrac{1}{4}\left(\dfrac{3U}{4}\right)^4+\dfrac{1}{4}\left(\dfrac{3U}{4}\right)^7+\dfrac{1}{4}\left(\dfrac{3U}{4}\right)^{10}+\dfrac{3}{2\times4^2}\left(\dfrac{3U}{4}\right)^{10}+\cdots$.

4.6 $\hat{X}=0.380224$, $\hat{e}=0.000001$.

4.7. The same answers as in Problem 4.5.

4.8 The same answers as in Problem 4.6.

4.9 As the sample size increases, the fit is better and the numbers become statistically representative.

4.10 A sample of 50 data points is not truly representative of the random behavior in annual sales.

4.11 $\mu_X=1.253$, $\sigma_X=0.655$. Generated Rayleigh random numbers have similar values, and the serial correlation coefficient is close to zero. A visual comparison between the theoretical density and the random-numbers frequency histogram seems reasonable:

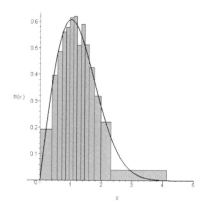

4.12

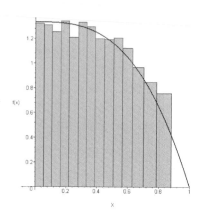

4.13

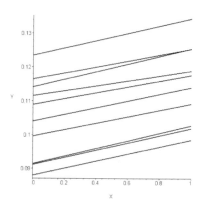

4.14 $\mu_Y=0.105$, $\sigma_Y=0.0100$.

4.15 $\mu_h=z+e^{-t^2/2}$, $\sigma_h^2 \equiv 0.505(1+e^{-2t^2})-e^{t^2}$.

4.16 $\mu_Y \approx aX+b\mu_Z e^{-aX}-b^2(\mu_Z^2+\sigma_Z^2)e^{-2aX}$.

Chapter 5
5.1

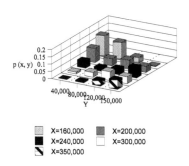

5.2 $P(X>240,000,\ Y>80,000)=0.209$.
5.3 $E\{T\}=\$320,050$.
5.4

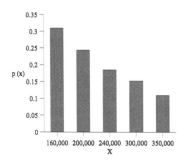   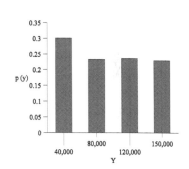

5.5 $P(X=240,000\,|\,Y=80,000)=0.103$.
5.6   (1) $f_{XY}(x,\ y)$ is a proper joint probability density function.
      (2)

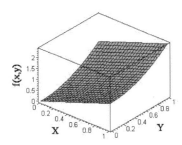

5.7 $P(0.25\le Y\le 0.75)=0.4625$.
5.8 $f_{X|Y}(x\,|\,y>0.8)=0.761x+0.619$.
5.9 $P(0.4<X<0.6\,|\,Y>0.8)=0.2$.

5.10 $\mu_X=\mu_Y=0.600$, $\sigma_X=0.271$, $\sigma_Y=0.283$, $C_{XY}=-0.010$, $\rho_{XY}=-0.131$. X and Y are not statistically independent. However, they are mildly correlated; about 13% of the value of Y is explained by that of X.

5.11 $T\sim N(\mu_T, \sigma_T)$, $\mu_T=80days$, $\sigma_T=5.254days$.

5.12 $P(-1\leq E_T\leq 1)=0.3859$.

Chapter 6

6.1 (1) $\bar{X}\sim N(59.125, 0.712)$. (2) $N=225$ times.

6.2 $\hat{\lambda}=\dfrac{1}{N}\sum_{i=1}^{N} k_i$.

6.3 Moments estimates: $\hat{\mu}_C=79.740^\circ C$, $\hat{\sigma}_C=0.911^\circ C$.
     Maximum Likelihood estimates: $\hat{\mu}_C=79.740^\circ C$, $\hat{\sigma}_C=0.864^\circ C$.

6.4 Same answers as in Example 6.4.

6.5 We are 95% confident that the mean strength is between $L=2941.1psi$, and $U=3061.4psi$.

6.6 We are 95% confident that the standard deviation is between $L=13.80$ and $U=28.20m/month$.

6.7 Same answers as in Example 6.7.

6.8 At the 5% level of significance, there is reason to believe that the mean is different from 13500 psi.

6.9 At the 5% level of significance, there is reason to believe that the workers mean salary is less than the industry-wide of $11.65 per hour and the company may be underpaying its employees.

6.10 At the 5% level of significance, there is reason to believe that the joints standard deviation is equal to 1.5cm.

Chapter 7

7.1 The histogram of the logarithms appears a little more symmetric than that of the arithmetic values. The hypothesis of Log-Normality could be further studied.

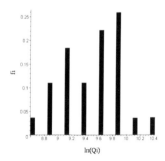

7.2 The Weibull distribution could be a good model, although the parameters, m and c, are not the optimal ones and must be estimated.

7.3 A slightly better fit is obtained. The Log-Normal distribution might be an appropriate model.

7.4 Plotting the empirical cumulative distribution function on Normal probability paper yields a cloud of points to which a straight line might be reasonably fitted. Extremely high values (i.e., those with return periods of about 25 days) may not be accurately described by a Normal distribution.

7.5 We accept H_0. We do not suspect the data has been rigged.

7.6 Cell limits coincide with sample values, and there are many cells with zero observed frequency. The choice of interval width is inappropriate and may lead to incorrect conclusions.

7.7 We accept H_0. We do not suspect the data has been rigged.

7.8 Initial running of the program shows that there are several cells with absolute frequencies of zero values. This would cause the rejection of the null hypothesis. After adjusting the corresponding width of such cells, the null hypothesis is accepted again.

7.9 We accept H_0. We do not suspect the data has been rigged.

7.10 We accept H_0. We do not suspect the data has been rigged.

7.11 We accept H_0. We do not suspect the data has been rigged.

Chapter 8

8.1 (1) $a=1.792 m/cm$, $b=7.768\ m$. (2) $r_{WL}=0.714$. (4)$L(3.5)=14.04\ m$.

8.2 (1) $a=1.182\ (m^3 \times 10^6)$, $b=0.240\ (m^3 \times 10^6)^{-year}$. (3) $G(30)=35.90\ (m^3 \times 10^6)$.

8.3 We are 95% confident that a is between 0.193 and 3.391.

8.4 Follow Example 8.4. Although a linear relationship is suggested, the cloud of points is scarce and disperse. The correlation coefficient is 0.71. The wide confidence band reflects these facts.

8.5 We are 90% confident that the intercept b is between $1.194m$ and $14.342m$.

8.6 Modify the program in Problem 8.2 according to the guidelines in Example 8.4.

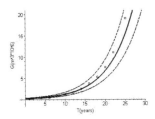

Chapter 9

9.1 (3) 0.135. (4) 1000$hours$.

9.2 $f_T(\tau\,|\,T>t)=\omega e^{-\omega(\tau-t)}$, $\tau > t$.

9.3 $R(t)=e^{-\omega t}$, $h(t)=\omega$.

9.4 $F_T(t)=t/10$, $R(t)=1-t/10$, $h(t)=1/(10-t)$.

9.5 $R_s=0.908$.

9.6 (1) $R_s = e^{-0.06t}$. (2) $E\{T_s\} = 16.667 hours$.

9.7 (1) $R_p = e^{-\omega_1 t} + e^{-\omega_2 t} - e^{-(\omega_1 + \omega_2)t}$. (2) $E\{T_p\} = \dfrac{1}{\omega_1} + \dfrac{1}{\omega_2} - \dfrac{1}{\omega_1 + \omega_2}$.

9.8 $Q_{25} = 39,321 m^3/s$.

9.9

N	$\hat{g}$	$\hat{c}$
50	8.598	7.788
100	9.498	7.971
500	10.364	8.079
1000	10.098	8.011

Accuracy improves with sample size.

Chapter 10

10.1 Same answers as those of Example 10.1.

10.2 $N=25$.

10.3 10 additional samples.

10.4 $N=22$.

10.5 $N=67$.

10.6 $N=16$.

Chapter 11

11.1 At the 5% level of significance there is evidence that $\mu_{X_2} < \mu_{X_1}$ and the government lab has a downward bias relative to the company lab.

11.2 At the 5% level of significance there is evidence against the statement that $\mu_{X_1} = \mu_{X_2}$. There is a "significant" difference in the two methods.

11.3 At the 5% level of significance there is no evidence against the statement $\sigma_{X_1}^2 = \sigma_{X_2}^2$.

11.4 We accept H_0. We say, at the 5% level of significance, there is reason to believe that the three means are equal.

11.5 Same answer as Problem 11.4.

11.6 Accept H_1. At the 1% level of significance at least one of the mean corrosion readings is different from the others.

11.7 Accept H_1. At the 5% level of significance there is reason to believe that at least one of the mean dissolved oxygen content is different.

Chapter 12

12.1

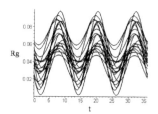
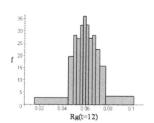

12.2

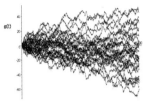

12.3
12.4
$$\mu_R = E\{R_g\} = \mu_A - \mu_B \cos\left(\frac{\pi(t-C)}{6}\right) \ . \ \ \sigma_R^2(t) = \sigma_A^2 + \sigma_B^2 \cos^2\left(\frac{\pi(t-C)}{6}\right) \ .$$

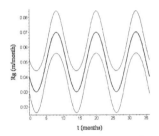

12.5 Curves are similar to those in Problem 12.4.
12.6 Curves are similar to those in Problem 12.4.
12.7

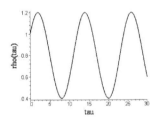

12.8.

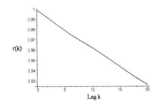

12.9 $S(\omega) = 1/(1+\omega^2)$.
12.10 The correlogram suggests an uncorrelated process consistent with a single
       probability distribution assumed in Example 7.7. The spectral density
       fluctuates widely around 1 suggesting that no single frequency dominates.

12.11

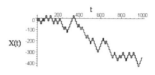

12.12

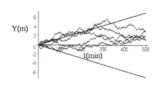

12.13

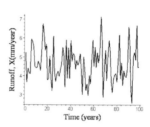

12.14

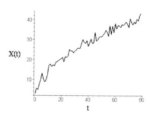

Chapter 13
13.1 $\sigma_q^2 = 2\sigma_T^2 \rho$
13.2 $\mu_Y = 0$, $\sigma_Y^2(t) = \sigma_X^2[t - \sin(\omega t)\cos(\omega t)/\omega]/2$
13.3 $\mu_C(t) = \mu_0 e^{-2}$, $\sigma_C^2(t) = \sigma_0^2 e^{-t}$
13.4

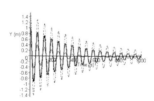

13.5 $\mu_z = 1.4142\left(e^{1.7071t} - e^{0.2929t}\right)$, $\sigma_z^2(t) = \left(e^{1.7071t} - e^{0.2929t}\right)^2$

13.6

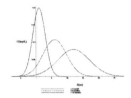

13.7 (1)

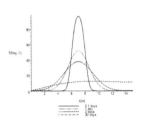

(2)

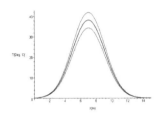

13.8 As in Example 13.15

13.9

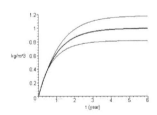

13.10

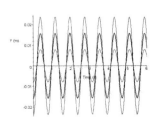

13.11 $\mu_y(t) = \dfrac{\mu_A e^{-bt}}{Mb^2 - \beta b + k}$, $\sigma_y^2(t) = \dfrac{\sigma_A^2 e^{-bt}}{Mb^2 - \beta b + k}$

13.12

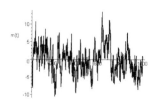

13.13   The flow field and velocity direction changes completely:

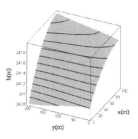

13.14   Higher transmissivity values translates into lower magnitude and lower amplitude in seasonal heads:

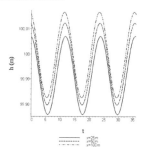

BIBLIOGRAPHY

Abbaoui, K., and Cherruault, Y., 1994. Convergence of Adomian's Method Applied to Differential Equations. Comp. Math. Applic., 28(5):103-109.

Adomian, G., 1994. Solving Frontier Problems of Physics: the Decomposition Method. Kluwer Academic, Dordrecht, The Netherlands

Adomian, G., 1991. A Review of the Decomposition Method and Some Recent Results for Nonlinear Equations. Comp. Math. Application, 21(5):101-127.

Adomian, G., 1986. Nonlinear Stochastic Operator Equations. Academic Press, San Diego, CA.

Adomian, G., 1983. Stochastic Systems. Academic Press, New York, NY.

Adomian, G., and Serrano, S.E., 1998. Stochastic Contaminant Transport Equation in Porous Media. Appl. Math. Letters, 1191):53-55.

Antony, J., 2003.Design of Experiments for Engineers and Scientists. Butterworth-Heinemann, Oxford, U.K.

Ayres, F., 1952. Differential Equations. Schaum's Outline Series. McGraw-Hill, New York, NY.

AMORC, 1995. Unto Thee I Grant. Supreme Grand Lodge of the of the Ancient and Mystical Order Rosae Crucis. San Jose, CA.

Barrentine, L. B., 1999. An Introduction to Design of Experiments: A Simplified Approach. American Society for Quality Press. Milwaukee, WI.

Barlow, R. , 1998. Engineering Reliability. SIAM, Philadelphia, PA.

Barlow, R., and Proschan, F., 1987. Mathematical Theory of Reliability. SIAM, Philadelphia, PA.

Barlow, R., and Proschan, F., 1975. Theory of Reliability and Life Testing. Holt, Rinehart & Winston, New York, NY.

Box, G.E.P., and Jenkins, G.M., 1976. Time Series Analysis: Forecasting and Control. Revised Edn, Holden Day, San Francisco, CA

Braun, M., 1983. Differential Equations and Their Applications. Springer-Verlag, New York, NY.

Burrus, C.S., Gopinath, R.A., and Guo, H., 1998. Introduction to Wavelets and Wavelet Transforms. Prentiche Hall, Upper Saddle River, NJ.

Cakmak, A.S., Botha, J.F., and Gray, W.G., 1987. Computational and Applied Mathematics for Engineering Analysis.Springer-Verlag, New York, NY.

Carslaw, H.S., and Jaeger, J.C., 1971. Conduction of Heat in Solids. Oxford University Press, New York, NY.

Chan, Y.T., 1995. Wavelet Basics. Kluwer Academic, Norwood, MA.

Chapra, S.C., and Canale, R.P., 1988. Numerical Methods for Engineers, 2nd ed. McGraw-Hill, New York, NY.

Cherruault, Y., 1989. Convergence of Adomian's Method. Kybernetes, 18(2):31-38.

Cherruault, Y., Saccomardi, G., and Some, B., 1992. New Results for Convergence of Adomian's Method Applied to Integral Equations. Math. Comput. Modelling, 16(2):85-93.

Cochran, W.G., 1977. Sampling Techniques. Third Ed., John Wiley & Sons, New York, NY.

Cooper, G.R., and McGillem, D.D., 1999. Probabilistic Methods of Signal and System Analysis. Oxford University Press, New York, NY.

Devore, J.L., 1995. Probability and Statistics for Engineering and the Sciences. Fourth Ed. Duxbury Press, Pacific Grove, CA,

Fourier, J., 1878. The Analytical Theory of Heat. Translated to English by Alexander Freeman, Cambridge University Press, London U.K.

Gabet, L., 1994. The Decomposition Method and Distributions. Computers Math. Applic., 27(3):41-49.

Gabet, L., 1993. The Decomposition Method and Linear Partial Differential Equations. Math. Comput. Modelling, 17(6):11-22.

Gabet, L., 1992. Equisse d'une Théorie Décompositionnelle et Application aux Equations aux Dérivées Partialles. Dissertation, Ecole Centrale de Paris, France.

Gardiner, C.W., 1985. Handbook of Stochastic Methods for Physics, Chemistry and the Natural Sciences. Springer-Verlag, New York, NY.

Gilbert, R.O., 1987. Statistical Methods for Environmental Pollution Monitoring. Van Nostrand Reinhold, New York, NY.

Guenther, W.C., 1964. Analysis of Variance. Prentice-Hall, Englewood Cliffs, NJ

Haan, C.T., 1977. Statistical Methods in Hydrology. The Iowa State University Press. Ames, IA.

Hicks, C.R., 1999. Fundamental Concepts in the Design of Experiments. Fifth Ed. Oxford University Press, Oxford, U.K.

Hornbeck, R.W., 1975. Numerical Methods. Quantum Publishers Inc., New York, NY.

Hutson, V., and Pym, J.S., 1980. Applications of Functional Analysis and Operator Theory. Academic Press, New York, NY.

Kottegoda, N.T., 1980. Stochastic Water Resources Technology. The Macmillan Press Ltd., London, U.K.

Ikeda, N., and Watanabe, S., 1989. Stochastic Differential Equations and Diffusion Processes. North Holland, New York, NY.

Jazwinski, A.H., 1970. Stochastic Processes and Filtering Theory. Academic press, New York, NY.

Johnson, R.A., 2000. Miller & Freund's Probability and Statistics for Engineers. Sixth Edition. Prentice Hall, Upper Saddle River, NJ.

Leemis, L. M., 2009. Reliability: Probabilistic Models and Statistical Methods. Lawrence Leemis, ISBN 978-0692000274

Li, J.C.R., 1961. Introduction to Statistical Inference. J.W. Edwards, Ann Arbor, MI.

Maurois, A., 2007. An Art of Living. Translation by S.E. Serrano. SpiralPress, Ambler PA.

Mendenhall, W., and Ott, L., 1976. Understanding Statistics, 2nd Ed. Duxbury Press, North Scituate, MA.

Mendenhall, W., 1968. Introduction to Linear Models and the Design and Analysis of Experiments. Wadsworth Publishing, Belmont, CA.

Milton, J.S., and Arnold, J.C., 1995. Introduction to Probability and Statistics.McGraw-Hill, New York, NY.

Montgomery, D.C., and Runger, G.C., 1994. Applied Statistics and Probability for Engineers. John Wiley & Sons Inc., New York, NY.

Moore, H.L., 1914. Economic Cycles: Their Law and Cause. Macmillan, New York, NY.

Myint-U, Tyn, 1987. Partial Differential Equations for Scientists and Engineers. North Holland, New York, NY.

Newland, D.E., 1993. Random Vibrations, Spectral and Wavelet Analysis, 3rd Edition. Addison Wesley Longman, Reading, MA.

O'Connor, P., 2002. Practical Reliability Engineering. Wiley, Hoboken, NJ.

Oden, J.T., 1979. Applied Functional Analysis. Prentice Hall, Upper Saddle River, NJ.

Oden, J.T., and Demkowicz, L., 2010. Applied Functional Analysis. Chapman & Hall/CRC Press.

Ogata, K., 1998. System Dynamics. Prentice Hall, Upper Saddle River, NJ.

Ogden, R.T., 1997. Essential wavelets for Statistical Application and data

Analysis. Birkhäuser, Boston, MA.

Ott, W.R., 1995. Environmental Statistics and Data Analysis. CRC Press LCC, Boca Raton, FL.

Papoulis, A., 1984. Probability, Random Variables, and Stochastic Processes. McGraw-Hill Book Company. New York, NY.

Parlange, J.Y., 1971. Theory of Water Movement in Soils: I. One-dimensional Absorption., Soil Sci., 111(2):134-137.

Philip, J.R., 1955. Numerical Solution of Equations of the Diffusion Type with Diffusivity Concentration-dependent. Trans. Faraday Soc.,51(7), 391.

Polibarinova-Kochina, P.Y., 1977. Theory of Groundwater Movement (in Russian). Nauka, Moscow, Russia.

Powers, D.L., 1979. Boundary Value Problems. Academic press, New York, NY.

Reinhard, H., 1987. Differential Equations. Foundations and Applications. Macmillan, New York, NY.

Roach, G.F., 1982. Green's Functions. Second Edition. Cambridge University Press, New York, NY.

Research and Education Association, 1982. Theory of Linear Systems. Research and Education Association, New York, NY.

Rosenkrantz, w.A., 1997. Introduction to Probability and Statistics for Scientists and Engineers. McGraw-Hill, NewYork, NY,

Ross, S., 1998. A First Course in Probability. Fifth Edition. Prentice-Hall, Upper Saddle River, NJ.

Serrano, S.E., 2010. Hydrology for Engineers, Geologists, and Environmental Professionals. An Integrated Treatment of Surface, Subsurface, and Contaminant Hydrology. HydroScience Inc., Ambler, PA.

Serrano, S.E., 2004. Modeling Infiltration with Approximate Solutions of Richard's Equation. ASCE Journal of Hydrologic Engineering,

9(5):421-432.

Serrano, S. E., 2003. Propagation of Nonlinear Reactive Contaminants in Porous Media. Water Resources Research, American Geophysical Union, 39(8):1228-1242.

Serrano, S. E., 1998. Analytical Decomposition of the Non-Linear Infiltration Equation. Water Resources Research, American Geophysical Union, 34(3):397-407.

Serrano, S.E., 1996. Hydrologic Theory of Dispersion in Heterogeneous Aquifers. ASCE J. Hydrologic Engr., 1(4):144-151.

Serrano, S.E., 1995(1). Analytical Solutions of the Nonlinear Groundwater Flow Equation in Unconfined Aquifers and the Effect of Heterogeneity. Water Resources Research, 31(11):2733-2742.

Serrano, S.E., 1995(2). Forecasting Scale-Dependent Dispersion from Spills in Heterogeneous Aquifers. Journal of Hydrology 169:151-169.

Serrano, S.E., 1992(1), Semianalytical Methods in Stochastic Groundwater Transport. Applied Mathematical Modelling, 16:181-191.

Serrano, S.E., 1992(2). Migration of Chloroform in Aquifers. ASCE Journal of Environmental Engr., 118(2):167-182.

Serrano, S.E., 1988(1). General Solution to Random Advective-Dispersive Transport Equation in Porous Media. 1: Stochasticity in the Sources and in the Boundaries. Stochastic Hydrology & Hydraulics 2(2):79-98.

Serrano, S.E., 1988(2). General Solution to Random Advective-Dispersive Transport Equation in Porous Media. 1: Stochasticity in the Parameters. Stochastic Hydrology & Hydraulics 2(2):99-112.

Serrano, S.E., and Adomian, G., 1996. New Contributions to the Solution of Transport Equations in Porous Media. Mathl. Comput. Modelling, 24(4):15-25.

Singpurwalla, N. D., 2006. Reliability and Risk: A Bayesian Perspective. Wiley, Hoboken, NJ.

Soong, T.T., 1973. Random Differential Equations in Science and Engineering. Academic Press, New York, NY.

Spiegel, M.R., 1980. Advanced Mathematics for Engineers and Scientists. Schaum's Outline Series. McGraw-Hill, New York, NY.

Spiegel, M.R., 1967. Applied Differential Equations. Prentice-Hall Englewood Cliffs, NJ.

Walpole, R.E., Myers, R.H., and Myers, S.L., 1998. Probability and Statistics for Engineers and Scientists. Sixth Ediiton. Prentice Hall, Upper Saddle River, NJ.

Wazwaz, A. M., and Gorguis A., 2004. Exact Solutions for Heat like and Wave like Equations with Variable Coefficients. Applied Math. and comp., 149, 15-29.

Wazwaz, A.-M., 2000. A New Algorthm for Calculating Adomian Polynomials for Nonlinear Operators. Appl. Math. Comput., 111:53-69.

Wiener, N., 1930. Generalized Harmonic Analysis. Acta Mathematica, 55:117-258.

Woods, R.L., and Lawrence, K.L., 1997. Modeling asn Simulation of Dynamic Systems. Prentice Hall, Upper Saddle River, NJ.

Zauderer, E., 1983. Partial Differential Equations of Applied Mathematics. John Wiley, New York, NY.

Ziemer, R.E., 1997. Elements of Engineering Probability & Statistics. Prentice-Hall, Upper Saddle River, NJ.

Ziemer, R.E., Tranter, W.H., and Fannin, D.R., 1993. Linear Systems: Continuous and Discrete. Prentice Hall, Upper Saddle River, NJ.

INDEX

HYDROLOGY
FOR ENGINEERS, GEOLOGISTS, AND ENVIRONMENTAL PROFESSIONALS

SECOND EDITION, COMPLETELY REVISED
An Integrated Treatment of Surface, Subsurface, and Contaminant Hydrology
Sergio E. Serrano, Ph.D.

ISBN 978-0-9655643-4-2, published by
HydroScience Inc., Ambler, PA

Hydrology for Engineers, Geologists and Environmental Professionals second edition is a new comprehensive hydrology treatise for today's research and consulting environmental professional. Hydrology for Engineers, Geologists and Environmental Professionals integrates the subjects of surface, subsurface, and contaminant hydrology. It prepares the reader to analyze today's environmental problems. The author presents the fundamental concepts of physical and contaminant hydrology of watersheds, rivers, lakes, soils, and aquifers in an easy and accessible manner to the environmental professional. Many practical examples and solved problems illustrate the concepts. State of the art research on groundwater and contaminant transport modeling is explained in a clear fashion. Recent research developments in nonlinear hydrologic science and simulation are included in this new edition. New solutions of nonlinear infiltration are presented with simple numerical applications. New developments in analytical decomposition are presented as simple and practical means to complex nonlinear hydrologic problems, such as regional groundwater flow modeling in homogeneous or heterogeneous media, regular or irregularly-shaped domains, steady or transient problems, multiple pumping wells, and nonlinear flow. For contaminant transport, new applications to the simulation of nonlinear decay, nonlinear sorption, and unsaturated-saturated zones contaminant propagation are presented along with simple programs.

"The presentations are clear and concise and the illustrative examples are well chosen and complement the text." *Transactions, American Geophysical Union*

"The chapter topics and organization of the book are presented in excellent format...The figure, table, paragraph, and chapter layouts are easily followed by the reader..." *Journal of the American Water Resources Association*

Features:
Consistent SI notation; 124 solved examples; 187 proposed problems; 152 illustrations; 71 tables; 46 short computer programs; answers to problems; extensive bibliography; index; 590 pages.